Analytic and Geometric Inequalities and Applications

Mathematics and Its Applications

Managing Editor:

M. HAZEWINKEL

Centre for Mathematics and Computer Science, Amsterdam, The Netherlands

Volume 478

Analytic and Geometric Inequalities and Applications

Edited by

Themistocles M. Rassias

Department of Mathematics,
National Technical University of Athens,
Zografou Campus,
Athens, Greece

and

Hari M. Srivastava

Department of Mathematics and Statistics,
University of Victoria,
Victoria, British Columbia, Canada

SPRINGER-SCIENCE+BUSINESS MEDIA, B.V.

A C.I.P. Catalogue record for this book is available from the Library of Congress.

ISBN 978-94-010-5938-1   ISBN 978-94-011-4577-0 (eBook)
DOI 10.1007/978-94-011-4577-0

Printed on acid-free paper

Table of Contents

PREFACE

Analytic and Geometric Inequalities and Applications is devoted to recent advances in a variety of inequalities of Mathematical Analysis and Geometry. Subjects dealt with in this volume include: Fractional order inequalities of Hardy type, differential and integral inequalities with initial time difference, multi-dimensional integral inequalities, Opial type inequalities, Gruss' inequality, Furuta inequality, Laguerre–Samuelson inequality with extensions and applications in statistics and matrix theory, distortion inequalities for analytic and univalent functions associated with certain fractional calculus and other linear operators, problem of infimum in the positive cone, alpha-quasi-convex functions defined by convolution with incomplete beta functions, Chebyshev polynomials with integer coefficients, extremal problems for polynomials, Bernstein's inequality and Gauss–Lucas theorem, numerical radii of some companion matrices and bounds for the zeros of polynomials, degree of convergence for a class of linear operators, open problems on eigenvalues of the Laplacian, fourth order obstacle boundary value problems, bounds on entropy measures for mixed populations as well as controlling the velocity of Brownian motion by its terminal value. A wealth of applications of the above is also included.

We wish to express our appreciation to the distinguished mathematicians who contributed to this volume. Finally, it is our pleasure to acknowledge the fine cooperation and assistance provided by the staff of Kluwer Academic Publishers.

June 1999

Themistocles M. Rassias
Hari M. Srivastava

PROBLEM OF INFIMUM IN THE POSITIVE CONE

TSUYOSHI ANDO

Faculty of Economics, Hokusei Gakuen University, Sapporo 004-8631, Japan

Abstract. It is known that for (bounded) self-adjoint operators A, B on a Hilbert space $\mathcal{H}$ the infimum $A \wedge B$, with respect to the order induced by the cone of positive (semi-definite) operators, exists only when A and B are comparable, that is, $A \geq B$ or $A \leq B$. In this paper we present a necessary and sufficient condition for that, given $A, B \geq 0$, the infimum considered in the positive cone exists.

1. Problem

The space of (bounded) self-adjoint operators on a Hilbert space $\mathcal{H}$ is provided with the order relation induced by positive semi-definiteness; for self-adjoint A, B the order relation $A \geq B$ means $A - B$ is *positive (semi-definite)*. In particular, $A \geq 0$ means positive semi-definiteness of A. With respect to this order, the space of self-adjoint operators does not become a lattice. More exactly, for self-adjoint A, B the infimum $A \wedge B$ exists only when A and B is comparable, that is, $A \geq B$ or $A \leq B$.

The situation is different if we take the *positive cone* $\mathcal{P} = \mathcal{P}(\mathcal{H})$, the cone of positive (semi-definite) operators, in place of the whole space of self-adjoint operators. For instance, it is well known that for any two orthoprojections P, Q the infimum $P \wedge Q$ in $\mathcal{P}$ always exists and is equal to the orthoprojection to the intersection of the range of P and that of Q.

In this paper we present a necessary and sufficient condition for that, given $A, B \geq 0$, the infimum $A \wedge B$ in $\mathcal{P}$ exists. This problem has been studied by several authors in mathematical physics (see [6], [7], [8] and [9]). Among others, Moreland and Gudder [9] have solved this problem in the finite dimensional case. We explain also how our result reduces to theirs in the finite dimensional case.

1991 *Mathematics Subject Classification.* Primary 47D20; Secondary 47A63
Key words and phrases. Positive cone; Lattice, Operator inequalities

T.M. Rassias and H.M. Srivastava (eds.), Analytic and Geometric Inequalities and Applications, 1–12.
© 1999 *Kluwer Academic Publishers.*

2. Reduction

To discuss existence of the infimum $A \wedge B$ in $\mathcal{P}$, we may assume $\ker(A + B) = \{0\}$. (Here ker denotes the *kernel*.) In fact, let P be the orthoprojection to the orthocomplement $\mathcal{K}$ of $\ker(A + B)$. Then the map

$$0 \leq X \longmapsto \Phi(X) = PX|_{\mathcal{K}}$$

gives an affine, order-isomorphism from $\{X ; 0 \leq X \leq A, B\}$ onto a subset $\mathcal{Q}$ of the positive cone $\mathcal{P}(\mathcal{K})$ satisfying that $Y \in \mathcal{Q}, 0 \leq Z \leq Y \implies Z \in \mathcal{Q}$, and $\ker(\Phi(A) + \Phi(B)) = \{0\}$.

When $\ker(A + B) = \{0\}$, there is an affine, order-isomorphism φ from the set $\{X ; 0 \leq X \leq A + B\}$ onto the set of positive contractions, that is, $\{Y ; 0 \leq Y \leq 1\}$, given by

$$X = (A + B)^{1/2} \cdot \varphi(X) \cdot (A + B)^{1/2}. \tag{1}$$

Here important is the fact that

$$\varphi(A) + \varphi(B) = 1,$$

in particular, $\varphi(A)$ and $\varphi(B)$ commute. Therefore we first discuss existence of the infimum $A \wedge B$ in $\mathcal{P}$ under the assumption $A + B = 1$.

3. Commuting Case

Suppose that

$$A, \ B \geq 0, \quad A + B = 1. \tag{2}$$

Then according to the spectral theorem there is uniquely an increasing family of orthoprojection $\{E(\lambda) ; 0 \leq \lambda \leq 1\}$ with $E(1) = 1$ such that

$$A = \int_0^1 \lambda dE(\lambda) \quad \text{and} \quad B = \int_0^1 (1 - \lambda)dE(\lambda). \tag{3}$$

Lemma 1. *Under the setting (3), if the infimum $A \wedge B$ in $\mathcal{P}$ exists, then*

$$A \wedge B = \int_0^1 \min(\lambda, 1 - \lambda)dE(\lambda).$$

Proof. Let

$$C \overset{\text{def}}{=} \int_0^1 \min(\lambda, 1 - \lambda)dE(\lambda).$$

Then obviously $0 \leq C \leq A, B$. We claim that C is a *maximal* element in the set $\{X ; 0 \leq X \leq A, B\}$ (cf. Ando [5]). To see this, take D such that

$$C \leq D \leq A, \ B.$$

Then
$$0 \leq D - C \leq A - C, \ B - C$$
so that there exist contractions X, Y such that
$$(D - C)^{1/2} = (A - C)^{1/2}X \quad \text{and} \quad (D - C)^{1/2} = (B - C)^{1/2}Y,$$
which implies
$$\text{ran}(D - C)^{1/2} \subset \text{ran}(A - C)^{1/2} \cap \text{ran}(B - C)^{1/2}.$$
Here ran denotes the *range*. Since by the spectral representation (3)
$$(A - C)^{1/2} = \int_0^1 \sqrt{\max(2\lambda - 1, 0)}dE(\lambda) = \int_{\frac{1}{2}+}^1 \sqrt{2\lambda - 1}dE(\lambda),$$
$$(B - C)^{1/2} = \int_0^1 \sqrt{\max(1 - 2\lambda, 0)}dE(\lambda) = \int_0^{\frac{1}{2}-} \sqrt{1 - 2\lambda}dE(\lambda),$$
we can infer that
$$\text{ran}(A - C)^{1/2} \cap \text{ran}(B - C)^{1/2} = \{0\},$$
which implies
$$\text{ran}(D - C)^{1/2} = \{0\},$$
or equivalently $D - C = 0$, that is, $D = C$. This proves our claim on maximality. Now if the infimum $A \wedge B$ in $\mathcal{P}$ exists, it must coincide with the maximial element C. This completes the proof. $\square$

Theorem 2. *Under the setting (3), the infimum $A \wedge B$ in $\mathcal{P}$ exists if and only if*
$$\int_{0+}^{1-} \lambda dE(\lambda) \geq \int_{0+}^{1-} (1 - \lambda)dE(\lambda)$$
or
$$\int_{0+}^{1-} \lambda dE(\lambda) \leq \int_{0+}^{1-} (1 - \lambda)dE(\lambda).$$
In each case, the infimum coincides with the smaller of the above two and is equal to
$$\int_0^1 \min(\lambda, 1 - \lambda)dE(\lambda).$$

Here remark
$$\int_{0+}^{1-} \lambda dE(\lambda) = A \cdot \{1 - E(\{0\}) - E(\{1\})\}, \tag{4}$$
$$\int_{0+}^{1-} (1 - \lambda)dE(\lambda) = B \cdot \{1 - E(\{0\}) - E(\{1\})\}. \tag{5}$$

Proof. A proof of the "if" part is easy. Suppose, for instance, via (4) and (5), that

$$A \cdot \{1 - E(\{0\}) - E(\{1\})\} \geq B \cdot \{1 - E(\{0\}) - E(\{1\})\},$$

and let

$$C_0 \stackrel{\mathrm{def}}{=} B \cdot \{1 - E(\{0\}) - E(\{1\})\}.$$

Then obviously $0 \leq C_0 \leq A, B$. Take D such that $0 \leq D \leq A, B$. Then since by (3)

$$A \cdot E(\{0\}) = 0 = B \cdot E(\{1\}),$$

the inequality $0 \leq D \leq A, B$ implies

$$D \cdot E(\{0\}) = D \cdot E(\{1\}) = 0$$

so that

$$\begin{aligned} D &= \{1 - E(\{0\}) - E(\{1\})\} \cdot D \cdot \{1 - E(\{0\}) - E(\{1\})\} \\ &\leq \{1 - E(\{0\}) - E(\{1\})\} \cdot B \cdot \{1 - E(\{0\}) - E(\{1\})\} = C_0. \end{aligned}$$

This proves that C_0 is the infimumu of A, B in $\mathcal{P}$.

To prove the "only if" part by contradiction, suppose that

$$\int_{0+}^{1-} \lambda dE(\lambda) \ngeq \int_{0+}^{1-} (1 - \lambda) dE(\lambda) \tag{6}$$

and

$$\int_{0+}^{1-} \lambda dE(\lambda) \nleq \int_{0+}^{1-} (1 - \lambda) dE(\lambda), \tag{7}$$

but A, B admits the infimum in $\mathcal{P}$. Then by Lemma 1 this infimum coincides with

$$C \stackrel{\mathrm{def}}{=} \int_0^1 \min(\lambda, 1 - \lambda) dE(\lambda).$$

By assumptions (6) and (7), there exist disjoint subsets $\Delta_1, \Delta_2 \subset (0,1)$ and $\epsilon > 0$ such that $E(\Delta_1) \neq 0, E(\Delta_2) \neq 0$ and

$$\begin{aligned} 1 - 3\epsilon &\geq \lambda \geq (1 - \lambda) + \epsilon & (\lambda \in \Delta_1), \tag{8} \\ 1 - 3\epsilon &\geq 1 - \lambda \geq \lambda + \epsilon & (\lambda \in \Delta_2). \tag{9} \end{aligned}$$

We may assume, without loss of generality, that

$$\dim(\mathrm{ran} E(\Delta_1)) \geq \dim(\mathrm{ran} E(\Delta_2)).$$

Then there is a partial isometry V with domain $\mathrm{ran} E(\Delta_2)$ and range in $\mathrm{ran} E(\Delta_1)$. Define an operator D by

$$\begin{aligned} D \stackrel{\mathrm{def}}{=} &B \cdot E(\Delta_1) - \epsilon E(\Delta_1) + A \cdot E(\Delta_2) \\ &- \epsilon E(\Delta_2) + \sqrt{2}\epsilon V \cdot E(\Delta_2) + \sqrt{2}\epsilon E(\Delta_2) \cdot V^*. \end{aligned}$$

We claim that A, $B \geq D \geq 0$. First let us show $D \geq 0$. Since by (8)

$$1 - \lambda \geq 3\epsilon \quad (\lambda \in \Delta_1)$$

it follows from (3) that

$$B \cdot E(\Delta_1) - \epsilon E(\Delta_1) \geq 2\epsilon E(\Delta_1),$$

and similarly from (9) and (3) that

$$A \cdot E(\Delta_2) - \epsilon E(\Delta_2) \geq 2\epsilon E(\Delta_2).$$

Further since by definition of V

$$E(\Delta_2) \cdot V^* V \cdot E(\Delta_2) = E(\Delta_2) \quad \text{and} \quad E(\Delta_1) \cdot V \cdot E(\Delta_2) = V \cdot E(\Delta_2)$$

we can conclude from the above that

$$
\begin{aligned}
D & \geq \sqrt{2}\epsilon\{E(\Delta_1) + E(\Delta_2) + V \cdot E(\Delta_2) + E(\Delta_2) \cdot V^*\} \\
& = \sqrt{2}\epsilon\{E(\Delta_1) + V \cdot E(\Delta_2)\}^*\{E(\Delta_1) + V \cdot E(\Delta_2)\} \geq 0
\end{aligned}
$$

Next let us turn to the proof of A, $B \geq D$. Since by (3) and (8)

$$(A - B) \cdot E(\Delta_1) \geq \epsilon E(\Delta_1)$$

it follows from the definitin of D and the property of V that

$$
\begin{aligned}
A - D & = A \cdot E((\Delta_1 \cup \Delta_2)^c) + (A - B) \cdot E(\Delta_1) + \epsilon E(\Delta_1) \\
& \quad + \epsilon E(\Delta_2) - \sqrt{2}\epsilon V \cdot E(\Delta_2) - \sqrt{2}\epsilon E(\Delta_2) \cdot V^* \\
& \geq \epsilon\{2E(\Delta_1) + E(\Delta_2) - \sqrt{2}V \cdot E(\Delta_2) - \sqrt{2}E(\Delta_2) \cdot V^*\} \\
& = \epsilon\{\sqrt{2}E(\Delta_1) - V \cdot E(\Delta_2)\}^*\{\sqrt{2}E(\Delta_1) - V \cdot E(\Delta_2)\} \geq 0.
\end{aligned}
$$

This proves $A \geq D$. The inequality $B \geq D$ is proved similarly.

Now since by Lemma 1 C is the infimum $A \wedge B$ in $\mathcal{P}$, we can conclude $C \geq D$. To see that this causes a contradiction, take a unit vector $x \in \mathrm{ran}\, E(\Delta_2)$ and let $y \stackrel{\mathrm{def}}{=} Vx$. Then y is a unit vector in $\mathrm{ran}\, E(\Delta_1)$. On the other hand, since by the definitions of C and D

$$C - D = C \cdot E((\Delta_1 \cup \Delta_2)^c) + \epsilon\{E(\Delta_1) + E(\Delta_2) - \sqrt{2}V \cdot E(\Delta_2) - \sqrt{2}E(\Delta_2) \cdot V^*\},$$

we have

$$
\begin{aligned}
\langle (C - D)(x + y), x + y \rangle & = \epsilon\{\langle x, x \rangle + \langle y, y \rangle - \sqrt{2}\langle y, y \rangle - \sqrt{2}\langle x, x \rangle\} \\
& = -2(\sqrt{2} - 1)\epsilon < 0,
\end{aligned}
$$

which contradicts $C - D \geq 0$. This proves the theorem. $\qquad \square$

4. Parallel sum and short

How can we recapture

$$\int_{0+}^{1-} \lambda dE(\lambda) \quad \text{and} \quad \int_{0+}^{1-} (1 - \lambda) dE(\lambda)$$

from A and B in the form (3) without appealing to the spectral representation ?

For this purpose, recall the definition of parallel sum operation $X : Y$ for $X, Y \geq 0$, introduced for matrices by Anderson and Duffin [1] and later for Hilbert space operators by Pekarev and Smul'yan [11].

As a positive quadratic form, *parallel sum* $X : Y$ is defined as

$$\langle (X : Y)a, a \rangle \overset{\text{def}}{=} \inf\{\langle Xb, b \rangle + \langle Yc, c \rangle \; ; \; b + c = a\}. \tag{10}$$

When $X + Y$ is invertible, this definition takes the form

$$X : Y = X - X(X + Y)^{-1}X = Y - Y(X + Y)^{-1}Y, \tag{11}$$

and when both X and Y are invertible,

$$X : Y = (X^{-1} + Y^{-1})^{-1}. \tag{12}$$

As a consequence of definition (10), parallel addition has the following properties; here X, Y, Z are all in $\mathcal{P}$,

(a) $0 \leq X : Y = Y : X \leq X, \; Y,$

(b) $(\alpha X) : (\alpha Y) = \alpha(X : Y) \quad (\alpha \geq 0),$

(c) $(X : Y) : Z = X : (Y : Z),$

(d) the map $(X, Y) \longmapsto X : Y$ is (jointly) *monotone* in the sense

$$X_1 \geq X_2, \; Y_1 \geq Y_2 \implies X_1 : Y_1 \geq X_2 : Y_2$$

(e) the map is (strongly) *continuous from above* in the sense:

$$X_n \downarrow X, \; Y_n \downarrow Y \implies X_n : Y_n \downarrow X : Y.$$

A little non-trivial is the following relation for ranges;

(f) $$\operatorname{ran}(A : B)^{1/2} = \operatorname{ran}(A^{1/2}) \cap \operatorname{ran}(B^{1/2}).$$

Recall the definition of short (operation) (see Anderson and Trapp [2]). Given a closed subspace $\mathcal{M} \subset \mathcal{H}$, for each $X \geq 0$ there is the maximum element in the set

$$\{Z \; ; \; 0 \leq Z \leq X, \; \operatorname{ran}(Z) \subset \mathcal{M}\}.$$

This maximum element is called the *short* of X to the subspace $\mathcal{M}$. By identifying a closed subspace $\mathcal{M}$ with the orthoprojection $P = P_{\mathcal{M}}$ to itself, we shall denote the short by $[P]X$. Since inclusion $\text{ran}(Z) \subset \mathcal{M}$ is equivalent to the requirement that there is $\gamma = \gamma(Z) \geq 0$ such that $Z \leq \gamma P$, the short $[P]X$ is defined by

$$[P]X = \max\{Z \ ; \ 0 \leq Z \leq X, \ \gamma P \text{ for some } \gamma \geq 0\} \tag{13}$$

Therefore if P commutes with X, then $[P]X = XP$. In general case, it is not difficult to see

$$[P]X = \lim_{n \to \infty} (nP) : X \quad \text{(strong convergence)}.$$

Motivated by this relation, Ando [4] introduced a notion of generalized short $[Y]X$ with $Y \geq 0$ (called the *Y-absolutely continuous part* of X) by

$$[Y]X \overset{\text{def}}{=} \lim_{n \to \infty} (nY) : X. \tag{14}$$

Here strong convergence on the right side is always guaranted because

$$0 \leq (nY) : X \leq ((n+1)Y) : X \leq X \quad (n = 1, 2, \ldots).$$

The following properties of generalized short are immediate; here all X, Y are in $\mathcal{P}$,

(A) $0 \leq [Y]X \leq X$ and $[Y]Y = Y$,

(B) $[Y](\alpha X) = \alpha[Y]X$ $(\alpha \geq 0)$,

(C) $\alpha Z \leq Y \leq \beta Z$ for some $\alpha, \beta > 0 \implies [Y]X = [Z]X$,

(D) $[Y]X = [X : Y]X$,

(E) $[P]1 = P$ for orthoprojection P,

(F) the map $(X, Y) \longmapsto [Y]X$ is (jointly) *monotone* in the sense:

$$X_1 \geq X_2, \ Y_1 \geq Y_2 \implies [Y_1]X_1 \geq [Y_2]X_2.$$

A little non-trivial is the following property whose proof is found in Nishio [8]:

(G) $[Y]X = [[X]Y]X$.

Lemma 3. *Under the setting* (3)

$$\int_{0+}^{1-} \lambda dE(\lambda) = [B]A \quad and \quad \int_{0+}^{1-} (1 - \lambda)dE(\lambda) = [A]B.$$

Proof. By (11)

$$(nB) : A = \int_0^1 f_n(\lambda)dE(\lambda)$$

where
$$f_n(\lambda) \stackrel{def}{=} \lambda - \frac{\lambda^2}{\lambda + n(1 - \lambda)}.$$
As $n \to \infty$, $f_n(\lambda)$ converges increasingly to λ for $0 < \lambda < 1$ while
$$f_n(1) = f_n(0) = 0 \quad (n = 1, 2, \ldots).$$
Therefore by definition (14)
$$[B]A = \lim_{n\to\infty} \int_0^1 f_n(\lambda) dE(\lambda) = \int_0^1 \lim_{n\to\infty} f_n(\lambda) dE(\lambda) = \int_{0+}^{1-} \lambda dE(\lambda),$$
and similarly
$$[A]B = \int_{0+}^{1-} (1 - \lambda) dE(\lambda).$$
This completes the proof. □

Now the content of Theorem 2 can be stated without use of spectral representation.

Theorem 4. *When $A, B \geq 0$ and $A + B = 1$, the infimum $A \wedge B$ in $\mathcal{P}$ exists if and only if $[A]B$ and $[B]A$ are comparable. In this case*
$$A \wedge B = \min([A]B, [B]A).$$

5. General case

Fix A, $B \geq 0$ with $\ker(A+B) = \{0\}$. By (1) there is an affine, order isomorphism $\varphi(\cdot)$ from the set $\{X \; ; \; 0 \leq X \leq A + B\}$ onto the set $\{Z \; ; \; 0 \leq Z \leq 1\}$, for which $\varphi(A) + \varphi(B) = 1$.

Lemma 5. *The map φ in (1) satisfies the condition*
$$\varphi([Y]X) = [\varphi(Y)]\varphi(X) \quad (0 \leq X, \; Y \leq A + B).$$

Proof. First let us establish the following identity
$$\varphi(X : Y) = \varphi(X) : \varphi(Y). \tag{15}$$
Take an arbitrary vector $a \in \mathcal{H}$. Since $\ker(A+B) = \{0\}$ implies that $\mathrm{ran}(A+B)^{1/2}$ is dense in $\mathcal{H}$, by (10) and (1)
$$\langle (\varphi(X) : \varphi(Y))(A + B)^{1/2}a, (A + B)^{1/2}a \rangle$$
$$= \inf\{ \langle \varphi(X)b', b' \rangle + \langle \varphi(Y)c', c' \rangle \; ; \; (A + B)^{1/2}a = b' + c' \}$$
$$= \inf\{ \langle \varphi(X)(A + B)^{1/2}b, (A + B)^{1/2}b \rangle + \langle \varphi(Y)(A + B)^{1/2}c, (A + B)^{1/2}c \rangle \; ;$$
$$a = b + c \}$$
$$= \inf\{ \langle Xb, b \rangle + \langle Yc, c \rangle \; ; \; a = b + c \} = \langle (X : Y)a, a \rangle$$
$$= \langle (A + B)^{1/2}\varphi(X : Y)(A + B)^{1/2}a, a \rangle$$
$$= \langle \varphi(X : Y)(A + B)^{1/2}a, (A + B)^{1/2}a \rangle,$$

which yields (15), because $(A + B)^{1/2}$ has dense range.

Again take an arbitrary vector $a \in \mathcal{H}$. Then by definition (1), (14) and (15)

$$
\begin{aligned}
\langle \varphi([Y]X) \cdot (A+B)^{1/2}a, (A+B)^{1/2}a \rangle &= \langle ([Y]X)a, a \rangle \\
&= \lim_{n \to \infty} \langle (((nY):X)a, a \rangle = \lim_{n \to \infty} \langle \varphi((nY):X)(A+B)^{1/2}a, (A+B)^{1/2}a \rangle \\
&= \lim_{n \to \infty} \langle (n\varphi(Y):\varphi(X)) \cdot (A+B)^{1/2}a, (A+B)^{1/2}a \rangle \\
&= \langle [\varphi(Y)]\varphi(X) \cdot (A+B)^{1/2}a, (A+B)^{1/2}a \rangle,
\end{aligned}
$$

which yields the assertion because $(A + B)^{1/2}$ has dense range. $\square$

Now we are in position to transform Theorem 4 to the general case.

Theorem 6. *Given A, $B \geq 0$, the infimum $A \wedge B$ in $\mathcal{P}$ exists if and only if $[B]A$ and $[A]B$ are comparable, that is,*

$$
[B]A \geq [A]B \quad \text{or} \quad [B]A \leq [A]B.
$$

In this case

$$
A \wedge B = \min([B]A, [A]B).
$$

Proof. We may assume that $\ker(A + B) = \{0\}$. Then since the map $\varphi(\cdot)$ in (1) is order-isomorphism between $\{X \; ; \; 0 \leq X \leq A, \; B\}$ and $\{Z \; ; \; 0 \leq Z \leq \varphi(A), \; \varphi(B)\}$, the infimum $A \wedge B$ in $\mathcal{P}$ exists if and only if the infimum $\varphi(A) \wedge \varphi(B)$ in $\mathcal{P}$ exists. Further in this case

$$
\varphi(A \wedge B) = \varphi(A) \wedge \varphi(B).
$$

By Theorem 4 the infimum $\varphi(A) \wedge \varphi(B)$ in $\mathcal{P}$ exists if and only if $[\varphi(B)]\varphi(A)$ and $[\varphi(A)]\varphi(B)$ are comparable, and in this case the infimum coincides with the smaller one of those two. Then since by Lemma 5

$$
[\varphi(B)]\varphi(A) = \varphi([B]A), \quad [\varphi(A)]\varphi(B) = \varphi([A]B)
$$

and $\varphi(\cdot)$ is order-isomorphism, the assertion follows. $\square$

6. Special cases

As pointed out in the begining, for every pair of orthoprojections P, Q the infimum $P \wedge Q$ in $\mathcal{P}$ always exists and is equal to the orthoprojection to $\operatorname{ran}(P) \cap \operatorname{ran}(Q)$. Let us discuss a connection of this fact with Theorem 6.

Lemma 7. *Let P be an orthoprojection. Then for every positive contraction $0 \leq A \leq 1$ the infimum $A \wedge P$ in $\mathcal{P}$ exists and is equal to the short $[P]A$.*

Proof. By property (E) and (F), it follows from $0 \leq A \leq 1$ that

$$
[P]A \leq [P]1 = P,
$$

which implies again by (F) and (G)

$$[P]A = [[P]A]([P]A) \leq [[P]A]P = [A]P.$$

Now the assertion follows from Theorem 6.   □

An immediate consequence of Lemma 7 is that for orthoprojections P, Q

$$[P]Q = [Q]P = P \wedge Q.$$

In this connection recall the identity (see Anderson and Schreiber [6])

$$2(P : Q) = P \wedge Q.$$

Lemma 8. *If $B \geq 0$ is of rank ≤ 1, then for every $A \geq 0$ the infimum $A \wedge B$ in $\mathcal{P}$ exists.*

Proof. $0 \leq [A]B \leq B$ implies that $[A]B$ is a non-negative scalar multiple of B. Similarly each $(nB) : A$ is a non-negative scalar multiple of B $(n = 1, 2, \ldots)$, so is $[B]A$ as the limit. Since any two scalar multiplies of B are comparable, $[B]A$ and $[A]B$ are comparable, and the infimum $A \wedge B$ in $\mathcal{P}$ exists by Theorem 6.   □

Theorem 9.
(i) *Let $0 \leq B \leq 1$. Then the infimum $A \wedge B$ in $\mathcal{P}$ exists for every $0 \leq A \leq 1$ if and only if B is of rank ≤ 1 or is an orthoprojection.*

(ii) *Let $B \geq 0$. Then the infimum $A \wedge B$ in $\mathcal{P}$ exists for every $A \geq 0$ if and only if B is of rank ≤ 1.*

Proof. (i) The "if" part is proved in Lemma 7 and Lemma 8. To prove the "only if" part by contradiction, suppose that B is of rank ≥ 2 and is not an orthoprojection. According to the spectral theorem, there are mutually annihilating non-zero orthoprojections P, Q, commuting with B, such that for some $0 < \epsilon \leq \gamma < 1$

$$BP \geq \gamma P \quad \text{and} \quad \gamma Q \geq BQ \geq \epsilon Q.$$

Let

$$A \overset{\text{def}}{=} \frac{\gamma + \epsilon}{2} P + \frac{1 + \gamma}{2} Q.$$

Then $0 \leq A \leq 1$, and A and $BP + BQ$ are not comparable. Since

$$\frac{1 + \gamma}{2\epsilon} B \geq \gamma P + \frac{1 + \gamma}{2} Q \geq A$$

there is α, $\beta > 0$ such that $\alpha(A : B) \leq A \leq \beta(A : B)$. Therefore by (A) and (C)

$$[B]A = [B : A]A = [A]A = A.$$

On the other hand, since

$$P + Q \geq A \geq \epsilon(P + Q)$$

and $P + Q$ is an orthoprojection commuting with B, we can see by (C)

$$[A]B = [P + Q]B = BP + BQ.$$

Therefore $[B]A$ and $[A]B$ are not comparable, and by Theorem 6 the infimum $A \wedge B$ in $\mathcal{P}$ does not exist. This proves (i).

(ii) The "if" part is proved in Lemma 8. To prove the "only if" part by contradiction, suppose that B is of rank ≥ 2. Via multiplication by a positive scalar we may assume that $0 \leq B \leq 1/2$. Then since B is not an orthoprojection, as in the proof of (i) there is $0 \leq A \leq 1$ such that the infimum $A \wedge B$ in $\mathcal{P}$ does not exist. This proves (ii). $\quad\square$

7. Finite dimensional case

In this section we assume $\dim(\mathcal{H}) < \infty$. Therefore every $A \geq 0$ has closed range and

$$\mathrm{ran}(A) = \mathrm{ran}(A^{1/2}).\tag{16}$$

We can consider the orthoprojection P_A onto $\mathrm{ran}(A)$. Since there are constants α, $\beta > 0$ such that

$$\alpha P_A \leq A \leq \beta P_A,$$

it follows from property (C) for generalized short

$$[A]B = [P_A]B.\tag{17}$$

Let $P_{A,B}$ be the orthoprojection to the intersection $\mathrm{ran}(A) \cap \mathrm{ran}(B)$. Since by (16) and (f)

$$\begin{aligned}\mathrm{ran}(A : B) &= \mathrm{ran}(A : B)^{1/2} = \mathrm{ran}(A^{1/2}) \cap \mathrm{ran}(B^{1/2})\\ &= \mathrm{ran}(A) \cap \mathrm{ran}(B),\end{aligned}$$

the following relation holds

$$P_{A,B} = P_{A:B}.\tag{18}$$

With a rather complicated method, Moreland and Gudder [9] established the following results (for the finite dimensional case).

(i) *For any orthoprojection P and $0 \leq A \leq 1$ the infimum $A \wedge P$ in $\mathcal{P}$ exists,*

(ii) *for $A, B \geq 0$, the infimum $A \wedge B$ in $\mathcal{P}$ exists if and only if $A \wedge P_{A,B}$ and $B \wedge P_{A,B}$ are comparable. In this case*

$$A \wedge B = \min(A \wedge P_{A,B}, B \wedge P_{A,B}).$$

The assertion (i) is just Lemma 7. Further by Lemma 7 it follows from (18), (17) and (D) that

$$A \wedge P_{A,B} = [P_{A,B}]A = [P_{A:B}]A = [A:B]A = [B]A$$

and similarly $B \wedge P_{A,B} = [A]B$. Therefore assertion (ii) coincides with Theorem 6.

Finally let us mention that Theorem 9 (for the finite dimensional case) is also pointed out in Moreland and Gudder [9].

References

1. W.N. Anderson, Jr. and R.J. Duffin, *Series and parallel addition of matrices*, J. Math. Anal. Appl. **26**(1969), 576-594.
2. W.N. Anderson, Jr. and M. Schreiber, *The infima of two projections*, Acta Sci. Math. (Szeged) **33**(1972), 165-168.
3. W.N. Anderson, Jr. and G.E. Trapp, *Shorted operators II*, SIAM J. Appl. Math. **28**(1975), 60 -71.
4. T. Ando, *Lebesgue-type decomposition of positive operators*, Acta Sci. Math. (Szeged), **38**(1976), 253-260.
5. ———, *Parametrization of minimal points of some convex sets of matrices*, Acta Sci. Math. (Szeged), **57**(1993), 3 -10.
6. S. Gudder, *Lattice properties of quantum effects*, J. Math. Phys. **37**(1996), 2637-2642.
7. ———, *Examples, problems, and results in effect algebras*, Int. J. Theor. Phys. **35**(1996), 2365-2376.
8. P. Lahti and M. Maczynski, *On the order structure of the set of effects in quantum mechanics*, J. Math. Phys. **36**(1995), 1673-1680.
9. T. Moreland and S. Gudder, *Infima of Hilbert space effects*, to appear in Linear Alg. Appl.
10. K. Nishio, *Characterization of Lebesgue-type decompositon of positive operators*, Acta Sci. Math. (Szeged), **42**(1980), 143-152.
11. E.L. Pekarev and Ju. L. Smul'yan, *Parallel addtion and parallel subtraction of operators*, Izv. Akad. Nauk SSSR, Ser. Mat. **40**(1976), 366-387.

Open Problems on Eigenvalues of the Laplacian[†]

Mark S. Ashbaugh[‡] (mark@math.missouri.edu)
Department of Mathematics, University of Missouri, Columbia, MO 65211-0001

Abstract. In this paper a number of open problems for the low eigenvalues of the Laplacian on a Euclidean domain are discussed. These include problems with Neumann boundary conditions as well as Dirichlet. Some problems for the biharmonic operator (with "clamped" boundary conditions) are also included, particularly the analogs of the Faber-Krahn result for the vibrating clamped plate and buckling problems. In many of the problems presented here, it is expected that the conjectured inequality saturates at a disk in $\mathbb{R}^2$ (or a ball in higher dimensions). For the most part, the problems listed complement those found in a prior paper of R. Benguria and the author [10], which, for example, discusses the Pólya conjectures for the Dirichlet and Neumann eigenvalues of a Euclidean domain. References to related problem lists and discussions due to L.E. Payne [32], [33], [34] and S.-T. Yau [49], [50], [51] are also given.

1991 Mathematics Subject Classification: Primary 35P15; Secondary 49R05, 49R10.

We shall consider the eigenvalues and eigenfunctions of a domain (= open connected set) Ω in Euclidean space $\mathbb{R}^n$. Throughout the discussion Ω will denote a bounded domain. We concentrate mainly on the *fixed membrane problem*,

$$-\Delta u = \lambda u \qquad \text{in } \Omega \subset \mathbb{R}^n, \qquad (1)$$

$$u = 0 \qquad \text{on } \partial\Omega, \qquad (2)$$

though later we shall also bring in several other classical eigenvalue problems associated with the Laplacian. Recall that for a bounded domain, problem (1)-(2) has a real and purely discrete spectrum $\{\lambda_i\}_{i=1}^{\infty}$ where

$$0 < \lambda_1 < \lambda_2 \leq \lambda_3 \leq \ldots \to \infty. \qquad (3)$$

[†] This paper was originally prepared in May 1996 in connection with the NSF-CBMS conference, "Advances in Inverse Spectral Geometry", featuring Carolyn Gordon as Principal Lecturer and held at Texas Tech University, Lubbock, Texas, June 24-29, 1996. It was made available to participants in this conference in substantially its present form at that time.

[‡] Partially supported by National Science Foundation grant DMS-9500968.

T.M. Rassias and H.M. Srivastava (eds.), Analytic and Geometric Inequalities and Applications, 13–28.
© 1999 *Kluwer Academic Publishers.*

Here we repeat eigenvalues according to their multiplicities. The associated sequence of real eigenfunctions will be denoted u_1, u_2, u_3, ..., assumed orthonormal. Our first two problems were suggested by Payne, Pólya, and Weinberger (PPW) in 1956.

1. For $\Omega \subset \mathbb{R}^2$ show that

$$\frac{\lambda_2 + \lambda_3}{\lambda_1} \le \left.\frac{\lambda_2 + \lambda_3}{\lambda_1}\right|_{\text{disk}} \approx 5.077. \tag{4}$$

In $\mathbb{R}^n$ this conjecture reads

$$\frac{\lambda_2 + \ldots + \lambda_{n+1}}{\lambda_1} \le \left.\frac{\lambda_2 + \ldots + \lambda_{n+1}}{\lambda_1}\right|_{n\text{-ball}} = \frac{n j_{p+1,1}^2}{j_{p,1}^2}, \tag{5}$$

where $p = (n-2)/2$ and $j_{\nu,k}$ denotes the kth positive zero of the Bessel function $J_\nu(t)$. For $(\lambda_2 + \lambda_3)/\lambda_1$, PPW [35] proved the bound 6, and subsequently others obtained the bounds $3 + \sqrt{7} \approx 5.646$ (Brands [16]), 5.622 (Hile and Protter [22]), $(15 + \sqrt{345})/6 \approx 5.5957$ (Marcellini [27]), and 5.525 (Ashbaugh and Benguria [9]). The best bound to date is 5.5066, also obtained by Ashbaugh and Benguria [11]. The last two bounds incorporated earlier work of deVries [20] and H.C. Yang [48], as well as that of PPW, Brands, and particularly Marcellini. For the corresponding developments for (5) (i.e., for dimensions larger than 2) and other background and discussion, the reader may consult [22], [5], [9], [10], [11]. In particular, [5] p. 1647, has a brief table of upper bounds for $(\lambda_2 + \lambda_3)/\lambda_1$ along with the conjectured optimal upper bound as given by (5). From this table the gap between the two values is seen to be quite small already for $n = 3, 4$ (and these could now be decreased a bit further based on the subsequent developments mentioned above).

2. For $\Omega \subset \mathbb{R}^2$ and all $m \ge 2$ show that

$$\frac{\lambda_{m+1}}{\lambda_m} < \left.\frac{\lambda_2}{\lambda_1}\right|_{\text{disk}} \approx 2.5387. \tag{6}$$

In $\mathbb{R}^n$ this conjecture reads

$$\frac{\lambda_{m+1}}{\lambda_m} < \left.\frac{\lambda_2}{\lambda_1}\right|_{n\text{-ball}} = \frac{j_{p+1,1}^2}{j_{p,1}^2} \tag{7}$$

where $p = (n-2)/2$. For $m = 1$ these inequalities hold as nonstrict inequalities (with equality if and only if Ω is a ball); see [35], [1],

[2], [7]. For $m = 2$ and 3 (6) and (7) are also known to hold [3], [4]. Payne Pólya, and Weinberger [35] showed that $\lambda_{m+1}/\lambda_m \leq 3$ for $\Omega \subset \mathbb{R}^2$. The straightforward extension of their argument to n dimensions yields the bound $\lambda_{m+1}/\lambda_m \leq 1 + 4/n$ (Thompson [45]). In fact, it would be interesting to obtain any upper bound better than $1 + 4/n$ in the cases $m \geq 4$.

Interestingly, the bounds (6) and (7) for $m = 2, 3$ follow from the stronger bound

$$\frac{\lambda_4}{\lambda_2} < \frac{\lambda_2}{\lambda_1}\bigg|_{n-\text{ball}} \tag{8}$$

proved in [4] using the bound $\lambda_2/\lambda_1 \leq \lambda_2/\lambda_1|_{n-\text{ball}}$ and Courant's nodal domain result for u_2 (u_2 has exactly two nodal domains). This suggests the stronger conjectures given in our next open problem.

3.   For $\Omega \subset \mathbb{R}^n$ and all $m \geq 1$ show that

$$\frac{\lambda_{2m}}{\lambda_m} \leq \frac{\lambda_2}{\lambda_1}\bigg|_{n-\text{ball}} \tag{9}$$

or even

$$\frac{\lambda_{km}}{\lambda_m} \leq \sup_{\Omega} \frac{\lambda_k}{\lambda_1}. \tag{10}$$

If these hold, they should probably hold with strict inequality for $m > 1$. The first of these, (9), holds for $m = 1, 2$ as already mentioned. The second also holds for $m = 1$ (trivially) and $m = 2$. For further background and remarks, consult [4], [8]. In particular, [8] shows that (9) holds "on average" in the sense that

$$\frac{\lambda_{2^k}}{\lambda_1} < \left(\frac{\lambda_2}{\lambda_1}\bigg|_{n-\text{ball}}\right)^k \quad \text{for } k = 2, 3, \ldots. \tag{11}$$

Also [4] proves weaker versions of (9) and (10) in which the m in λ_{2m} and λ_{km} is replaced by the number of nodal domains of u_m.

4.   For $\Omega \subset \mathbb{R}^n$ find the optimal bound on λ_3/λ_1 and the shape of domain that saturates it (possibly in a limiting sense). In $\mathbb{R}^2$ it is known that

$$3.1818 \approx \frac{35}{11} \leq \sup_{\Omega} \frac{\lambda_3}{\lambda_1} \leq 3.83103. \tag{12}$$

Hence the optimizing domain is definitely not the disk (which has $\lambda_3 = \lambda_2$ and therefore $\lambda_3/\lambda_1 \approx 2.5387$). The upper bound in (12) is proved in [11] based in part on prior work of H.C. Yang [48] and Marcellini [27], while the lower bound is λ_3/λ_1 for a $\sqrt{8} \times \sqrt{3}$ rectangle (which is where λ_3/λ_1 is maximized among rectangles). Previous upper bounds for λ_3/λ_1 in two dimensions include 5 (and implicitly 13/3) of PPW [35], $(7 + 2\sqrt{7})/3 \approx 4.097$ of Brands [16], 4.014 of Hile and Protter [22], $(15 + \sqrt{345})/6 \approx 3.9170$ of Marcellini [27], 3.90514 of Ashbaugh and Benguria [9], and 3.89804 of H.C. Yang [48]. For other discussion and references, see [5], [9], [10], [11]. In particular, [5], p. 1647, has a small table of upper bounds for λ_3/λ_1 in higher dimensions (which could now be improved somewhat, based on the subsequent developments mentioned above).

5. For $\Omega \subset \mathbb{R}^n$ show that

$$\frac{\lambda_{n+2}}{\lambda_1} \leq \left.\frac{\lambda_{n+2}}{\lambda_1}\right|_{n-\text{ball}} = \frac{j_{p+2,1}^2}{j_{p,1}^2} \tag{13}$$

where $p = (n-2)/2$. A brief discussion of this problem appears in [4] (see also (11) above). In two dimensions we know

$$4.56 \approx \frac{j_{2,1}^2}{j_{0,1}^2} \leq \sup_\Omega \frac{\lambda_4}{\lambda_1} \leq \frac{j_{1,1}^4}{j_{0,1}^4} \approx 6.445. \tag{14}$$

It may also be true that in two dimensions

$$\frac{\lambda_4 + \lambda_5}{\lambda_1} \leq \left.\frac{\lambda_4 + \lambda_5}{\lambda_1}\right|_{\text{disk}} = \frac{2j_{2,1}^2}{j_{0,1}^2} \approx 9.121 \tag{15}$$

and in general dimension n

$$\frac{\lambda_{n+2} + \ldots + \lambda_{n(n+3)/2}}{\lambda_1} \leq \left.\frac{\lambda_{n+2} + \ldots + \lambda_{n(n+3)/2}}{\lambda_1}\right|_{n-\text{ball}}$$
$$= \frac{(n-1)(n+2)}{2} \frac{j_{p+2,1}^2}{j_{p,1}^2} \tag{16}$$

but not much has been done regarding these. The indices here reflect the fact that for an n-ball the $(n+2)$th through $\left[\frac{1}{2}n(n+3)\right]$th eigenvalues are degenerate (all are equal to $j_{p+2,1}^2/R^2$ where $p = (n-2)/2$ and R is the radius of the ball), while the $\left[\frac{1}{2}(n+1)(n+2)\right]$th eigenvalue is strictly larger. Multiplicities (degeneracies) are determined by the dimension of the space of homogeneous harmonic polynomials (i.e., spherical harmonics) of degree ℓ in $\mathbb{R}^n$. Of relevance here are the cases $\ell = 0, 1$, and 2, where the corresponding dimensions are 1, n, and $\frac{1}{2}(n-1)(n+2)$, respectively. That the eigenvalues are ordered as stated follows from a cycling property of zeros of Bessel functions: $j_{\nu,k} < j_{\nu+1,k} < j_{\nu+2,k} < j_{\nu,k+1}$ for $\nu > -1$ and k a positive integer (see Ince [24], p. 252, probs. 8-9).

To this point our list has dealt exclusively with problems about eigenvalue ratios. We state one more open problem concerning eigenvalue ratios for the fixed membrane problem, and then move on to other topics.

6.   For $\Omega \subset \mathbb{R}^2$ show that

$$\frac{\lambda_1}{\lambda_2 - \lambda_1} + \frac{\lambda_1}{\lambda_3 - \lambda_1} \geq \left(\frac{\lambda_1}{\lambda_2 - \lambda_1} + \frac{\lambda_1}{\lambda_3 - \lambda_1}\right)\Big|_{\text{disk}} \approx 1.2998 \quad (17)$$

and, more generally, for $\Omega \subset \mathbb{R}^n$ show that

$$\sum_{i=1}^{n} \frac{\lambda_1}{\lambda_{i+1} - \lambda_1} \geq \left(\sum_{i=1}^{n} \frac{\lambda_1}{\lambda_{i+1} - \lambda_1}\right)\Big|_{n-\text{ball}} = \frac{n}{j_{p+1,1}^2 / j_{p,1}^2 - 1} \quad (18)$$

where $p = (n-2)/2$. This problem is related to Problem 1 above and, in fact, (17) and (18) would be implied by (4) and (5), respectively. In [5] weaker bounds than (17) and (18) are proved, based on a rearrangement approach introduced by Chiti [18], [19]. In particular, a bound of the form (17) is proved but with right-hand side 1.2611. More generally, a bound of the form (18) is proved with right-hand side $[2j_{p,1}^2 + n(n-4)]/6$ where $p = (n-2)/2$. As discussed in [5], this right-hand side has a relative error with respect to the right-hand side of (18) which is $O(1/n)$ as n goes to infinity.

We now turn to conjectured inequalities which take their inspiration from the Faber-Krahn inequality [21], [25], [26]. This inequality is

$$A\lambda_1 \geq (A\lambda_1)|_{\text{disk}} = \pi j_{0,1}^2 \quad (19)$$

in two dimensions, and

$$|\Omega|^{2/n} \lambda_1 \geq (|\Omega|^{2/n} \lambda_1)|_{n-\text{ball}} = C_n^{2/n} j_{p,1}^2, \quad (20)$$

in n dimensions, where A denotes the area of $\Omega \subset \mathbb{R}^2$, $|\Omega|$ denotes the n-dimensional volume of $\Omega \subset \mathbb{R}^n$, $C_n = \pi^{n/2}/\Gamma\left(\frac{n}{2}+1\right) =$ volume of the unit ball in $\mathbb{R}^n$, and $p = (n-2)/2$. Krahn [26] also observed that by means of Courant's nodal domain result for u_2 there follow from (19), (20)

$$A\lambda_2 > 2\pi j_{0,1}^2 \tag{21}$$

and

$$|\Omega|^{2/n}\lambda_2 > 2^{2/n} C_n^{2/n} j_{p,1}^2. \tag{22}$$

These inequalities saturate when Ω breaks into two disks (n-balls) of equal area (volume). Inequality (21) was also noted by Peter Szego somewhat later (see Pólya [37], p. 336). For higher eigenvalues we have the following conjectures.

   7.   For $\Omega \subset \mathbb{R}^2$

$$A\lambda_3 \geq (A\lambda_3)|_{\text{disk}} = \pi j_{1,1}^2, \tag{23}$$

and for $\Omega \subset \mathbb{R}^n$

$$|\Omega|^{2/n}\lambda_{n+1} \geq (|\Omega|^{2/n}\lambda_{n+1})|_{n-\text{ball}} = C_n^{2/n} j_{p+1,1}^2 \tag{24}$$

where $p = (n-2)/2$. See Wolf and Keller [47] for some support for (23).

   8.   For $\Omega \subset \mathbb{R}^2$

$$A(\lambda_2 + \lambda_3) \geq [A(\lambda_2 + \lambda_3)]|_{\text{disk}} = 2\pi j_{1,1}^2, \tag{25}$$

and for $\Omega \subset \mathbb{R}^n$

$$|\Omega|^{2/n}(\lambda_2 + \ldots + \lambda_{n+1}) \geq [|\Omega|^{2/n}(\lambda_2 + \ldots + \lambda_{n+1})]|_{n-\text{ball}}$$
$$= n C_n^{2/n} j_{p+1,1}^2 \tag{26}$$

where $p = (n-2)/2$. Inequalities (25) and (26) are related somewhat to (4) and (5) since in combination these would imply the Faber-Krahn inequality, (19) and (20).

   Not much is known toward Problems 7 and 8. Analogous, but even more tenuous conjectures would be

$$A\lambda_5 \geq (A\lambda_5)|_{\text{disk}} = \pi j_{2,1}^2 \tag{27}$$

and

$$A(\lambda_4 + \lambda_5) \geq [A(\lambda_4 + \lambda_5)]|_{\text{disk}} = 2\pi j_{2,1}^2 \tag{28}$$

for $\Omega \subset \mathbb{R}^2$; and

$$|\Omega|^{2/n}\lambda_{n(n+3)/2} \geq (|\Omega|^{2/n}\lambda_{n(n+3)/2})|_{n-\text{ball}} = C_n^{2/n} j_{p+2,1}^2$$

$$\tag{29}$$

and

$$|\Omega|^{2/n}(\lambda_{n+2} + \ldots + \lambda_{n(n+3)/2}) \geq [|\Omega|^{2/n}(\lambda_{n+2} + \ldots + \lambda_{n(n+3)/2})]|_{n-\text{ball}}$$
$$= nC_n^{2/n} j_{p+2,1}^2$$

$$\tag{30}$$

for $\Omega \subset \mathbb{R}^n$. In connection with these inequalities it might be noted that Krahn [26], based on his results (21)-(22), suggested that $|\Omega|^{2/n}\lambda_k$ might always take its minimum when Ω approaches k n-balls of equal volume. This is not true for $k = 3$ if $n = 2$, for example (see also [36]). Thus, in general it is an open problem to find the optimal lower bound to $|\Omega|^{2/n}\lambda_k$ for $k \geq 3$ and to discover the shape of domain that saturates it. Some interesting results in this direction were proved recently by Wolf and Keller [47]. In particular, they give some useful general facts helping to characterize the optimal lower bound to $|\Omega|^{2/n}\lambda_k$ and the domain(s) that saturate it.

We next consider the *free membrane problem*

$$-\Delta v = \mu v \qquad \text{in} \quad \Omega \subset \mathbb{R}^n, \tag{31}$$

$$\frac{\partial v}{\partial n} = 0 \qquad \text{on} \quad \partial\Omega. \tag{32}$$

Here $\partial/\partial n$ denotes the outward normal derivative on $\partial\Omega$. Assuming Ω is sufficiently smooth, (31)-(32) has spectrum $\{\mu_i\}_{i=0}^{\infty}$ where

$$0 = \mu_0 < \mu_1 \leq \mu_2 \leq \ldots \to \infty \tag{33}$$

(eigenvalues are repeated here according to their multiplicities). For μ_1 we have the well-known Szegö-Weinberger bound [42], [46]

$$A\mu_1 \leq (A\mu_1)|_{\text{disk}} \tag{34}$$

in two dimensions, and

$$|\Omega|^{2/n}\mu_1 \leq (|\Omega|^{2/n}\mu_1)|_{n-\text{ball}} = C_n^{2/n}p_{n/2,1}^2 \qquad (35)$$

in n dimensions, where $p_{\nu,k}$ denotes the kth positive zero of the derivative of $t^{1-\nu}J_\nu(t)$. For simply connected domains in $\mathbb{R}^2$ Szegö and Weinberger also noted (see [46]) that

$$\frac{1}{A}\left(\frac{1}{\mu_1} + \frac{1}{\mu_2}\right) \geq \left[\frac{1}{A}\left(\frac{1}{\mu_1} + \frac{1}{\mu_2}\right)\right]\bigg|_{\text{disk}} = \frac{2}{\pi p_{1,1}^2}, \qquad (36)$$

which follows from Szegö's proof of (34) in that case. As an open problem we therefore have:

9.   Show that (36) holds for an arbitrary smooth bounded domain $\Omega \subset \mathbb{R}^2$, and, more generally, show

$$\frac{1}{|\Omega|^{2/n}}\left(\frac{1}{\mu_1} + \cdots + \frac{1}{\mu_n}\right) \geq \left[\frac{1}{|\Omega|^{2/n}}\left(\frac{1}{\mu_1} + \cdots + \frac{1}{\mu_n}\right)\right]\bigg|_{n-\text{ball}}$$
$$= \frac{n}{C_n^{2/n}p_{n/2,1}^2} \qquad (37)$$

for an arbitrary smooth bounded domain in $\mathbb{R}^n$. Weaker variants of these bounds are known [6]. In particular, (36) holds if its right-hand side is replaced by $1/2\pi$ (we note in this connection that $p_{1,1} \approx 1.84118 < 2$), and (37) holds with right-hand side replaced by $n/[(n+2)C_n^{2/n}]$. Inequalities (36) and (37) are in a sense Neumann analogs of (17) and (18); see [6] for explanation.

10.   For $\Omega \subset \mathbb{R}^n$ and convex, show

$$\mu_2/\mu_1 \leq 4. \qquad (38)$$

This would say that μ_2/μ_1 is maximized in any number of dimensions by an essentially one-dimensional example, since saturation would occur for any rectangular parallelopiped having all but one of its dimensions tiny. This bound is probably not true without the convexity hypothesis. See also [34].

Finally, we come to two fourth order eigenvalue problems, the *vibrating clamped plate problem*

$$\Delta^2 w = \Gamma w \qquad\qquad \text{in}\quad \Omega \subset \mathbb{R}^n, \qquad (39)$$

$$w = 0 = \frac{\partial w}{\partial n} \qquad\qquad \text{on}\quad \partial\Omega, \qquad (40)$$

and the *buckling problem* for a clamped plate

$$\Delta^2 v = -\Lambda \Delta v \qquad \text{in} \quad \Omega \subset \mathbb{R}^n, \qquad (41)$$

$$v = 0 = \frac{\partial v}{\partial n} \qquad \text{on} \quad \partial \Omega. \qquad (42)$$

Analogs of the Faber-Krahn inequality for these problems were conjectured by Rayleigh [39] and Pólya and Szegö [38], respectively. For the clamped plate problem, partial results have been obtained beginning with Szegö [41] (see also [43] for some corrections), but it was not until 1992 that Nadirashvili [28], [29], [30], building on earlier work of Talenti [44], proved the two-dimensional case of this inequality. Subsequently, Ashbaugh and Benguria [12] gave a proof for the three-dimensional case, following the ideas of Nadirashvili. We therefore are left with the following problem.

   11.   For $\Omega \subset \mathbb{R}^n$ with $n \geq 4$ prove that the lowest eigenvalue Γ_1 of (39)-(40) satisfies

$$|\Omega|^{4/n}\Gamma_1 \geq \left(|\Omega|^{4/n}\Gamma_1 \right)|_{n-\text{ball}}. \qquad (43)$$

Bounds of this form but with lower right-hand sides have been proved by Talenti [44] and Ashbaugh and Laugesen [14]. See [12], [13], and [14] for more extensive discussions.

   Similarly, for the buckling problem we have the Faber-Krahn analog, as conjectured by Pólya and Szegö.

   12.   For $\Omega \subset \mathbb{R}^n$ with $n \geq 2$ prove that the lowest eigenvalue Λ_1 of (41)-(42) satisfies

$$|\Omega|^{2/n}\Lambda_1 \geq \left(|\Omega|^{2/n}\Lambda_1 \right)|_{n-\text{ball}}. \qquad (44)$$

Again Szegö [41], [43] had a partial result in this direction, as did Bramble and Payne [15]. However, as yet (44) is unproved in all dimensions $n \geq 2$. Bounds of the form of (44) but with lower right-hand sides were noted by Bramble and Payne and recently were investigated more extensively by Ashbaugh and Laugesen in [14]. See [13] and [14] for more discussion.

   It turns out that the best weaker version of (44) yet known depends upon the known (isoperimetric) bound of Payne [31] (with equality if and only if Ω is a ball)

$$\Lambda_1(\Omega) \geq \lambda_2(\Omega), \qquad (45)$$

where $\lambda_2(\Omega)$ denotes the second fixed membrane eigenvalue (=second Dirichlet eigenvalue) of $\Omega \subset \mathbb{R}^n$. This together with Krahn's bound (22) yields the best bound of the form (44) yet known.

It is interesting to note that if (45) could be improved to

$$\Lambda_1(\Omega) \geq \frac{1}{n}(\lambda_2(\Omega) + \cdots + \lambda_{n+1}(\Omega)), \tag{46}$$

then this together with (26) (Problem 8 above) would prove (44). However, these two inequalities might well be harder to prove than the original inequality.

Another conjecture that suggests itself in the light of the discussion above (in particular, (45)) and the bound

$$\Gamma_1(\Omega) \geq \lambda_1(\Omega)\Lambda_1(\Omega) \tag{47}$$

(easily proved from the variational characterizations of these three principal eigenvalues) is

$$|\Omega|^{4/n}\lambda_1(\Omega)\Lambda_1(\Omega) \geq [|\Omega|^{4/n}\lambda_1(\Omega)\Lambda_1(\Omega)]|_{n-\text{ball}} = C_n^{4/n}j_{p,1}^2 j_{p+1,1}^2. \tag{48}$$

If this were true then we could base a bound of the form (43) (but with weaker right-hand side) upon it.

With our current state of knowledge, from (47) we can only prove

$$\Gamma_1(\Omega) \geq \lambda_1(\Omega)\Lambda_1(\Omega) \geq \lambda_1(\Omega)\lambda_2(\Omega) > |\Omega|^{-4/n}2^{2/n}C_n^{4/n}j_{p,1}^4 \tag{49}$$

where $p = (n-2)/2$, using the Faber and Krahn results, (20) and (22). This is weaker than Ashbaugh and Laugesen's bound of this form [14], which has an additional $2^{2/n}$ on the right.

Other problems suggested by the line of development above would be to find the best lower bound for $|\Omega|^{4/n}\lambda_1(\Omega)\lambda_2(\Omega)$, or for $|\Omega|^{2/n}\lambda_2(\Omega)$ when Ω is convex. These might lead to improvements in the best lower bounds yet known for $|\Omega|^{4/n}\Gamma_1(\Omega)$ (for $n \geq 4$) and $|\Omega|^{2/n}\Lambda_1(\Omega)$ (for $n \geq 2$) using the ideas laid out above.

Concluding Remarks

It might be noted that in all our inequalities above we could dispense with the factors of A, $|\Omega|^{2/n}$, and $|\Omega|^{4/n}$ if we had stated them for domains of equal area/volume (or even for domains of unit area/volume). These factors can be thought of simply as scale factors, making the quantities we consider dimensionless. Another, rather elegant, formulation of those bounds which are expected to saturate for a disk/n-ball is to state them as comparisons of eigenvalues of Ω with those of $\Omega^\star \subset \mathbb{R}^n$, defined as an n-ball of the same volume as Ω (i.e., $\Omega^\star$ is an n-ball with $|\Omega^\star| = |\Omega|$). Note that this builds in an equal area/volume condition. Then, for example, the Faber-Krahn inequality becomes simply

$$\lambda_1(\Omega) \geq \lambda_1(\Omega^\star). \tag{50}$$

Many of our other inequalities would simplify similarly.

One could also investigate combinations of eigenvalues of the clamped plate and buckling problems, paralleling the combinations set forth above for the fixed membrane problem. In fact, here one should begin with the ratios Γ_2/Γ_1 and Λ_2/Λ_1, and only later consider ratios Γ_{m+1}/Γ_m, Λ_{m+1}/Λ_m, $(\Gamma_2+\Gamma_3)/\Gamma_1$, or $(\Lambda_2+\Lambda_3)/\Lambda_1$. Various results are known for Γ_2/Γ_1 and Λ_2/Λ_1, and even Γ_{m+1}/Γ_m and Λ_{m+1}/Λ_m, but the natural conjectures that Γ_2/Γ_1 and Λ_2/Λ_1 should be maximized when Ω is an n-ball are as yet unproved. Early results on these topics are due to Payne, Pólya, and Weinberger [35] and Hile and Yeh [23]. References to the more recent literature can be found in Section 2 of [10].

Finally, we mention that there are many other interesting conjectures and questions concerning eigenvalues (and eigenfunctions) of the Laplacian and bi-Laplacian that are not discussed here. Perhaps the most compelling of these are Pólya's conjectures for the Dirichlet and Neumann eigenvalues of the Laplacian. These state that λ_k is always greater than its large k (Weyl) asymptotic behavior, while μ_k is always less than its Weyl asymptotic behavior. For $n = 2$, these statements are expressed concisely by

$$\mu_k < \frac{4\pi k}{A} < \lambda_k \quad \text{for } k = 1, 2, \ldots \tag{51}$$

(recall that our convention for the μ_k's is that the first Neumann eigenvalue is $\mu_0 = 0$). These problems, and progress on them, are discussed extensively and references given, in the last section of [10], in which a few other open problems are also discussed. Other directions that have been explored somewhat, though to a lesser degree than the Euclidean case, are the parallel developments for the Laplacian (=Laplace-Beltrami operator) on a compact manifold or on a bounded domain on a manifold, with suitable boundary conditions imposed on $\partial\Omega$. For compact manifolds there is a parallel with the Euclidean case with Neumann boundary conditions, partly because the first eigenvalue is always 0 (with constant eigenfunction), and hence attention shifts to the second eigenvalue. We leave aside further discussion or references, referring the reader instead to Chavel's book [17].

For other problems and conjectures of a similar nature, one might consult the review papers of Payne [32], [33], [34] and the recent problem lists [49], [51] and expositions [50], [40] of Yau (respectively, Schoen and Yau for [40]). Items [49], [50], and [51] are reprinted in [40].

References

1. M.S. Ashbaugh and R.D. Benguria, Proof of the Payne-Pólya-Weinberger conjecture, Bull. Amer. Math. Soc. **25** (1991), 19-29.
2. M.S. Ashbaugh and R.D. Benguria, A sharp bound for the ratio of the first two eigenvalues of Dirichlet Laplacians and extensions, Annals of Mathematics **135** (1992), 601-628.
3. M.S. Ashbaugh and R.D. Benguria, Isoperimetric bound for λ_3/λ_2 for the membrane problem, Duke Math. J. **63** (1991), 333-341.
4. M.S. Ashbaugh and R.D. Benguria, Isoperimetric bounds for higher eigenvalue ratios for the n-dimensional fixed membrane problem, Proc. Royal Soc. Edinburgh **123A** (1993), 977-985.
5. M.S. Ashbaugh and R.D. Benguria, More bounds on eigenvalue ratios for Dirichlet Laplacians in n dimensions, SIAM J. Math. Anal. **24** (1993), 1622-1651.
6. M.S. Ashbaugh and R.D. Benguria, Universal bounds for the low eigenvalues of Neumann Laplacians in n dimensions, SIAM J. Math. Anal. **24** (1993), 557-570.
7. M.S. Ashbaugh and R.D. Benguria, A second proof of the Payne-Pólya-Weinberger conjecture, Commun. Math. Phys. **147** (1992), 181-190.
8. M.S. Ashbaugh and R.D. Benguria, Bounds for ratios of eigenvalues of the Dirichlet Laplacian, Proc. Amer. Math. Soc. **121** (1994), 145-150.
9. M.S. Ashbaugh and R.D. Benguria, The range of values of λ_2/λ_1 and λ_3/λ_1 for the fixed membrane problem, Rev. in Math. Phys. **6**, No.

5a (1994), 999-1009 [also appearing in *The State of Matter: A Volume Dedicated to E.H. Lieb*, M. Aizenman and H. Araki, editors, Advanced Series in Mathematical Physics, vol. **20**, World Scientific, Singapore, 1994, pp. 167-181].

10. M.S. Ashbaugh and R.D. Benguria, Isoperimetric inequalities for eigenvalue ratios, *Partial Differential Equations of Elliptic Type, Cortona, 1992*, A. Alvino, E. Fabes, and G. Talenti, editors, Symposia Mathematica, vol. **35**, Cambridge University Press, Cambridge, 1994, pp. 1-36.

11. M.S. Ashbaugh and R.D. Benguria, Bounds for ratios of the first, second, and third membrane eigenvalues, *Nonlinear Problems in Applied Mathematics: In Honor of Ivar Stakgold on his Seventieth Birthday*, T.S. Angell, L. Pamela Cook, R.E. Kleinman, and W.E. Olmstead, editors, Society for Industrial and Applied Mathematics, Philadelphia, 1996.

12. M.S. Ashbaugh and R.D. Benguria, On Rayleigh's conjecture for the clamped plate and its generalization to three dimensions, Duke Math. J. **78** (1995), 1-17.

13. M.S Ashbaugh, R.D. Benguria, and R.S. Laugesen, Inequalities for the first eigenvalues of the clamped plate and buckling problems, *General Inequalities 7*, C. Bandle, W.N. Everitt, L. Losonczi, and W. Walter, editors, International Series of Numerical Mathematics, vol. **123**, Birkhäuser, Basel, 1997, pp. 95-110.

14. M.S. Ashbaugh and R.S. Laugesen, Fundamental tones and buckling loads of clamped plates, Ann. Scuola Norm. Sup. Pisa Cl. Sci. (Ser. IV) **23** (1996), 383-402.

15. J.H. Bramble and L.E. Payne, Pointwise bounds in the first biharmonic boundary value problem, J. Math. and Phys. **42** (1963), 278-286.

16. J.J.A.M. Brands, Bounds for the ratios of the first three membrane eigenvalues, Arch. Rational Mech. Anal. **16** (1964), 265-268.

17. I. Chavel, *Eigenvalues in Riemannian Geometry*, Academic, New York, 1984.

18. G. Chiti, Inequalities for the first three membrane eigenvalues, Boll. Un. Mat. Ital. (5) **18-A** (1981), 144-148.

19. G. Chiti, A bound for the ratio of the first two eigenvalues of a membrane, SIAM J. Math. Anal. **14** (1983), 1163-1167.

20. H.L. de Vries, On the upper bound for the ratio of the first two membrane eigenvalues, Zeitschrift für Naturforschung **22A** (1967), 152-153.

21. G. Faber, Beweis, dass unter allen homogenen Membranen von gleicher Fläche und gleicher Spannung die kreisförmige den tiefsten Grundton gibt, Sitzungberichte der mathematisch-physikalischen Klasse der Bayerischen Akademie der Wissenschaften zu München Jahrgang, 1923, pp. 169-172.

22. G.N. Hile and M.H. Protter, Inequalities for eigenvalues of the Laplacian, Indiana Univ. Math. J. **29** (1980), 523-538.

23. G.N. Hile and R.Z. Yeh, Inequalities for eigenvalues of the biharmonic operator, Pac. J. Math. **112** (1984), 115-133.

24. E.L. Ince, *Ordinary Differential Equations*, Longmans, Green and Co., London, 1927.

25. E. Krahn, Über eine von Rayleigh formulierte Minimaleigenschaft des Kreises, Math. Ann. **94** (1925), 97-100.

26. E. Krahn, Über Minimaleigenschaften der Kugel in drei und mehr Dimensionen, Acta Comm. Univ. Tartu (Dorpat) **A9** (1926), 1-44. [English translation: Minimal properties of the sphere in three and more dimensions, *Edgar Krahn 1894-1961: A Centenary Volume*, Ü. Lumiste and J. Peetre, editors, IOS Press, Amsterdam, 1994, Chapter 11, pp. 139-174.]

27. P. Marcellini, Bounds for the third membrane eigenvalue, J. Diff. Eqs. **37** (1980), 438-443.

28. N.S. Nadirashvili, An isoperimetric inequality for the principal frequency of a clamped plate, Dokl. Akad. Nauk **332** (1993), 436-439 (in Russian) [English translation in Phys. Dokl. **38** (1993), 419-421].

29. N.S. Nadirashvili, New isoperimetric inequalities in mathematical physics, *Partial Differential Equations of Elliptic Type, Cortona, 1992*, A. Alvino, E. Fabes, and G. Talenti, editors, Cambridge University Press, Symposia Mathematica, vol. **35**, Cambridge, 1994, pp. 197-203.

30. N.S. Nadirashvili, Rayleigh's conjecture on the principal frequency of the clamped plate, Arch. Rational Mech. Anal. **129** (1995), 1-10.

31. L.E. Payne, Inequalities for eigenvalues of membranes and plates, J. Rational Mech. Anal. **4** (1955), 517-529.

32. L.E. Payne, Isoperimetric inequalities for eigenvalues and their applications, Autovalori e autosoluzioni, Centro Internazionale Matematico Estivo (C.I.M.E.) 2° Ciclo, Chieti, 1962, pp. 1-58.

33. L.E. Payne, Isoperimetric inequalities and their applications, SIAM Review **9** (1967), 453-488.

34. L.E. Payne, Some comments on the past fifty years of isoperimetric inequalities, *Inequalities: Fifty Years On from Hardy, Littlewood, and Pólya*, W.N. Everitt, editor, Marcel Dekker, New York, 1991, pp. 143-161.

35. L.E. Payne, G. Pólya, and H.F. Weinberger, On the ratio of consecutive eigenvalues, J. Math. and Phys. **35** (1956), 289-298 [reprinted in *George Pólya: Collected Papers, vol. III: Analysis*, J. Hersch and G.-C. Rota, editors, MIT Press, Cambridge, Massachusetts, 1984, pp. 420-429, with comments by J. Hersch on pp. 521-522].

36. J. Peetre, Faber-Krahn inequality and related topics, *Edgar Krahn 1894-1961: A Centenary Volume*, Ü. Lumiste and J. Peetre, editors, IOS Press, Amsterdam, 1994, Chapter 6, pp. 81-106.

37. G. Pólya, On the characteristic frequencies of a symmetric membrane, Math. Z. **63** (1955), 331-337 [reprinted in *George Pólya: Collected Papers, vol. III: Analysis*, J. Hersch and G.-C. Rota, editors, MIT Press, Cambridge, Massachusetts, 1984, pp. 413-419, with comments by J. Hersch on pp. 519-521].

38. G. Pólya and G. Szegö, *Isoperimetric Inequalities in Mathematical Physics*, Annals of Mathematics Studies, Number **27**, Princeton Unversity Press, Princeton, 1951.

39. J.W.S. Rayleigh, *The Theory of Sound*, second edition revised and enlarged (in 2 volumes), Dover Publications, New York, 1945 (republication of the 1894/96 edition).

40. R. Schoen and S.-T. Yau, *Lectures on Differential Geometry*, Conference Proceedings and Lecture Notes in Geometry and Topology, vol. 1, International Press, Boston, 1994. [This is an expanded English translation of S.-T. Yau and R. Schoen, *Differential Geometry*, Scientific Publication, Beijing, 1988 (in Chinese).]

41. G. Szegö, On membranes and plates, Proc. Nat. Acad. Sci. **36** (1950), 210-216 [reprinted in *Gabor Szegö: Collected Papers, vol. 3: 1945-1972*, R. Askey, editor, Birkhäuser, Boston, 1982, pp. 187-193, with comments by R. Askey on p. 194; see also [43] for corrections to Sections III and IV].

42. G. Szegö, Inequalities for certain eigenvalues of a membrane of given area, J. Rational Mech. Anal. **3** (1954), 343-356 [reprinted in *Gabor Szegö: Collected Papers, vol. 3: 1945-1972*, R. Askey, editor, Birkhäuser, Boston, 1982, pp. 373-386, with comments by R. Askey on p. 387].

43. G. Szegö, Note to my paper "On membranes and plates", Proc. Nat. Acad. Sci. **44** (1958), 314-316 [reprinted in *Gabor Szegö: Collected Papers, vol. 3: 1945-1972*, R. Askey, editor, Birkhäuser, Boston, 1982, pp. 481-483].

44. G. Talenti, On the first eigenvalue of the clamped plate, Ann. Mat. Pura Appl. (Ser. 4) **129** (1981), 265-280.

45. C.J. Thompson, On the ratio of consecutive eigenvalues in n-dimensions, Stud. Appl. Math. **48** (1969), 281-283.

46. H.F. Weinberger, An isoperimetric inequality for the n-dimensional free membrane problem, J. Rational Mech. Anal. **5** (1956), 633-636.

47. S.A. Wolf and J.B. Keller, Range of the first two eigenvalues of the Laplacian, Proc. Royal Soc. London **A 447** (1994), 397-412.

48. H.C. Yang, Estimates of the difference between consecutive eigenvalues, 1995 preprint (revision of International Centre for Theoretical Physics preprint IC/91/60, Trieste, Italy, April, 1991).

49. S.-T. Yau, Problem section, *Seminar on Differential Geometry*, S.-T. Yau, editor, Annals of Mathematics Studies, Number **102**, Princeton University Press, Princeton, 1982, pp. 669-706 [reprinted as pp. 277-314 of [40]].

50. S.-T. Yau, *Nonlinear Analysis in Geometry*, L'Enseignement Mathématique, Universitè de Genève, Genève, 1986 [reprinted as pp. 315-364 of [40]].

51. S.-T. Yau, Open problems in geometry, *Differential Geometry: Partial Differential Equations on Manifolds*, Proceedings of Symposia in Pure Mathematics, vol. **54**, part 1, R. Greene and S.-T. Yau, editors, Amer-

28

ican Mathematical Society, Providence, Rhode Island, 1993, pp. 1-28 [reprinted as pp. 365-409 of [40]].

ON AN INEQUALITY OF S. BERNSTEIN AND THE GAUSS-LUCAS THEOREM

A. AZIZ and N.A. RATHER
Post-Graduate Department of Mathematics and Statistics
University of Kashmir
Hazratbal, Srinagar 19006
India

Abstract. In this paper we first present an interesting generalization of the Gauss-Lucas Theorem. Next we use this result as a basic tool to prove certain compact generalizations of the well-known inequalities of S. Bernstein, P. Erdös and P.D. Lax, N.C. Ankeny and T.J. Rivlin, and others.

1. Introduction and the Main Results

The relative position of the real zeros and the critical points of a real differentiable function $f(z)$ is described in the well-known Theorem of Rolle which states that between any two zeros of $f(z)$ lies at least one zero of its derivative $f'(z)$. However, Rolle's Theorem is not true, in general, for analytic functions of a complex variable, as is shown by the function $f(z) = e^{iz} - 1$, which vanishes at $z = 0$ and at $z = 2\pi$, but $f'(z) = ie^{iz}$ never vanishes. This leads to a question as to what analogues of Rolle's Theorem are valid for at least a restrictive class of analytic functions such as polynomials. As a first step in this direction, we have the following celebrated result known as the Gauss-Lucas Theorem.

Theorem A (Gauss-Lucas Theorem). *If all the zeros of the polynomial $P(z)$ of degree $n \geq 2$ lie in the disk $D = \{z : |z - c| \leq r\}$, then all the zeros of its derivative $P'(z)$ also lie in D.*

This theorem has been rather thoroughly investigated [6] and further developed in several ways. The aim of this paper is to present an interesting generalization of Theorem A, which has a variety of useful applications. In fact, we first prove

Theorem 1. *If all the zeros of a polynomial $P(z)$ of degree $n \geq 2$ lie in the disk $D = \{z : |z - c| \leq r\}$, then, for every real or complex number β with $|\beta| \leq 1$ and $R \geq 1$, all the zeros of the polynomial $P(Rz - c(R - 1)) - \beta P(z)$ also lie in D.*

Remark 1. For $\beta = 1$, it follows from Theorem 1 that, if all the zeros of $P(z)$ lie in the disk $D = \{z : |z - c| \leq r\}$, then all the zeros of its derivative

$$P'(z) = \lim_{R \to 1} \left\{ \frac{P(Rz - c(R - 1)) - P(z)}{(R - 1)(z - c)} \right\}$$

29

T.M. Rassias and H.M. Srivastava (eds.), Analytic and Geometric Inequalities and Applications, 29–35.
© 1999 *Kluwer Academic Publishers.*

also lie in D. This is precisely the conclusion of the Gauss-Lucas Theorem.

The following corollary, which plays a key rôle in the rest of the paper, immediately follows from Theorem 1 by taking $c = 0$.

Corollary 1. *If all the zeros of the nth degree polynomial $P(z)$ lie in $|z| \leq r$, then so do the zeros of $P(Rz) - \beta P(z)$ for every real or complex number β with $|\beta| \leq 1$ and $R > 1$.*

If $P(z)$ is a polynomial of degree n, then we have

$$\max_{|z|=1} |P'(z)| \leq n \max_{|z|=1} |P(z)| \tag{1}$$

and

$$\max_{|z|=R>1} |P(z)| \leq R^n \max_{|z|=1} |P(z)| . \tag{2}$$

Inequality (1) is a well-known theorem of S. Bernstein (see [7, p. 53]). Inequality (2) is a simple deduction from the maximum modulus principle (see [9, p. 346] or [8, Vol. I, p. 137]). In both (1) and (2), equality holds only for $P(z) = \alpha z^n$, $\alpha \neq 0$.

If we restrict ourselves to the class of polynomials having no zero in $|z| < 1$, then (1) and (2) can be sharpened. In fact, if $P(z) \neq 0$ in $|z| < 1$, then (1) and (2) can, respectively, be replaced by

$$\max_{|z|=1} |P'(z)| \leq \frac{n}{2} \max_{|z|=1} |P(z)| \tag{3}$$

and

$$\max_{|z|=R>1} |P(z)| \leq \left(\frac{R^n + 1}{2}\right) \max_{|z|=1} |P(z)| . \tag{4}$$

Inequality (3) was conjectured by Erdös and later verified by Lax [5] (see also [2]). Ankeny and Rivlin [3] used (3) to prove the inequality (4). Here we use Corollary 1 to present some interesting generalizations of the inequalities (1), (2), (3), and (4). The following result is a compact generalization of the inequalities (1) and (2).

Theorem 2. *If $P(z)$ is a polynomial of degree n, then, for every real or complex number β with $|\beta| \leq 1$ and $R \geq 1$,*

$$|P(Rz) - \beta P(z)| \leq |R^n - \beta| \, |z|^n \max_{|z|=1} |P(z)| \quad for \quad |z| \geq 1. \tag{5}$$

The result is best possible and equality in (5) holds for $P(z) = \alpha z^n$, $\alpha \neq 0$.

Remark 2. Dividing the two sides of (5) by $R - 1$ and taking the limits as $R \to 1$ with $\beta = 1$, we get the inequality (1). For $\beta = 0$, the inequality (5) reduces to (2).

Next we use Theorem 2 to prove the following interesting result.

Theorem 3. *If $P(z)$ is a polynomial of degree n and $Q(z) = z^n \overline{P(1/\overline{z})}$, then, for every real or complex number β with $|\beta| \leq 1$ and $R \geq 1$,*

$$|P(Rz) - \beta P(z)| + |Q(Rz) - \beta Q(z)|$$

$$\leq \{|R^n - \beta| |z|^n + |1 - \beta|\} \max_{|z|=1} |P(z)| \quad \text{for} \quad |z| \geq 1. \tag{6}$$

The result is sharp and equality in (6) *holds for $P(z) = z^n + 1$.*

Remark 3. For $\beta = 1$ and $|z| = 1$, the inequality (6) reduces to a result due to Aziz [1, Lemma 2]. Dividing the two sides of (6) by $R - 1$ with $\beta = 1$ and letting $R \to 1$, we obtain a result due to Govil and Rahman [4] (see also [1]).

Finally, we use Theorem 3 to prove the following interesting result which includes the inequalities (3) and (4) as its special cases.

Theorem 4. *If $P(z)$ is a polynomial of degree n which does not vanish in $|z| < 1$, then, for every real or complex number β with $|\beta| \leq 1$ and $R > 1$,*

$$|P(Rz) - \beta P(z)| \leq \left\{ \frac{|R^n - \beta| |z|^n + |1 - \beta|}{2} \right\} \max_{|z|=1} |P(z)| \quad \text{for} \quad |z| \geq 1. \tag{7}$$

The result is best possible and equality in (7) *holds for $P(z) = z^n + 1$.*

Remark 4. Dividing the two sides of (7) by $R-1$ and taking the limits as $R \to 1$ with $\beta = 1$, we obtain (3). For $\beta = 0$, the inequality (7) reduces to the inequality (4) for $|z| = 1$.

2. A Set of Lemmas

For the proofs of Theorems 1 to 4, we need the following lemmas.

Lemma 1. *If $P(z)$ is a polynomial of degree n having all its zeros in $|z| \leq k$ ($k \leq 1$), then, for every $R > 1$,*

$$|P(Rz)| \geq \left(\frac{R+k}{1+k} \right)^n |P(z)| \qquad (|z| = 1).$$

Proof of Lemma 1. Since all the zeros of the polynomial $P(z)$ lie in $|z| \leq k \leq 1$, we write

$$P(z) = c \prod_{j=1}^{n} (z - r_j e^{i\theta_j}) \qquad (r_j \leq k; \ j = 1, \cdots, n),$$

so that, for each θ ($0 \leq \theta < 2\pi$) and each $R > 1$, it can be easily verified that

$$\left| \frac{P(Re^{i\theta})}{P(e^{i\theta})} \right| = \prod_{j=1}^{n} \left| \frac{Re^{i\theta} - r_j e^{i\theta_j}}{e^{i\theta} - r_j e^{i\theta_j}} \right|$$

$$= \prod_{j=1}^{n} \left(\frac{R + r_j}{1 + r_j} \right) \geq \prod_{j=1}^{n} \left(\frac{R + k}{1 + k} \right)$$

$$= \left(\frac{R+k}{1+k} \right)^n.$$

This implies that

$$|P(Re^{i\theta})| \geq \left(\frac{R+k}{1+k}\right)^n |P(e^{i\theta})| \qquad (R > 1; \ 0 \leq \theta \leq 2\pi).$$

Hence

$$|P(Rz)| \geq \left(\frac{R+k}{1+k}\right)^n |P(z)| \qquad (R > 1; \ |z| = 1).$$

This completes the proof of Lemma 1.

Next we need

Lemma 2. *If $P(z)$ is a polynomial of degree n which does not vanish in $|z| < 1$, then, for every real or complex number β with $|\beta| \leq 1$ and $R \geq 1$,*

$$|P(Rz) - \beta P(z)| \leq |Q(Rz) - \beta Q(z)| \qquad (|z| \geq 1),$$

where $Q(z) = z^n \overline{P(1/\overline{z})}$.

Proof of Lemma 2. The result is obvious for $R = 1$. Henceforth we assume that $R > 1$. Since all the zeros of the polynomial $P(z)$ lie in $|z| \geq 1$, therefore, for every complex number λ with $|\lambda| > 1$, the polynomial $f(z) = P(z) - \lambda Q(z)$, where $Q(z) = z^n \overline{P(1/\overline{z})}$, has all its zeros in $|z| \leq 1$. Applying Lemma 1 to the polynomial $f(z)$ with $k = 1$, we get

$$|f(Rz)| \geq \left(\frac{R+1}{2}\right)^n |f(z)| \qquad (|z| = 1; \ R > 1). \tag{8}$$

Clearly, $f(Re^{i\theta}) \neq 0$ for every $R > 1$ and $0 \leq \theta < 2\pi$, which implies that $|f(Rz)| > 0$ for $|z| = 1$ and $R > 1$. Now, for points $e^{i\theta}$ ($0 \leq \theta < 2\pi$), which are not the zeros of $f(z)$, we find from (8) that

$$|f(Re^{i\theta})| > |f(e^{i\theta})| \quad \text{for every } R > 1. \tag{9}$$

Since, by (8), the inequality (9) is obviously true for those points $e^{i\theta}$ ($0 \leq \theta < 2\pi$) which are the zeros of $f(z)$, it follows that

$$|f(z)| < |f(Rz)| \qquad (|z| = 1; \ R > 1).$$

If β is any real or complex number with $|\beta| \leq 1$, then, by using Rouche's theorem and noting that all the zeros of $f(Rz)$ lie in $|z| \leq (1/R) < 1$, we conclude that the polynomial

$$g(z) = f(Rz) - \beta f(z) = P(Rz) - \beta P(z) - \lambda \{Q(Rz) - \beta Q(z)\} \tag{10}$$

has all its zeros in $|z| < 1$ for every λ with $|\lambda| > 1$ and $R > 1$. This implies that

$$|P(Rz) - \beta P(z)| \leq |Q(Rz) - \beta Q(z)| \qquad (|z| \geq 1; \ R > 1). \tag{11}$$

If the inequality (11) is not true, then there is a point $z = z_0$ with $|z_0| \geq 1$ such that

$$|P(Rz_0) - \beta P(z_0)| > |Q(Rz_0) - \beta Q(z_0)| \qquad (R > 1).$$

Since all the zeros of $Q(z)$ lie in $|z| \leq 1$, it follows (just as in the case of $f(z)$) that all the zeros of $Q(Rz) - \beta Q(z)$ lie in $|z| < 1$ for every β with $|\beta| \leq 1$ and $R > 1$. Hence $Q(Rz_0) - \beta Q(z_0) \neq 0$ with $|z_0| \geq 1$. We take

$$\lambda = \frac{P(Rz_0) - \beta P(z_0)}{Q(Rz_0) - \beta Q(z_0)},$$

so that λ is a well-defined real or complex number with $|\lambda| > 1$, and with this choice of λ, from (10) we get

$$g(z_0) = 0, \quad \text{where } |z_0| \geq 1.$$

This is clearly a contradiction to the fact that all the zeros of $g(z)$ lie in $|z| < 1$. Thus

$$|P(Rz) - \beta P(z)| \leq |Q(Rz) - \beta Q(z)| \qquad (|z| \geq 1; \ R > 1),$$

which proves Lemma 2.

3. Proofs of Theorems 1 to 4

Proof of Theorem 1. The result is obviously true for $R = 1$. Henceforth we assume that $R > 1$. Since all the zeros of the polynomial $P(z)$ of degree $n \geq 2$ lie in disk $D = \{z : |z - c| \leq r\}$, therefore, all the zeros of the polynomial $G(z) = P(rz + c)$ lie in $|z| \leq 1$. Using Lemma 1 with $k = 1$, we have, for every $R > 1$,

$$|G(z)| < |G(Rz)| \qquad (|z| = 1).$$

Since all the zeros of $G(Rz)$ lie in $|z| \leq (1/R) < 1$, therefore, by the maximum modulus principle, it follows that

$$|G(z)| < |G(Rz)| \qquad (|z| \geq 1; \ R > 1).$$

This implies that, for every real or complex number β with $|\beta| \leq 1$ and $R > 1$, the polynomial $G(Rz) - \beta G(z)$ has all its zeros in $|z| < 1$. Because, if this is not true, then there is some point $z = z_0$ with $|z_0| \geq 1$ such that $G(Rz_0) - \beta G(z_0) = 0$. This gives

$$|G(Rz_0)| = |\beta G(z_0)| \leq |G(z_0)| \qquad (|z_0| \geq 1; \ R > 1),$$

which clearly contradicts (12). Hence, for every real or complex number β with $|\beta| \leq 1$ and $R > 1$, the polynomial

$$G(Rz) - \beta G(z) = P(Rrz + c) - \beta P(rz + c)$$

has all its zeros in $|z| < 1$. Replacing z by $(z - c)/r$, we conclude that, for every real or complex number β with $|\beta| \leq 1$ and $R > 1$, the polynomial

$$P(Rz - c(R - 1)) - \beta P(z)$$

has all its zeros in the disk $D = \{z : |z - c| < r\}$. This proves the desired result.

Proof of Theorem 2. For $R = 1$, the result is obvious. Henceforth we assume that $R > 1$. If

$$\max_{|z|=1} |P(z)| = M,$$

then $|P(z)| < M$ for $|z| = 1$. Using Rouche's theorem, it follows that all the zeros of the polynomial $F(z) = P(z) + \lambda z^n M$ lie in $|z| < 1$ for every λ with $|\lambda| > 1$. If β is any complex number with $|\beta| \leq 1$, then an application of Corollary 1 with $r = 1$ shows that all the zeros of the polynomial

$$F(Rz) - \beta F(z) = P(Rz) - \beta P(z) + \lambda(R^n - \beta) z^n M \tag{13}$$

also lie in $|z| < 1$ for every $R > 1$ and $|\lambda| > 1$. This implies that

$$|P(Rz) - \beta P(z)| \leq |R^n - \beta| |z|^n M \qquad (|z| \geq 1; \ R > 1). \tag{14}$$

For, if (14) is not true, then there is some point $z = z_0$ with $|z_0| \geq 1$ such that

$$|P(Rz_0) - \beta P(z_0)| > |R^n - \beta| |z_0|^n M \qquad (R > 1).$$

If we take

$$\lambda = -\frac{P(Rz_0) - \beta P(z_0)}{(R^n - \beta) z_0^n M},$$

then $|\lambda| > 1$, and we have $F(Rz_0) - \beta F(z_0) = 0$ with $|z_0| \geq 1$ and $R > 1$, which is clearly a contradiction to (13). Hence (14) is established and the proof of Theorem 2 is thus completed.

Proof of Theorem 3. Let $M = \max_{|z|=1} |P(z)|$, then $|P(z)| \leq M$ for $|z| = 1$. Using Rouche's theorem, it follows that, for every real or complex number α with $|\alpha| > 1$, the polynomial $F(z) = P(z) + \alpha M$ does not vanish in $|z| < 1$. Applying Lemma 2 to the polynomial $F(z)$, we get, for every real or complex number β with $|\beta| \leq 1$,

$$\begin{aligned} |P(Rz) - \beta P(z) + \alpha(1 - \beta)M| \\ \leq |Q(Rz) - \beta Q(z) + \alpha(R^n - \beta)z^n M| \qquad (|z| \geq 1; \ R > 1), \end{aligned} \tag{15}$$

where $Q(z) = z^n \overline{P(1/\bar{z})}$. Choosing the argument of α in the right side of (15) such that

$$|Q(Rz) - \beta Q(z) + \alpha(R^n - \beta)z^n M|$$

$$= |\alpha| |R^n - \beta| |z|^n M - |Q(Rz) - \beta Q(z)|,$$

(which is possible by Theorem 10), we obtain, for every β with $|\beta| \leq 1$ and $R > 1$,

$$|P(Rz) - \beta P(z)| - |\alpha|\,|1 - \beta|\,M$$

$$\leq |\alpha|\,|R^n - \beta|\,|z|^n - |Q(Rz) - \beta Q(z)| \qquad (|z| \geq 1)\,.$$

Equivalently, for $|z| \geq 1$ and $R > 1$,

$$|P(Rz) - \beta P(z)| + |Q(Rz) - \beta Q(z)| \leq |\alpha|\,\{|R^n - \beta|\,|z|^n + |1 - \beta|\}\,M.$$

Finally, letting $|\alpha| \to 1$, we get the desired result and the proof of Theorem 3 is thus completed.

Proof of Theorem 4. Combining Lemma 2 with the conclusion of Theorem 3, the desired result follows immediately.

Remark 5. A polynomial $P(z)$ of degree n is said to be self-inverse if $P(z) = uQ(z)$, where $Q(z) = z^n\,\overline{P(1/\bar{z})}$ and $|u| = 1$. It can now be easily seen that Theorem 4 holds for self-inverse polynomials as well.

References

[1] A. Aziz, Inequalities for the derivative of a polynomial, *Proc. Amer. Math. Soc.* **89**(1983), 259–266.

[2] A. Aziz and Q.G. Mohammad, Simple proof of a theorem of Erdös and Lax, *Proc. Amer. Math. Soc.* **80**(1980), 119–122.

[3] N.C. Ankeny and T.J. Rivlin, On a theorem of S. Bernstein, *Pacific J. Math.* **5**(1955), 849–852.

[4] N.K. Govil and Q.I. Rahman, Functions of exponential type not vanishing in a half plane and related polynomials, *Trans. Amer. Math. Soc.* **137**(1969), 501–517.

[5] P.D. Lax, Proof of a conjecture of P. Erdös on the derivative of a polynomial, *Bull. Amer. Math. Soc.* (*N.S.*) **50**(1944), 509–513.

[6] M. Marden, *Geometry of Polynomials*, Second edition, Math. Surveys **3**, Amer. Math. Soc., Providence, Rhode Island, 1966.

[7] G.V. Milovanović, D.S. Mitrinović, and Th. M. Rassias, *Topics in Polynomials: Extremal Properties, Inequalities, Zeros*, World Scientific Publishing Company, Singapore, 1994.

[8] G. Pólya and G. Szegö, *Aufgaben and Lehrsätze aus der Analysis*, Springer-Verlag, Berlin, 1925.

[9] M. Riesz, Über einen Satz des Herrn Serge Bernstein, *Acta Math.* **40**(1916), 337–347.

SECOND ORDER OPIAL INEQUALITIES IN $\mathcal{L}^p$ SPACES AND APPLICATIONS

RICHARD BROWN

Mathematics Department, University of Alabama,
Tuscaloosa, AL 35487.

VICTOR BURENKOV

School of Mathematics, University of Wales Cardiff,
23 Senghennydd Road, Cardiff CF B4YH, United Kingdom.

STEVE CLARK

Mathematics Department, University of Missouri Rolla,
Rolla, Missouri 65409.

DON HINTON

Department of Mathematics, University of Tennessee,
Knoxville, TN 37996.

Abstract. We determine the best constant and extremals of a second order Opial-type inequality in L^p spaces under zero boundary conditions. For $p = 2$ we investigate other boundary conditions as well. We apply our results to determine lower bounds for the disconjugacy interval for fourth-order differential equations.

1. Introduction

For $1 < p \leq \infty$ we consider the problem of determining the best constant and extremals of the second order Opial-type inequality

$$(1.1) \qquad \int_a^b |yy'|\, dx \leq K \left(\int_a^b |y''|^p\, dx \right)^{2/p}$$

for $y \in BC_0$ where

$$BC_0 = \{y : y \in D,\ y(a) = y'(a) = y(b) = y'(b) = 0\},$$
$$D = \{y : y \text{ is real},\ y, y' \in AC[a,b],\ y'' \in \mathcal{L}^p[a,b]\}.$$

Here $AC[a,b]$ and $\mathcal{L}^p[a,b]$ are respectively the absolutely continuous and Lebesgue p integrable functions on $[a,b]$.

It will be convenient to reduce the Opial inequality (1.1) to the solution of a certain minimization problem since the best constant K of (1.1) is just $K_0(a,b)^{-1}$ where

$$(1.2) \qquad K_0(a,b) := \inf\{J(y) : y \in BC_0\}$$

1991 *Mathematics Subject Classification.* Primary: 26D10; Secondary: 34C10, 34L05.
Key words and phrases. Opial inequality, extremals, eigenvalue problems, disconjugacy.

T.M. Rassias and H.M. Srivastava (eds.), Analytic and Geometric Inequalities and Applications, 37–52.
© 1999 *Kluwer Academic Publishers.*

38

and

$$J(y) = \frac{\left(\int_a^b |y''|^p \, dx\right)^{2/p}}{\int_a^b |yy'| \, dx}.$$

Moreover $y_0 \in BC_0$ clearly gives equality in (1.1) if and only if it is an extremal of (1.2).

For $p = 2$ second order Opial inequalities an as well as the nth-order generalization where $y^{(n)}$ replaces y'' in (1.1) or (1.2) have been considered by FitzGerald [2]. However, there is a point in FitzGerald's proof with which we have difficulty. It deals with an approximation procedure. In FitzGerald's proof, as well as ours below, a critical step is to reduce the problem to the analysis of functions for which y' can change sign only once in the interval $[a, b]$. To achieve this, FitzGerald starts with a polynomial $y \in BC_0$ which is an approximate extremal and uses y to construct a function $\hat{y}$ defined by

$$\hat{y}(x) = \begin{cases} \int_a^x |y'| \, dx, & a \leq x \leq c, \\ \int_x^b |y'| \, dx, & c \leq x \leq b, \end{cases}$$

where c is chosen so that $\int_a^c |y'| dx = \int_c^b |y'| dx$, Then $\hat{y}(a) = \hat{y}'(a) = \hat{y}(b) = \hat{y}'(b) = 0$, $|\hat{y}'(x)| = |y'(x)|$ a.e., and $\hat{y}(x) \geq |y(x)|$. However $\hat{y}'$ may be discontinuous at $x = c$ in which case $\hat{y} \notin BC_0$. The proof in [2] then continues by modifying $\hat{y}$ to a function $\tilde{y} \in BC_0$. We discuss now what we see as a difficulty in the procedure by which this is done. It is sufficient to take $[a, b] = [-1, 1]$, $c = 0$, and to assume $\hat{y}'(0^+) - \hat{y}'(0^-) = k \neq 0$. Suppose there is a sequence $\{z_\epsilon\}$ in BC_0 such that

$$\lim_{\epsilon \to 0} \|\hat{y}' - z_\epsilon'\|_2 = 0, \quad \lim_{\epsilon \to 0} \|z_\epsilon''\|_2 = M$$

where the norm is in $\mathcal{L}^2(-1, 1)$, and

$$M = \int_{-1}^{0^-} (\hat{y}'')^2 \, dx + \int_{0^+}^1 (\hat{y}'')^2 \, dx = \int_{-1}^{0^-} (y'')^2 \, dx + \int_{0^+}^1 (y'')^2 \, dx.$$

Since $\|z_\epsilon''\|_2$ is bounded, we have from

$$|z_\epsilon'(t) - z_\epsilon'(s)| \leq \left| \int_s^t z_\epsilon'' \, dx \right| \leq \left(\int_s^t (z_\epsilon'')^2 \, dx \right)^{\frac{1}{2}} (t - s)^{\frac{1}{2}},$$

that $\{z_\epsilon'\}$ is equicontinous and uniformly bounded (set $s = -1$). From the convergence of $\{z_\epsilon'\}$ in $\mathcal{L}^2(-1, 1)$ to $\hat{y}'$, we conclude that $\{z_\epsilon'\}$ converges uniformly to $\hat{y}'$. The uniform convergence of the continuous functions z_ϵ' to $\hat{y}'$ implies $\hat{y}'$ is continuous which is contrary to our assumption. The argument shows that if $\hat{y}'$ is "smoothed" in the sense of $\|\hat{y}' - z_\epsilon'\|_2 \to 0$ as $\epsilon \to 0$ by functions $z_\epsilon \in BC_0$, then $\|z_\epsilon''\|_2 \to \infty$ as $\epsilon \to 0$. By a similar argument it follows that if $\hat{y}$ is "smoothed" in the sense of $\|\hat{y} - z_\epsilon\|_2 \to 0$ as $\epsilon \to 0$ by functions $z_\epsilon \in BC_0$, then $\|z_\epsilon''\|_2 \to \infty$ as $\epsilon \to 0$.

In this paper we present an alternate construction which avoids this problem and works for all $1 < p \leq \infty$. As noted by FitzGerald, it is sufficient to make a convenient choice of $[a, b]$ since by translation and scaling the general problem is reduced to a specific case. It can be shown by a scaling argument that

$$K_0(a, b) = \frac{K_0}{(b-a)^{4-2/p}}$$

where $K_0 = K_0(0, 1)$. Consequently, we can take $[a, b] = [0, 1]$ throughout. In the first five sections of the paper we take $p = 2$ since this case has valuable applications and the proofs are paradigmatic for general p. Section 2 derives necessary conditions for an extremal of (1.2) and shows that an extremal exists for (1.2) as well as for more general problems. In section 3 we prove that $K_0 = 192$ and that there exists a family of extremals consisting of nonzero scalar multiples of a cubic spline. These results agree with FitzGerald. In section 4 we point out several problems which can be solved by similar methods and give extremals and best constants for them. In section 5 we apply the Opial inequalities we have derived in two ways. The first application gives optimal lower bounds for the smallest eigenvalue for the transversal vibrations of a nonhomogeneous rod under various end conditions; in the second, we determine lower bounds on the length of the disconjugacy interval for fourth-order differential equations. Finally, in section 6 we show how the arguments solving (1.1) or (1.2) for $p = 2$ will apply with minor changes to general p. This version of the second order Opial inequality is believed to be new.

2. Existence and Properties of Extremals when $p = 2$

First we see what necessary conditions an extremal of (1.2) must satisfy. If y_0 is an extremal, then computing for $h \in BC_0$,

$$\frac{d}{d\epsilon} J(y_0 + \epsilon h)|_{\epsilon=0} = 0,$$

yields that,

$$(2.1) \qquad \int_0^1 2y_0'' h'' \, dx \int_0^1 |y_0' y_0| \, dx - \int_0^1 (y_0'')^2 \, dx \int_0^1 [\text{sgn}(y_0 y_0')](y_0 h)' \, dx = 0$$

where $\text{sgn}(\cdot)$ is the sign function. Using the fact that

$$\int_0^1 (y_0'')^2 \, dx = K_0 \int_0^1 |y_0 y_0'| \, dx,$$

(2.1) reduces to

$$(2.2) \qquad \int_0^1 2y_0'' h'' \, dx - K_0 \int_0^1 [\text{sgn}(y_0 y_0')](y_0 h)' \, dx = 0.$$

On an interval (α, β) on which $y_0 y_0'$ is of constant sign, we may use standard calculus of variation arguments for smooth test functions $h(x)$ with support $[\alpha, \beta]$ to conclude that $\int_\alpha^\beta y_0'' h'' \, dx = \int_\alpha^\beta y_0^{(iv)} h \, dx = 0$. From this we have $y^{(iv)}(x) = 0$ a.e. on (α, β), i.e., $y(x)$ is a cubic polynomial on (α, β). To see what natural boundary conditions to expect, suppose $y_0 y_0'$ is of constant sign on some $(0, \delta)$. With smooth test functions $h(x)$ with support $[0, \delta]$, two integration by parts, and using $y^{(iv)}(x) = 0$ a.e. gives from (2.2) that

$$(2.3) \qquad -y_0''(0)h'(0) + y_0'''(0)h(0) + (K_0/2)\text{sgn}(y_0 y_0')(0)[y_0(0)h(0)] = 0.$$

Moreover if $y_0 y_0'$ is of constant sign on some $(1 - \delta, 1)$, then we conclude that (2.3) holds with function evaluations at 1 instead of 0. In our particular minimization problem $y_0, h \in BC_0$, so that no natural boundary conditions are imposed at 0. However in section 4 we will consider more general problems which will require examination of (2.3) to determine the natural boundary conditions satisfied by the extremal y_0.

Theorem 1. *If $p = 2$ the infimum K_0 of (1.2) is positive, and an extremal y_0 exists.*

Proof. Let $\{y_n\}$ be a sequence in BC_0 such that $J(y_n) \to K_0$ as $n \to \infty$. Without loss of generality, we may assume $\int_0^1 |y_n y_n'| \, dx = 1$. Then since $\{\|y_n''\|_2\}$ is a bounded sequence, $\{y_n''\}$ has a weakly convergent subsequence in $\mathcal{L}^2(0, 1)$, say $y_n'' \rightharpoonup g$ as $n \to \infty$. From $y_n'(t) = \int_0^t y_n'' \, dx$, we have that

$$(2.4) \qquad |y_n'(t) - y_n'(s)|^2 = \left| \int_s^t y_n'' \, dx \right|^2 \le |t - s| \int_0^1 (y_n'')^2 \, dx.$$

Inequality (2.4) implies $\{y_n'\}$ is equicontinuous and uniformly bounded (set $s = 0$). Thus $\{y_n'\}$ has a uniformly convergent subsequence; without loss, assume $y_n' \to h$ uniformly as $n \to \infty$. Since $y_n'' \rightharpoonup g$ as $n \to \infty$, $y_n' \to h$ uniformly as $n \to \infty$, and $y_n'(0) = 0$ for all n, we conclude that $h(t) = \int_0^t g(x) \, dx$; furthermore $y_n'(1) = 0$ implies that $h(1) = 0$. Set $y(t) = \int_0^t h(x) \, dx$. Then $y(0) = 0$, $y'(0) = h(0) = 0$, $y'(1) = h(1) = 0$, and

$$y(1) = \int_0^1 h \, dx = \lim_{n \to \infty} \int_0^1 y_n' \, dx = \lim_{n \to \infty} [y_n(1) - y_n(0)] = 0.$$

Thus $y \in BC_0$ with $y'' = h' = g$. The uniform convergence of y_n' to h implies $\{y_n\}$ conveges uniformly to $y(x)$; hence

$$\int_0^1 |yy'| \, dx = \lim_{n \to \infty} \int_0^1 |y_n y_n'| \, dx = 1.$$

The weak convergence of y_n'' to $g = y''$ implies that

$$\|g\|_2^2 = \|y''\|_2^2 \le \liminf_{n \to \infty} \|y_n''\|_2^2 = K_0.$$

On the other hand $\|y''\|_2^2 \ge K_0$ since $y \in BC_0$ and $\int_0^1 |yy'| \, dx = 1$. Thus $\|y''\|_2^2 = K_0$ and y is an extremal for (1.1). Finally, since $\int_0^1 |yy'| \, dx = 1$, y is nonzero on a set of positive measure. The initial conditions $y(0) = y'(0) = 0$ imply that y cannot be a nontrivial linear function. Therefore $K_0 > 0$.

An examination of this proof shows that it can be adopted to the more general functional

$$J_{p,w}(y) = \frac{\int_0^1 p(x) y''(x)^2 \, dx}{\int_0^1 w(x) |(yy')(x)| \, dx}$$

where $p(x) > 0, w(x) > 0$ are measurable functions such that $\int_0^1 \frac{dx}{p(x)} < \infty$ and $\int_0^1 w(x) \, dx < \infty$. The boundary conditions $y(0) = y'(0) = y(1) = y'(1) = 0$ may also be replaced by other boundary conditions. We return to this point in section 4.

3. Determination of K_0 for $p = 2$

First we prove a lemma on polynomial approximation of an extremal using the density of the polynomials in $\mathcal{L}^2(0, 1)$.

Lemma 1. *Let $y \in BC_0$ and let $\{p_n\}$ be a sequence of polynomials such that $\|y'' - p_n\|_2 \to 0$ as $n \to \infty$. Define*

$$(3.1) \qquad q_n(x) := \int_0^x (x - s)p_n(s)\, ds + a_n x^2 + b_n x^3,$$

where a_n, b_n are determined by the requirement $q_n(1) = q_n'(1) = 0$. Then $q_n \in BC_0$, $q_n \to y$ and $q_n' \to y'$ uniformly as $n \to \infty$, and $\|y'' - q_n''\|_2 \to 0$ as $n \to \infty$.

Proof. Since $y \in BC_0$ we have that

$$y(x) = \int_0^x (x - s)y''(s)\, ds, \; y(1) = \int_0^1 (1 - s)y''(s)\, ds\ = 0,$$

$$y'(1) = \int_0^1 y''(s)\, ds = 0.$$

Thus a_n, b_n satisfy the equations,

$$(3.2) \qquad q_n(1) = 0 = \int_0^1 (1 - s)[p_n(s) - y''(s)]\, ds + a_n + b_n,$$

$$(3.3) \qquad q_n'(1) = 0 = \int_0^1 [p_n(s) - y''(s)]\, ds + 2a_n + 3b_n.$$

From $\|p_n - y''\|_2 \to 0$ as $n \to \infty$, it is clear from (3.2) and (3.3) that a_n, $b_n \to 0$ as $n \to \infty$. The uniform convergence of q_n to y and q_n' to y' now follows from the equations

$$q_n(x) - y(x) = \int_0^x (x - s)[p_n(s) - y''(s)]\, ds + a_n x^2 + b_n x^3,$$

$$q_n'(x) - y'(x) = \int_0^x [p_n(s) - y''(s)]\, ds + 2a_n x + 3b_n x^2.$$

For $y \in BC_0$, we define the following set:

$$\mathcal{C}(y) := \{s \in (0, 1) : y'(s) = 0 \text{ and } y' \text{ changes sign at } s\}.$$

and let $|\mathcal{C}(y)|$ denote the cardinality of $\mathcal{C}(y)$.

Lemma 2. *Suppose $p \in BC_0$ is a nonzero polynomial. Let*

$$\mathcal{R}(p) := \{y \in BC_0 : J(y) \leq J(p), |\mathcal{C}(y)| \leq |\mathcal{C}(p)|\}.$$

Then there exists $y \in \mathcal{R}(p)$ such that $\mathcal{C}(y)$ consists of a single point.

42

Proof. Since $p \in \mathcal{R}(p)$ and $\mathcal{C}(p)$ is finite, there is a member $\hat{y} \in \mathcal{R}(p)$ such that the number of elements in $\mathcal{C}(\hat{y})$ is minimal. Suppose $\mathcal{C}(\hat{y})$ contains two elements s_1, s_2 which we take to be consecutive with $s_1 < s_2$. Since $\mathcal{C}(\hat{y}) = \mathcal{C}(-\hat{y})$, we can assume s_1 is the first positive member of $\mathcal{C}(\hat{y})$ and that $\hat{y}(x) > 0$ on $(0, s_1)$. Thus $\hat{y}'$ switches from positive to negative at s_1.

Case (i): $\hat{y}(s_2) \geq 0$. Since $\hat{y}'$ switches from negative to positive at s_2 and $y(s_2) \geq 0$, there is a first point $s > s_2$ at which $s \in \mathcal{C}(\hat{y})$ (since $\hat{y}(1) = 0$); call it s_3. Define the function z on $[s_1, s_3]$ by:

$$(3.4) \qquad z(x) := \hat{y}(s_1) + \hat{y}(s_3) - \hat{y}(-x + s_1 + s_3).$$

Then $z(s_1) = \hat{y}(s_1)$, $z(s_3) = \hat{y}(s_3)$, and $\hat{y}(x) \leq \max[\hat{y}(s_1), \hat{y}(s_3)]$ on $[s_1, s_3]$ implies that $z(x) \geq 0$ on $[s_1, s_3]$. Also $z'(s_1) = \hat{y}'(s_3) = 0$, $z'(s_3) = \hat{y}'(s_1) = 0$. The graph of z is that of $\hat{y}$ on $[s_1, s_3]$ rotated about the midpoint, then flipped over, and finally translated vertically so as to leave initial and terminal values unchanged. Note also that $z'(s^*) = 0$ and z is nondecreasing and nonincreasing respectively on (s_1, s^*) and (s^*, s_3) where $s^* = s_1 + s_3 - s_2$. We now show that

$$(3.5) \qquad \int_{s_1}^{s_3} (z'')^2 \, dx = \int_{s_1}^{s_3} (\hat{y}'')^2 \, dx,$$

$$(3.6) \qquad \int_{s_1}^{s_3} |zz'| \, dx > \int_{s_1}^{s_3} |\hat{y}\hat{y}'| \, dx.$$

Equation (3.5) is is a consequence of the change of variable $u = -x + s_1 + s_3$ since

$$\int_{s_1}^{s_3} z''(x)^2 \, dx = \int_{s_1}^{s_3} \hat{y}''(-x + s_1 + s_3)^2 \, dx = \int_{s_1}^{s_3} \hat{y}''(u)^2 \, du.$$

For the first integral of (3.6), using $z(x) \geq 0$, we have,

$$(3.7) \qquad \int_{s_1}^{s_3} |z'(x)z(x)| \, dx = \int_{s_1}^{s_3} |\hat{y}'(u)|[\hat{y}(s_1) + \hat{y}(s_3) - \hat{y}(u)] \, du$$

$$= \int_{s_1}^{s_2} -\hat{y}'(u)[\hat{y}(s_1) + \hat{y}(s_3) - \hat{y}(u)] \, du$$

$$+ \int_{s_2}^{s_3} \hat{y}'(u)[\hat{y}(s_1) + \hat{y}(s_3) - \hat{y}(u)] \, du$$

$$= [\hat{y}(s_1) + \hat{y}(s_3)][\hat{y}(s_1) - 2\hat{y}(s_2) + \hat{y}(s_3)]$$

$$+ \frac{1}{2}[-\hat{y}^2(s_1) + 2\hat{y}^2(s_2) - \hat{y}^2(s_3)]$$

after simplification. The second integral is

$$\int_{s_1}^{s_3} |\hat{y}'(u)\hat{y}(u)| \, du = \int_{s_1}^{s_2} -\hat{y}'(u)\hat{y}(u) \, du + \int_{s_2}^{s_3} \hat{y}'(u)\hat{y}(u) \, du$$

$$= \frac{1}{2}[\hat{y}^2(s_1) - 2\hat{y}^2(s_2)^2 + \hat{y}^2(s_3)].$$

Using these two evaluations and simplifying yields that

$$\int_{s_1}^{s_3} |z'(x)z(x)|\, dx - \int_{s_1}^{s_3} |\hat{y}'(u)\hat{y}(u)|\, du$$
$$= 2[\hat{y}(s_1) - \hat{y}(s_2)][\hat{y}(s_3) - \hat{y}(s_2)] > 0;$$

hence (3.6) holds. Then the function $\tilde{y}$ defined by :

$$(3.8) \qquad \tilde{y}(x) = \begin{cases} \hat{y}(x), & x \notin (s_1, s_3), \\ z(x), & x \in (s_1, s_3), \end{cases}$$

belongs to $\mathcal{R}(p)$ and $\mathcal{C}(\tilde{y}) = s^* \cup \mathcal{C}(\hat{y}) \setminus \{s_1, s_2, s_3\}$ since $z'(x) = -\hat{y}'(-x+s_1+s_3)$ will be nonnegative to the right of s_1 and nonpositive to the left of s_3. This contradicts the minimality of $\mathcal{C}(\hat{y})$.

Case (ii): $\hat{y}(s_2) < 0$. Let $s_3 = 1$ if $\mathcal{C}(\hat{y})$ contains no member to the right of s_2; otherwise let s_3 be the first member of $\mathcal{C}(\hat{y}) > s_2$.

Subcase (1): $\hat{y}(s_3) \geq 0$. Define z on $[s_1, s_3]$ as in (3.4). Then as in Case (i), (3.5) holds. Also as in Case (i), $z(x) \geq 0$, so the calculation (3.7) is still valid. For the calculation of $\int_{s_1}^{s_3} |\hat{y}(u)\hat{y}'(u)|\, du$, define the intermediate points α, β, $s_1 < \alpha < s_2 < \beta \leq s_3$ by $\hat{y}(\alpha) = \hat{y}(\beta) = 0$ and $\hat{y}(x) < 0$ on (α, β). Then

$$(3.9) \qquad \int_{s_1}^{s_3} |\hat{y}(u)\hat{y}'(u)|\, du = \int_{s_1}^{\alpha} -\hat{y}(u)\hat{y}'(u)\, du + \int_{\alpha}^{s_2} \hat{y}(u)\hat{y}'(u)\, du$$
$$- \int_{s_2}^{\beta} \hat{y}(u)\hat{y}'(u)\, du + \int_{\beta}^{s_3} \hat{y}(u)\hat{y}'(u)\, du$$
$$= \frac{1}{2}[\hat{y}^2(s_1) + 2\hat{y}^2(s_2) + \hat{y}^2(s_3)].$$

From (3.7) and (3.9) we compute after simplifying that

$$(3.10) \qquad \int_{s_1}^{s_3} |z'(x)z(x)|\, dx - \int_{s_1}^{s_3} |\hat{y}(u)\hat{y}'(u)|\, du$$
$$= 2(-\hat{y}(s_1)\hat{y}(s_2) + \hat{y}(s_1)\hat{y}(s_3) - \hat{y}(s_2)\hat{y}(s_3)).$$

The equation (3.10) is positive since $\hat{y}(s_1) > 0$, $\hat{y}(s_2) < 0$, and $\hat{y}(s_3) \geq 0$. Defining $\tilde{y}$ as in (3.8) leads to a contradiction as in Case (i).

Subcase (2): $\hat{y}(s_3) < 0$. Then $s_3 < 1$ and $\hat{y}(1) = 0$ implies there is a first member s_4 in $\mathcal{C}(\hat{y}) > s_3$. The function $-\hat{y}$ now satisfies the condition of Case (i) with s_2, s_3, s_4 replacing s_1, s_2, s_3 in that argument. Thus the proof of Lemma 2 is complete.

Theorem 2. *When $p = 2$ the constant $K_0 = 192$ in (1.2) with $[a, b] = [0, 1]$. Moreover there are extremals y such that $|\mathcal{C}(y)| = 1$ which are non-zero multiples of a cubic spline with a knot at $1/2$. Thus $y = Ay_0$, where A is a real number, $A \neq 0$, and*

$$(3.11) \qquad y_0(x) = \begin{cases} 12x^2 - 16x^3, & 0 \leq x < 1/2, \\ 16x^3 - 36x^2 + 24x - 4, & 1/2 \leq x \leq 1. \end{cases}$$

Proof. By Lemmas 1 and 2 we can choose a sequence $\{y_n\}$ in BC_0 such that $J(y_n) \to K_0$ as $n \to \infty$, and y_n has only one point c_n where y_n' changes sign; further we can normalize y_n so that $\int_0^1 |y_n y_n'| \, dx = 1$. The proof of Theorem 1 shows that $\{y_n\}$ has a subsequence which converges uniformly on $[0, 1]$ to an extremal y_0. Thus y_0 can have only one point where y_0' changes sign; denote this point by c. We take $y_0(x)$ to be positive on $(0, 1)$. From $y_0(0) = y_0'(0) = y_0'(c) = 0$, we get that

$$(3.12) \qquad y_0(c) = \int_0^c (c - u) y_0''(u) \, du, \qquad 0 = y_0'(c) = \int_0^c y_0''(u) \, du.$$

Hence for all ξ,

$$y_0(c) = \int_0^c (\xi - u) y_0''(u) \, du.$$

Since y_0 is monotone on $[0, c]$ and on $[c, 1]$,

$$(3.13) \qquad \int_0^1 |y_0 y_0'| \, du = \int_0^c y_0(u) y_0'(u) \, du + \int_c^1 -y_0(u) y_0'(u) \, du$$
$$= y_0^2(c).$$

On the other hand, by Cauchy-Schwarz and (3.12),

$$(3.14) \qquad y_0^2(c) \leq \int_0^c (\xi - u)^2 \, du \int_0^c y_0''(u)^2 \, du.$$

Choose $\xi = c/2$ in (3.14); then

$$(3.15) \qquad y_0^2(c) \leq \frac{c^3}{12} \int_0^c y_0''(u)^2 \, du.$$

Similarly,

$$y_0^2(c) = \left(\int_c^1 (u - \xi) y_0''(u) \, du \right)^2$$
$$\leq \int_c^1 (u - \xi)^2 \, du \int_c^1 y_0''(u)^2 \, du$$
$$(3.16) \qquad = \frac{(1 - c)^3}{12} \int_c^1 y_0''(u)^2 \, du$$

with $\xi = (1 + c)/2$. Combining (3.15) and (3.16) and subsitution into (3.13) yields that

$$\int_0^1 |y_0 y_0'| \, dx = y_0^2(c) \leq \frac{1}{12} \left(\frac{1}{c^3} + \frac{1}{(1 - c)^3} \right)^{-1} \int_0^1 (y_0'')^2 \, du$$
$$\leq \frac{1}{12} \sup_{0 < c < 1} \left(\frac{1}{c^3} + \frac{1}{(1 - c)^3} \right)^{-1} \int_0^1 (y_0'')^2 \, du$$
$$(3.17) \qquad = \frac{1}{192} \int_0^1 (y_0'')^2 \, du,$$

where a calculus argument in (3.17) shows the supremum occurs at $c = 1/2$. Equation (3.17) gives $192 \leq K_0$. On the other hand, the condition for equality in the Cauchy-Schwartz inequality used above and equality in (3.17) gives that

$c = 1/2$, on $[0, 1/2]$, $y_0''(u)$ is a multiple of $u - 1/4$, and on $[1/2, 1]$, $y_0''(u)$ is a multiple of $u - 3/4$. These requirements, together with the boundary conditions $y_0(0) = y_0'(0) = y_0(1) = y_0'(1) = 0$, uniquely determine y_0 when $y_0(1/2)$ is given. For example with $y_0(1/2) = 1$, y_0 is given by (3.11). A calculation shows y_0 is a cubic spline since y_0, y_0', y_0'' are continuous at $x = 1/2$.

We remark that the inequality

$$y(c)^2 \le \frac{1}{192} \int_0^1 (y'')^2 \, du$$

given by (3.17) may also be obtained by Green's function techniques, e.g., see Reid [3]. However it is still necessary to determine conditions of equality.

4. Other Boundary Conditions

In this section we consider classes of functions other than BC_0. Define

$$BC_1 = \{y \in D : y(0) = y(1) = 0\},$$
$$BC_2 = \{y \in D : y(0) = y'(0) = y(1) = 0\},$$
$$BC_3 = \{y \in D : y(0) = y'(0) = y'(1) = 0\},$$
$$BC_4 = \{y \in D : y(0) = y'(1) = 0\},$$
$$BC_5 = \{y \in D : y(0) = y'(0) = 0\},$$

and let for $i = 1, \cdots, 5$

$$K_i = \inf\{J(y) : y \in BC_i\}.$$

In problems BC_3, BC_4, and BC_5 it is easy to see that an extremal is monotone on $[0, 1]$. This is a result of the fact that if $y \in BC_i$, $i = 3, 4,$ or 5, then $y^\#(x) := \int_0^x |y'(s)| \, ds$ is also in BC_i, and $J(y^\#) \le J(y)$. Returning to the necessary conditions of section 2, we see that an extremal y_i must be a cubic polynomial and satisfy the boundary conditions of BC_i as well as natural boundary conditions imposed by (see (2.3)) either or both of the conditions

$$y_i''(0)h'(0) - y_i'''(0)h(0) - (K_i/2)\text{sgn}(y_i y_i')(0)[y_i(0)h(0)] = 0,$$
$$y_i''(1)h'(1) - y_i'''(1)h(1) - (K_i/2)\text{sgn}(y_i y_i')(1)[y_i(1)h(1)] = 0.$$

For BC_i, $i = 1, \cdots, 5$, the existence proof of section 2 may be modified to yield the existence of an extremal. For problems BC_1 and BC_2, it is necessary to use Lemmal 2 and a modified version of Lemmal 1 to prove that there is an extremal y_i with exactly one sign change in y_i'. With p_n as in Lemma 1, a polynomial q_n approximating $y \in BC_i$ is constructed having the form

$$q_n(x) = y'(0)x + \int_0^x (x - s)p_n(s) \, ds + a_n x^2$$

for BC_1 (note that $q_n(1) = \int_0^1 (1 - s)[p_n(s) - y''(s)] \, ds + a_n$) and with the form

$$q_n(x) = \int_0^x (x - s)p_n(s) \, ds + a_n x^2$$

for BC_2 (note that $q_n(1) = \int_0^1 (1 - s)[p_n(s) - y''(s)] \, ds + a_n$). In BC_1 and BC_2 one imposes continuity of y_i at the point c where y_i' changes sign as well as the condition

$y_i'(c) = 0$. These conditions plus the natural boundary conditions and boundary conditions of BC_i, $i = 1, 2$, again determine the extremals. For problems BC_1 and BC_2 an argument similar to that of Theorem 2 can be given to determine the value of c where the extremal changes sign. As an alternate method for determining c and the best constant, one can construct all cubic splines with a knot at c and satisfying the end conditions plus natural boundary conditions. This can be accomplished by software such as MAPLE, and we have checked our solution to problem BC_2 in this way.

Theorem 3. *When $p = 2$ and $[a, b] = [0, 1]$ the constants K_i together with corresponding extremals y_i[1] with $|C(y_i)| \leq 1$ and all boundary conditions, including natural boundary conditions, for problems BC_i are given below.*

BC_1: $K_1 = 48$, $y_1(0) = y_1(1) = y_1'(0) = y_1''(1) = 0$, $y_1(c) = 1$, $y_1'(c) = 0, c = 1/2,$

$$y_1(x) = \begin{cases} \frac{3}{2}x - 2x^3, & 0 \leq x < \frac{1}{2}, \\ \frac{3}{2}(1 - x) - 2(1 - x)^3, & \frac{1}{2} \leq x \leq 1. \end{cases}$$

BC_2: $K_2 = 3(1 + \sqrt{2})^4$, $y_2(0) = y_2'(0) = y_2(1) = y_2''(0) = 0$, $y_2(c) = 1$, $y_2'(c) = 0$, $c = \sqrt{2}/(1 + \sqrt{2}),$

$$y_2(x) = \begin{cases} 3\left(\frac{x}{c}\right)^2 - 2\left(\frac{x}{c}\right)^3, & 0 \leq x < c, \\ \frac{3}{2}\left(\frac{1-x}{1-c}\right) - \frac{1}{2}\left(\frac{1-x}{1-c}\right)^3, & c \leq x \leq 1. \end{cases}$$

BC_3: $K_3 = 24, y_3(0) = y_3'(0) = y_3'(1) = 0$, $y_3(1) = 1$, $y_3'''(1) + K_3/2 = 0,$

$$y_3(x) = 3x^2 - 2x^3, \quad 0 \leq x \leq 1.$$

BC_4: $K_4 = 6, y_4(0) = y_4''(0) = y_4'(1) = 0$, $y_4(1) = 1$, $y_4'''(1) + K_4/2 = 0,$

$$y_4(x) = \frac{3}{2}x - \frac{1}{2}x^3, \quad 0 \leq x \leq 1.$$

BC_5: $K_5 = 6, y_5(0) = y_5'(0) = 0$, $y_5''(1) = 0$, $y_5(1) = 1$, $y_5'''(1) + K_5/2 = 0,$

$$y_5(x) = \frac{3}{2}x^2 - \frac{1}{2}x^3, \quad 0 \leq x \leq 1.$$

The problem (1.2) may also be considered over a larger class of functions with just one boundary condition, for example

$$BC_7 = \{y \in D : y(0) = 0\} \quad \text{or} \quad BC_8 = \{y \in D : y'(0) = 0\}.$$

In both the cases the infimum of (1.2) is 0. For example in BC_7, $y_7(x) = x$ is an extremal with $J(y_7) = 0$, and in BC_8, $y_n(x) := 1 + x^2/n \in BC_8$ and satisfies $J(y_n) \to 0$ as $n \to \infty$.

5. Applications

---

[1]For BC_1 and BC_2 we normalize the extremal y_i by the condition $y_i(c) = 1$. For BC_i, $i = 3, 4, 5$, we normalize by $y_i(1) = 1$.

As a first application of the Opial inequalities we consider the transverse vibrations of a rod on $0 \leq x \leq 1$ with density $\rho(x)$. The eigenvalue problem is,

$$(5.1) \qquad y^{(iv)} = \lambda \rho(x)y, \qquad 0 \leq x \leq 1,$$

where boundary conditions are imposed at $x = 0, 1$. For simplicity, we have normalized Young's modulus E and the moment of inertia I of the rod so that $EI = 1$. First we consider both ends clamped so that

$$(5.2) \qquad y(0) = y'(0) = y(1) = y'(1) = 0,$$

and suppose the smallest eigenvalue of (5.1)–(5.2) is λ_0 (thus the lowest angular frequency of oscillation is $\sqrt{\lambda_0}$). Multiplying (5.1) by y and integrating by parts twice, we obtain that

$$(5.3) \qquad \int_0^1 y''(x)^2 \, dx = \lambda_0 \int_0^1 \rho(x)y(x)^2 \, dx.$$

Set $P(x) = \int_0^x \rho(u) \, du - (1/2) \int_0^1 \rho(u) \, du$. Then a further integration by parts of the right side of (5.3) gives us

$$(5.4) \qquad \int_0^1 y''(x)^2 \, dx = \lambda_0 \int_0^1 P'(x)y(x)^2 \, dx$$

$$= -2\lambda_0 \int_0^1 P(x)y(x)y'(x) \, dx$$

$$\leq \lambda_0 \left(\int_0^1 \rho(x) \, dx \right) \int_0^1 |y(x)y'(x)| \, dx$$

since $|P(x)| \leq 1/2 \int_0^1 \rho(x) \, dx$ as $\rho(x) > 0$. Hence (5.4) combined with (3.17) gives

$$(5.5) \qquad \lambda_0 \geq \frac{\int_0^1 y''(x)^2 \, dx}{\int_0^1 \rho(x) \, dx \int_0^1 |y(x)y'(x)| \, dx} \geq \frac{192}{\int_0^1 \rho(x) \, dx}.$$

We now show why inequality (5.5) is best possible. Let $M = \int_0^1 \rho(x) \, dx$ be a fixed mass and set for ϵ small,

$$\rho_\epsilon(x) = \begin{cases} \epsilon, & x \notin [\frac{1}{2} - \epsilon, \frac{1}{2} + \epsilon], \\ \frac{M}{2\epsilon} - (\frac{1}{2} - \epsilon), & x \in [\frac{1}{2} - \epsilon, \frac{1}{2} + \epsilon]. \end{cases}$$

Then $\int_0^1 \rho_\epsilon(x) \, dx = M$. Let $\lambda_0(\rho_\epsilon)$ be the smallest eigenvalue of

$$y^{(iv)} = \lambda \rho_\epsilon y, \qquad y(0) = y'(0) = y(1) = y'(1) = 0.$$

By the Rayleigh quotient characterization of $\lambda_0(\rho_\epsilon)$,

$$\lambda_0(\rho_\epsilon) = \inf_{y \in BC_0} \frac{\int_0^1 y''(x)^2 \, dx}{\int_0^1 \rho_\epsilon(x)y(x)^2 \, dx}.$$

In particular for $y(x)$ given by the cubic spline (3.11),

$$\lambda_0(\rho_\epsilon) \leq \frac{\int_0^1 y_0''(x)^2 \, dx}{\int_0^1 \rho_\epsilon(x)y_0(x)^2 \, dx},$$

and thus

$$\limsup_{\epsilon \to 0} \lambda_0(\rho_\epsilon) \leq \frac{\int_0^1 y_0''(x)^2 \, dx}{M y_0(1/2)^2} = \frac{192}{M}$$

which shows the constant 192 cannot be increased. For $\rho(x) \equiv 1$, a computation shows that the value of $\lambda_0(1)$ is given by the smallest positive root of $\cos(\lambda^{1/4}) \cosh(\lambda^{1/4}) = 1$ which is $\lambda_0(1) \approx 500.5$.

As a second example consider the rod with one end clamped and the other end supported. This leads to the eigenvalue problem

$$y^{(iv)} = \mu \rho(x) y, \quad y(0) = y'(0) = y(1) = y''(1) = 0.$$

Following the analysis in the previous example and using the best constant K_2 of BC_2 leads to the inequality for the first eigenvalue,

$$\mu_0 \geq \frac{K_2}{\int_0^1 \rho(x) \, dx} = \frac{3(1 + \sqrt{2})^4}{\int_0^1 \rho(x) \, dx},$$

which again can be shown to be the best possible inequality, i.e., the constant K_2 cannot be replaced by a larger constant.

Finally, the rod with two supported ends is associated with the eigenvalue problem

$$y^{(iv)} = \eta \rho(x) y, \quad y(0) = y''(0) = y(1) = y''(1) = 0.$$

and consequently to the best inequality for the first eigenvalue,

$$\eta_0 \geq \frac{K_1}{\int_0^1 \rho(x) \, dx} = \frac{48}{\int_0^1 \rho(x) \, dx},$$

As another application we apply 2nd order Opial inequalities to disconjugacy theory. Applications of 1st order Opial inequalities may be found in [1]. Suppose $q(x)$ is a real, locally $\mathcal{L}^1$ function on $[a, b]$. b is said to be a conjugate point of a if there is a nontrivial solution y to the boundary value problem

$$(5.6) \qquad y^{(iv)} + q(x)y = 0, \quad y(0) = y'(a) = y(b) = y'(b) = 0.$$

The absence of conjugate points plays an important role in control theory as it leads to the existence of solutions to Riccati equations and optimal controls. We now show how the Opial inequality (1.1) leads to a lower bound on the length of the interval $[a, b]$ in terms of the function $q(x)$. We begin by multiplying (5.6) by y and integrating by parts; this yields, for an antiderivative $Q(x)$ of $q(x)$, that

$$(5.7) \qquad \int_a^b y''(x)^2 \, dx = -\int_a^b q(x)y(x)^2 \, dx = 2 \int_a^b Q(x)y(x)y'(x) \, dx$$

$$\leq 2 \max_{a \leq x \leq b} |Q(x)| \int_a^b |yy'| \, dx$$

$$\leq 2 \max_{a \leq x \leq b} |Q(x)| \frac{(b-a)^3}{192} \int_a^b y''(x)^2 \, dx$$

by Theorem 2. If m, M are defined by

$$m = \min_{a \leq x \leq b} \int_a^x q(\xi) \, d\xi, \qquad M = \max_{a \leq x \leq b} \int_a^x q(\xi) \, d\xi,$$

then the choice of $Q(x) = \int_a^x q(\xi) \, d\xi - (M+m)/2$ shows there exists x_1, x_2 in $[a, b]$ such that

(5.8)
$$\max_{a \leq x \leq b} |Q(x)| = \frac{1}{2} \left| \int_{x_1}^{x_2} q(\xi) \, d\xi \right|$$

since $Q(x)$ will have a maximum of $(M - m)/2$ and a minimum of $(m - M)/2$. Substitution of (5.8) in to (5.7) gives the Liapunov-type inequality

(5.9)
$$\frac{192}{(b-a)^3} \leq \left| \int_{x_1}^{x_2} q(\xi) \, d\xi \right|,$$

improving a result of Reid [3, Theorem 3.2 and §4] who showed that

$$\frac{192}{(b-a)^3} \leq \left| \int_a^b q_+(\xi) \, d\xi \right|$$

where q_+ denotes the nonnegative part of q.

Further (5.9) is a strict inequality since equality in Theorem 2 implies y is a cubic spline which is contrary to y being a solution of (5.6) as solutions of (5.6) require that y, y', y'', y''' be absolutely continuous.

In a similar way each of the inequalities of section 4, BC_i, $i = 1, \cdots, 5$, leads to an equality of type (5.9). The anti-derivative $Q(x)$ must be $\int_x^b q(\xi) \, d\xi$ in cases where $y(b) \neq 0$. For example, suppose y is a nontrivial solution of

(5.10)
$$y^{(iv)} + q(x)y = 0, \quad y(a) = y'(a) = y''(b) = y'''(b) = 0.$$

Then with $Q(x) = \int_x^b q(\xi) \, d\xi$, the first two lines of (5.7) and the Opial inequality for BC_5 leads to

$$\frac{3}{(b-a)^3} \leq \max_{a \leq x \leq b} \left| \int_x^b q(\xi) \, d\xi \right|.$$

Again the inequality is strict since an extremal y_5 for BC_5 satisfies $y_5'''(b) \neq 0$, contrary to (5.10).

6. A version of the second order Opial inequality in $\mathcal{L}^p(0, 1)$

We consider now the case of general $1 < p \leq \infty$, $p \neq 2$, in (1.1) and (1.2). To show that $K_0 > 0$ and that an extremal of (1.2) exists, we repeat the proof of Theorem 2 with suitable changes. Let $\{y_n\}$ be a sequence such that $\lim_{n \to \infty} J(y_n) = K_0$. We can assume as before that $\int_0^1 |y_n y_n'| \, dx = 1$. By Alaoglu's theorem and the fact that $\mathcal{L}^p(0, 1)$ is reflexive if $1 < p < \infty$, the bounded set $\{y_n''\}$ has a weakly convergent subsequence to a function $g \in \mathcal{L}^p(0, 1)$. If $p = \infty$ there is a weak-star convergent subsequence. In either case we replace (2.4) by

$$|y_n'(t) - y_n'(s)|^2 \leq |t - s|^{2/q} \left(\int_0^1 |y_n''|^p \, dx \right)^{2/p}$$

where $1/q + 1/p = 1$ and conclude that $\{y_n'\}$ has a uniformly convergent subsequence. The remaining details of the proof do not differ conceptually from Theorem 1 and are omitted.

50

Similarly Lemmas 1 and 2 in no way depend on the fact that $p = 2$ and may be repeated nearly verbatim.

Theorem 4. *If $1 < p < \infty$ and $q = p/(p-1)$ then $K_0 = 64(q+1)^{2/q}$. If $p = \infty$ and $q = 1$ then $K_0 = 256$. Moreover there are extremals y such that $|\mathcal{C}(y)| = 1$ which are non-zero multiples of the function*

$$y_0(x) = \int_0^1 G(x, u) f(u) \, du$$

where

(6.1)
$$G(x, u) = \chi_{[0,1/2)}(x)(x - u)_+ + \chi_{[1/2,1]}(x)(x - u)_-$$

$$f(u) = \begin{cases} \operatorname{sgn}(1/4 - u)|1/4 - u|^{q/p} & \text{if } 0 < u < 1/2, \\ \operatorname{sgn}(u - 3/4)|u - 3/4|^{q/p} & \text{if } 1/2 \le u < 1 \end{cases}$$

if $1 < p < \infty$ and

$$f(u) = \begin{cases} \operatorname{sgn}(1/4 - u) & \text{if } 0 < u < 1/2, \\ \operatorname{sgn}(u - 3/4) & \text{if } 1/2 \le u < 1 \end{cases}$$

if $p = \infty$. Here

$$(x - u)_+ = \begin{cases} (x - u) & \text{if } x > u, \\ 0 & \text{otherwise} \end{cases}$$

and $(x - u)_- = |x - u| - (x - u)_+$. Furthermore when $1 < p < \infty$ y_0, y_0', and y_0'' are continuous on $[0,1]$ while y_0''' has a jump at $1/2$.

Proof. As in the proof of Theorem 2 using the revised versions of Lemmas 1 and 2 we can obtain a sequence $\{y_n\}$ in BC_0 with the properties that y_n has only one point c_n where y_n' changes sign, $\int_0^1 |y_n y_n'| \, dx = 1$, $J_p(y_n) \to K_{0,p}$, $\{y_n\}$ converges uniformly to an extremal y_0 as $n \to \infty$, and $\{y''\}$ converges weakly or in the weak-star sense (when $p = \infty$). Hence y_0 also has only one point c where y_0' changes sign. We take $y_0(x)$ to be positive on $(0,1)$. As in the $p = 2$ case it follows since $y_0 \in BC_0$ that for all ξ,

(6.2)
$$y_0(c) = \int_0^c (\xi - u) y_0''(u) \, du$$

(6.3)
$$= \int_c^1 (u - \xi) y_0''(u) \, du.$$

Also since y_0 is monotone on $[0, c]$ and on $[c, 1]$,

(6.4)
$$\int_0^1 |y_0 y_0'| \, du = y_0^2(c).$$

From (6.2), (6.3), (6.4), and Hölder's inequality we get that

(6.5)
$$y_0^2(c) \le \left(\int_0^c |\xi - u|^q \, du \right)^{2/q} \left(\int_0^c |y_0''(u)|^p \, du \right)^{2/p}$$

(6.6)
$$y_0^2(c) \le \left(\int_c^1 |u - \xi|^q \, du \right)^{2/q} \left(\int_c^1 y_0''(u)^2 \, du \right)^{2/p}.$$

We choose $\xi = c/2$ in (6.5) and $\xi = (1+c)/2$ in (6.6). Then a calculation and substitution into (6.4) gives the estimate

$$\int_0^1 |y_0 y_0'| \, dx \leq \left[\left(\frac{c^{q+1}}{(q+1)2^q}\right)^{-p/q} + \left(\frac{(1-c)^{q+1}}{(q+1)2^q}\right)^{-p/q}\right]^{-2/p}$$

$$\times \left(\int_0^1 |y_0''|^p \, dx\right)^{2/p}$$

$$\leq \left(\frac{1}{4(q+1)^{2/q}}\right) \sup_{c \in (0,1)} \left[\frac{1}{c^{(q+1)(p/q)}} + \frac{1}{(1-c)^{(q+1)(p/q)}}\right]^{-2/p}$$

$$(6.7) \qquad\qquad \times \left(\int_0^1 |y_0''|^p \, dx\right)^{2/p}$$

when $1 < p < \infty$ or

$$(6.8) \qquad \int_0^1 |y_0 y_0'| \, dx \leq \left(\frac{1}{8}\right) \sup_{c \in (0,1)} \left[\frac{1}{c^4} + \frac{1}{(1-c)^4}\right]^{-1} \|y_0''\|_{\infty,(0,1)}^2$$

if $p = \infty$. The optimal value of c in either (6.7) or (6.8) is $c = 1/2$ and gives after calculation the inequalities

$$(6.9) \qquad \int_0^1 |y_0 y_0'| \, dx \leq \frac{1}{64(q+1)^{2/q}} \left(\int_0^1 |y_0''|^p \, dx\right)^{2/p}$$

for $1 < p < \infty$ or

$$(6.10) \qquad \int_0^1 |y_0 y_0'| \, dx \leq \frac{1}{256} \|y_0''\|_{\infty,[0,1]}$$

if $p = \infty$.

We now show that this estimate of K_0 is sharp by constructing y_0. The conditions for equality in the Hölder inequalities (6.5) and (6.6) with the choice $c = 1/2$ imply that up to multiplication by a positive constant

$$(6.11) \qquad y_0''(u) = \begin{cases} \text{sgn}(1/4 - u)|1/4 - u|^{q/p} & \text{if } 0 < u < 1/2, \\ \text{sgn}(u - 3/4)|u - 3/4|^{q/p} & \text{if } 1/2 \leq u < 1 \end{cases}$$

for $1 < p < \infty$. On the other hand, if $p = \infty$ there are equalities in

$$y_0(1/2) = \int_0^{1/2} (1/4 - u) y_0''(u) \, du \leq \left(\int_0^{1/2} |1/4 - u| \, du\right) \|y_0''\|_{\infty,(0,1/2)}$$

$$y_0(1/2) = \int_{1/2}^1 (u - 3/4) y_0''(u) \, du \leq \left(\int_{1/2}^1 |u - 3/4| \, du\right) \|y_0''\|_{\infty,(1/2,1)}$$

if and only if (again up to multiplication by a positive constant)

$$(6.12) \qquad y_0''(u) = \text{sgn}(1/4 - u)\chi_{(0,1/2)} + \text{sgn}(u - 3/4)\chi_{(1/2,1)}.$$

Let f be given by (6.11) or (6.12) and define y_0 by

$$y_0(x) = \begin{cases} \int_0^x (x - u) f(u) \, du & \text{if } 0 \leq x < 1/2, \\ \int_x^1 (u - x) f(u) \, du & \text{if } 1/2 \leq x \leq 1. \end{cases}$$

Equivalently,

$$y_0(x) = \int_0^1 G(x,u)f(u)\,du$$

where $G(x,u)$ is the Green's function defined by (6.1).

A calculation verifies that $y_0 \in BC_0$, y_0, y_0', y_0'' are continuous when $1 < p < \infty$, and that there is equality in (6.9) and (6.10).

REMARK. If $p = 1$ the analysis leading to Theorem 4 will give a sequence of approximate extremals $\{y_n\}$ with y_n, y_n' converging uniformly to a function y_0. But although y_0 satisfies the boundary conditions y_0' need not be absolutely continuous so that $y_0 \notin BC_0$. The problem is that $\{y_n''\}$ need not converge weakly to an $\mathcal{L}^1$ function. Since each of the approximate extremals y_n can be assumed to have one sign change in y_n' at some point $c_n \in (0,1)$ we can still however apply the analysis following (6.2) to find that $K_0 = 64$.

References

1. R. C. Brown and D. B. Hinton, *Opial's inequality and oscillation of 2nd order equations*, Proc. Amer. Math. Soc. **125** (1997), 1123–1129.
2. C. H. FitzGerald, *Opial-Type inequalities that involve higher derivatives*, International Series of Numerical Mathematics, Vol. 71 (1984), Bikerhauser Verlag, Basel.
3. W. T. Reid, *A generalized Liapunov inequality*, J. Differential Equations **13** (1973), 182–196.

ON MULTI-DIMENSIONAL INTEGRAL INEQUALITIES AND APPLICATIONS

WING-SUM CHEUNG

Department of Mathematics, The University of Hong Kong, Pokfulam Road, Hong Kong

THEMISTOCLES M. RASSIAS

Department of Mathematics, National Technical University of Athens, Zografou Campus, Greece

Abstract. In this article some new multi-dimensional integral inequalities of the Poincaré- and Sobolev-types are established. The methods used are systematic and algorithmic, and are more efficient than the usual eigenvalue-problem techniques. Furthermore, the results can serve as generators of other interesting multi-dimensional integral inequalites.

1. Introduction

As we are all aware of, analytic inequalities play an essential and indispensable role in the study of qualitative as well as quantitative properties of solutions of differential and integral equations. In fact, they are important in virtually every area of analysis. As an important feature of C^1 functions of several variables is the gradient, it is not surprising to observe that amongst all multi-dimensional analytic inequalities, those involving the gradient of C^1 functions are most commonly used and in fact are most important.

Two major types of multi-dimensional analytic inequalities involving gradients are the so-called Poincaré-type and Sobolev-type integral inequalities. *Poincaré's inequality* is the integral inequality [17, 23]

$$(1) \qquad \lambda_0 \int_\Omega f^2 dx \leq \int_\Omega |\nabla f|^2 dx \, ,$$

where Ω is a bounded region in $\mathbb{R}^2$ or $\mathbb{R}^3$, $f \in C^1(\Omega)$, $f = 0$ on the boundary $\partial\Omega$ of Ω, and $\lambda_0 \in \mathbb{R}$ is the smallest eigenvalue of the problem

$$\begin{cases} \Delta f + \lambda f = 0 & \text{in } \Omega \\ f = 0 & \text{on } \partial\Omega \, . \end{cases}$$

Exhibiting an estimate of the average of f^2 on Ω by that of $|\nabla f|^2$ on Ω, Poincaré's inequality is one of the few most important multi-dimensional integral inequalities. Its importance does not only lie on the fact that it gives an effective and sensible estimate of the $\mathcal{L}^2$-norm of f for our disposal in the study differential and integral equations, not only lie on the fact that it is venerable, but also lies on the fact that it is extremely inspiring. In fact, a good part of multi-dimensional integral inequalities are originated from it. Some of these may bear other names, but they

T.M. Rassias and H.M. Srivastava (eds.), Analytic and Geometric Inequalities and Applications, 53–67.
© 1999 *Kluwer Academic Publishers.*

all share the same spirit as Poincaré's inequality does. In general, such variations of Poincaré's inequality are collectively known as *Poincaré-type integral inequalities*. A brief account of such inequalities can be found in, say, Beckenbach-Bellman [1], Hardy-Littlewood-Pólya [11], Mitrinović [17], Nirenberg [18], and more recently Horgan et al [13, 14, 15], Milovanović-Mitrinović-Rassias [16], Pachpatte [19, 21], Rassias [24, 25], and Cheung [3, 6, 7].

Another important type of multi-dimensional integral inequalities, which can be regarded as a variation of Poincaré-type integral inequalities, is the so-called *Sobolev-type integral inequalities*. A typical Sobolev-type integral inequality is [eg. 15]

$$(2) \qquad \int_\Omega f^4 dx \le k \left(\int_\Omega f^2 dx \right) \left(\int_\Omega |\nabla f|^2 dx \right) ,$$

where Ω is a bounded region in $\mathbb{R}^2$, $f \in C^1(\Omega)$, $f = 0$ on $\partial\Omega$, and k is a constant. (2) and its variations are also important in the study of finite element analysis as well as differential and integral equations. Various results on such type of integral inequalities can be found in, say, Beckenbach-Bellman [1], Friedman [10], Horgan et al [12, 13, 15], Nirenberg [18], Pachpatte [20, 21, 22], and Cheung [2, 6].

It is the purpose of this article to establish new multi-dimensional Poincaré-type and Sobolev-type integral inequalities which improve some existing results of inequalities of such types in the literature. The technique used here is along the line with those used in [2-9, 19-22] and is simpler than the customary eigenvalue-problem technique. Furthermore, an important feature of the technique used here is that it is rather algorithmic and is easy to apply, and the results here can serve as generators of a whole bunch of integral inequalities of Poincaré-type, Sobolev-type, and further variations of them.

2. Notations and Preliminaries

In the sequel $m \ge 2$ and $n \ge 1$ will denote two fixed integers. For consistency we shall always denote indices running from 1 to m by Greek alphabets $\alpha, \beta, \gamma, \ldots$ and those from 1 to n by English letters $i, j, k, \ldots$. Let $\Omega = \prod_{i=1}^{n} [a_i, b_i] \subset \mathbb{R}^n$ be a fixed rectangular region. As customary, a general point in $\mathbb{R}^n$ will be denoted as $x = (x_1, \ldots, x_n)$ and the volume form on $\mathbb{R}^n$ as $dx = dx_1 \cdots dx_n$. $C^1(\Omega)$ will be the space of all continuously differentiable functions on Ω and $C_0^1(\Omega)$ the subspace of $C(\Omega)$ consisting of all those functions in $C^1(\Omega)$ which vanish on the boundary $\partial\Omega$ of Ω. For any $f \in C^1(\Omega)$, partial derivatives of f will be denoted as f_i and the gradient of f by ∇f.

Since all summations and products in the sequel will be either over $\alpha, \beta, \gamma, \ldots$ from 1 to m or over $i, j, k, \ldots$ from 1 to n, for the sake of convenience we shall simply drop the terminals of all summations and products and write $\sum$ for $\sum_{\alpha=1}^{m}$, $\prod$ for $\prod_{i=1}^{n}$, etc.

The following elementary inequalities which are easy consequences of the geometric-arithmetic mean inequality, will be needed in the sequel.

Lemma A [11, 17]. *For any real numbers p_α, q_α, $c_\alpha > 0$ with $\sum_\alpha q_\alpha/p_\alpha = 1$,*

$$\prod_\alpha c_\alpha^{q_\alpha} \le \sum_\alpha \frac{q_\alpha}{p_\alpha} c_\alpha^{p_\alpha} \ ,$$

where the inequality holds if and only if $c_1 = \cdots = c_m$.

Lemma B [11, 17]. *For any $r_i \ge 0$ and $s \ge 0$,*

$$\left(\sum r_i\right)^s \le c(s,n) \sum r_i^s \ ,$$

where

$$c(s,n) = \begin{cases} n^{s-1} & \text{if } s \ge 1 \\ 1 & \text{if } 0 \le s < 1 \ . \end{cases}$$

3. Poincaré-type Integral Inequalities

Let $M := \max\{b_i - a_i : i = 1, \ldots, n\}$ be the maximum of the length of the sides of the rectangular region $\Omega \subset \mathbb{R}^n$.

Theorem 3.1. *For any $f^\alpha \in C_0^1(\Omega)$, any real numbers $p_\alpha \ge 2$, $q_\alpha \ge 0$ with $\sum_\alpha q_\alpha/p_\alpha = 1$, and any $C_\alpha > 0$,*

$$\int_\Omega \prod_\alpha |f^\alpha|^{q_\alpha} \le \frac{C}{n} \sum_\alpha \frac{q_\alpha}{p_\alpha} \left(\frac{MC_\alpha}{2}\right)^{p_\alpha} \int_\Omega |\nabla f|^{p_\alpha} \ ,$$

where $C := \prod_\beta C_\beta^{-q_\beta}$.

Theorem 3.1 generalizes and improves a number of results of Poincaré-type integral inequalities in the literature [3, 6, 7, 15, 17, 19, 21]. For example, the following are simple consequences of Theorem 3.1.

Corollary 3.1 [7]. *For any $f^\alpha \in C_0^1(\Omega)$ and any real numbers $p_\alpha \ge 2$, $q_\alpha \ge 0$ with $\sum_\alpha q_\alpha/p_\alpha = 1$,*

$$\int_\Omega \prod_\alpha |f^\alpha|^{q_\alpha} \le \frac{1}{n} \sum_\alpha \frac{q_\alpha}{p_\alpha} \left(\frac{M}{2}\right)^{p_\alpha} \int_\Omega |\nabla f^\alpha|^{p_\alpha} \ .$$

Proof. This follows immediately from Theorem 3.1 by letting $C_\alpha = 1$ for all α. Q.E.D.

Corollary 3.2. *For any $f^\alpha \in C_0^1(\Omega)$ and any real numbers $p_\alpha \geq 2$, $q_\alpha > 0$ with $\sum\limits_\alpha q_\alpha/p_\alpha = 1$,*

$$\int_\Omega \prod_\alpha |f^\alpha|^{q_\alpha} \leq \frac{C}{n} \sum_\alpha \int_\Omega |\nabla f|^{p_\alpha} ,$$

where

$$C := \prod_\alpha \left[\left(\frac{M}{2}\right)^{q_\alpha} \left(\frac{q_\alpha}{p_\alpha}\right)^{\frac{q_\alpha}{p_\alpha}} \right] .$$

Proof. This follows immediately from Theorem 3.1 by letting

$$C_\alpha = \frac{2}{M} \left(\frac{p_\alpha}{q_\alpha}\right)^{1/p_\alpha}$$

for all α. Q.E.D.

Corollary 3.3 [3, 6, 7]. *For any $f^\alpha \in C_0^1(\Omega)$ and any real numbers $p_\alpha \geq 2$ with $\sum\limits_\alpha 1/p_\alpha = 1$,*

$$\int_\Omega \prod_\alpha |f^\alpha| \leq \frac{1}{n} \sum_\alpha \frac{1}{p_\alpha} \left(\frac{M}{2}\right)^{p_\alpha} \int_\Omega |\nabla f^\alpha|^{p_\alpha} .$$

Proof. This is immediate from Corollary 3.1 by letting $q_\alpha = 1$ for all α. Q.E.D.

Corollary 3.4 [3, 6, 7]. *For any $f^\alpha \in C_0^1(\Omega)$ and any real numbers $q_\alpha \geq 0$ with $q := \sum\limits_\alpha q_\alpha \geq 2$,*

$$\int_\Omega \prod_\alpha |f^\alpha|^{q_\alpha} \leq \frac{1}{n} \left(\frac{M}{2}\right)^q \sum_\alpha \frac{q_\alpha}{q} \int_\Omega |\nabla f^\alpha|^q .$$

Proof. This is immediate from Corollary 3.1 by letting $p_\alpha = q$ for all α. Q.E.D.

Corollary 3.5 [3, 6, 7]. *For any $f^\alpha \in C_0^1(\Omega)$,*

$$\int_\Omega \prod_\alpha |f^\alpha| \leq \frac{1}{nm} \left(\frac{M}{2}\right)^m \sum_\alpha \int_\Omega |\nabla f^\alpha|^m .$$

Proof. This is immediate from Corollary 3.3 with $p_\alpha = m$ for all α or from Corollary 3.4 with $q_\alpha = 1$ for all α. Q.E.D.

To prove Theorem 3.1 we need the following basic lemma.

Lemma 3.1. *If $f \in C_0^1(\Omega)$, then for any $t \in \Omega$,*

$$|f(t)| \leq \frac{1}{2n} \sum_i \int_{a_i}^{b_i} |f_i(t_1, \ldots, t_{i-1}, u_i, t_{i+1}, \ldots, t_n)| du_i .$$

Proof. Since $f = 0$ on $\partial\Omega$, for each $i = 1, \ldots, n$, we have

$$f(t) = \int_{a_i}^{t_i} f_i(t_1, \ldots, t_{t-i}, u_i, t_{i+1}, \ldots, t_n) du_i$$

and

$$f(t) = -\int_{t_i}^{b_i} f_i(t_1, \ldots, t_{i-1}, u_i, t_{i+1}, \ldots, t_n) du_i .$$

Taking absolute value of each of these equations and adding them up with respect to i, we conclude that

$$2n|f(t)| \leq \sum_i \int_{a_i}^{b_i} |f_i(t_1, \ldots, t_{i-1}, u_i, t_{i+1}, \ldots, t_n)| du_i . \qquad \text{Q.E.D.}$$

Proof of Theorem 3.1. By Lemma A, Lemma 3.1 and Lemma B, we have

$$\prod_\alpha |f^\alpha(t)|^{q_\alpha} = \left(\prod_\beta C_\beta^{-q_\beta}\right) \prod_\alpha |C_\alpha f^\alpha(t)|^{q_\alpha}$$

$$\leq C \sum_\alpha \frac{q_\alpha}{p_\alpha} C_\alpha^{p_\alpha} |f^\alpha(t)|^{p_\alpha}$$

$$\leq C \sum_\alpha \frac{q_\alpha}{p_\alpha} C_\alpha^{p_\alpha} \left[\frac{1}{2n} \sum_i \int_{a_i}^{b_i} |f_i^\alpha(t_1, \ldots, u_i, \ldots, t_n)| du_i\right]^{p_\alpha}$$

$$\leq C \sum_\alpha \frac{q_\alpha}{p_\alpha} C_\alpha^{p_\alpha} \left(\frac{1}{2n}\right)^{p_\alpha} c(p_\alpha, n)$$

$$\times \sum_i \left(\int_{a_i}^{b_i} |f_i^\alpha(t_1, \ldots, u_i, \ldots, t_n)| du_i\right)^{p_\alpha}$$

for all $t \in \Omega$. Since $p_\alpha \geq 1$, $c(p_\alpha, n) = n^{p_\alpha - 1}$ and so

$$\prod_\alpha |f^\alpha(t)|^{q_\alpha} \leq \frac{C}{n} \sum_\alpha \frac{q_\alpha}{p_\alpha} \left(\frac{C_\alpha}{2}\right)^{p_\alpha} \sum_i \left(\int_{a_i}^{b_i} |f_i^\alpha(t_1, \ldots, u_i, \ldots, t_n)| du_i\right)^{p_\alpha} .$$

By Hölder's Inequality, we have

$$\prod_\alpha |f^\alpha(t)|^{q_\alpha}$$

$$\leq \frac{C}{n} \sum_\alpha \frac{q_\alpha}{p_\alpha} \left(\frac{C_\alpha}{2}\right)^{p_\alpha} \sum_i \left[(b_i - a_i)^{\frac{p_\alpha - 1}{p_\alpha}} \left(\int_{a_i}^{b_i} |f_i^\alpha(t_1, \ldots, u_i, \ldots, t_n)|^{p_\alpha} du_i\right)^{\frac{1}{p_\alpha}}\right]^{p_\alpha}$$

$$= \frac{C}{n} \sum_\alpha \frac{q_\alpha}{p_\alpha} \left(\frac{C_\alpha}{2}\right)^{p_\alpha} M^{p_\alpha - 1} \sum_i \int_{a_i}^{b_i} |f_i^\alpha(t_1, \ldots, u_i, \ldots, t_n)|^{p_\alpha} du_i .$$

Now since

$$\sum_i \int_\Omega \int_{a_i}^{b_i} |f_i^\alpha(t_1, \ldots, u_i, \ldots, t_n)|^{p_\alpha} \, du_i \, dt$$

$$= \sum_i \int_{a_i}^{b_i} \int_\Omega |f_i^\alpha(t)|^{p_\alpha} \, dt \, du_i$$

$$= \sum_i \left(\int_\Omega |f_i^\alpha(t)|^{p_\alpha} \, dt \right)(b_i - a_i)$$

$$\leq M \int_\Omega \sum_i |f_i^\alpha(t)|^{p_\alpha} \, dt \, ,$$

we have

$$\int_\Omega \prod_\alpha |f^\alpha(t)|^{q_\alpha} \, dt$$

$$\leq \frac{C}{n} \sum_\alpha \frac{q_\alpha}{p_\alpha} \left(\frac{C_\alpha}{2}\right)^{p_\alpha} M^{p_\alpha - 1} \sum_i \int_\Omega \int_{a_i}^{b_i} |f_i^\alpha(t_1, \ldots, u_i, \ldots, t_n)|^{p_\alpha} \, du_i \, dt$$

$$\leq \frac{C}{n} \sum_\alpha \frac{q_\alpha}{p_\alpha} \left(\frac{C_\alpha}{2}\right)^{p_\alpha} M^{p_\alpha} \int_\Omega \sum_i |f_i^\alpha(t)|^{p_\alpha} \, dt$$

$$= \frac{C}{n} \sum_\alpha \frac{q_\alpha}{p_\alpha} \left(\frac{MC_\alpha}{2}\right)^{p_\alpha} \int_\Omega \left[\left(\sum_i |f_i^\alpha(t)|^{p_\alpha} \right)^{2/p_\alpha} \right]^{p_\alpha/2} dt$$

$$\leq \frac{C}{n} \sum_\alpha \frac{q_\alpha}{p_\alpha} \left(\frac{MC_\alpha}{2}\right)^{p_\alpha} \int_\Omega \left[c\left(\frac{2}{p_\alpha}, n\right) \sum_i (|f_i^\alpha(t)|^{p_\alpha})^{2/p_\alpha} \right]^{p_\alpha/2} dt$$

by Lemma B. Since $p_\alpha \geq 2$, $c\left(\dfrac{2}{p_\alpha}, n\right) = 1$ for all α and so

$$\int_\Omega \prod_\alpha |f^\alpha(t)|^{q_\alpha} \, dt \leq \frac{C}{n} \sum_\alpha \frac{q_\alpha}{p_\alpha} \left(\frac{MC_\alpha}{2}\right)^{p_\alpha} \int_\Omega |\nabla f^\alpha(t)|^{p_\alpha} \, dt \, . \qquad \text{Q.E.D.}$$

Note that from the preceding results, Poincaré-type integral inequalities involving only 1 function (the case $m = 1$) can be obtained easily for free. For instance, we have

Corollary 3.6. *For any* $f \in C_0^1(\Omega)$, *and any real number* $q \geq 2$,

$$\int_\Omega |f|^q \leq \frac{1}{n} \left(\frac{M}{2}\right)^q \int_\Omega |\nabla f|^q \, .$$

Proof. This follows from Corollary 3.4 by letting $f^\alpha = f$ for all α. Q.E.D.

Remark. Note that when $m = n = 2$ and Ω is a square, the Poincaré-type integral inequality obtained in Corollary 3.6 above is sharper than that in [17]. In fact, here our constant is $M^2/8$, while in [17] the constant is $7M^2/12$.

4. Sobolev-type Integral Inequalities

Again we let $M = \max\{b_i - a_i : i = 1, \ldots, n\}$.

Theorem 4.1. *For any $f^\alpha \in C_0^1(\Omega)$, any real numbers $p_\alpha \geq 1$, $q_\alpha \geq 0$ with $\sum\limits_\alpha q_\alpha/p_\alpha = 1$, and any $C_\alpha > 0$,*

$$\int_\Omega \prod_\alpha |f^\alpha|^{q_\alpha} \leq \frac{CM}{2\sqrt{n}} \sum_\alpha q_\alpha C_\alpha^{p_\alpha} \left(\int_\Omega |f^\alpha|^{2(p_\alpha - 1)} \right)^{1/2} \left(\int_\Omega |\nabla f^\alpha|^2 \right)^{1/2} ,$$

where

$$C := \prod_\beta C_\beta^{-q_\beta} .$$

Theorem 4.1 generalizes and improves some existing results of Sobolev-type integral inequalities in the literature [2, 6, 17, 20, 21, 22]. For instance, we have the following consequences.

Corollary 4.1. [2, 6]. *For any $f^\alpha \in C_0^1(\Omega)$ and any real numbers $p_\alpha \geq 1$, $q_\alpha \geq 0$ with $\sum\limits_\alpha q_\alpha/p_\alpha = 1$,*

$$\int_\Omega \prod_\alpha |f^\alpha|^{q_\alpha} \leq \frac{M}{2\sqrt{n}} \sum_\alpha q_\alpha \left(\int_\Omega |f^\alpha|^{2(p_\alpha - 1)} \right)^{1/2} \left(\int_\Omega |\nabla f^\alpha|^2 \right)^{1/2} .$$

Proof. This follows immediately from Theorem 4.1 by setting $C_\alpha = 1$ for all α.   Q.E.D.

Corollary 4.2. *For any $f^\alpha \in C_0^1(\Omega)$ and any real numbers $p_\alpha \geq 1$, $q_\alpha > 0$ with $\sum\limits_\alpha q_\alpha/p_\alpha = 1$,*

$$\int_\Omega \prod_\alpha |f^\alpha|^{q_\alpha} \leq \frac{CM}{2\sqrt{n}} \sum_\alpha \left(\int_\Omega |f^\alpha|^{2(p_\alpha - 1)} \right)^{1/2} \left(\int_\Omega |\nabla f^\alpha|^2 \right)^{1/2} ,$$

where

$$C := \prod_\alpha (q_\alpha)^{q_\alpha/p_\alpha} .$$

Proof. This is immediate from Theorem 4.1 by letting $C_\alpha = q_\alpha^{-1/p_\alpha}$ for all α.   Q.E.D.

Corollary 4.3 [2, 6]. *For any $f^\alpha \in C_0^1(\Omega)$ and any real numbers $p_\alpha \geq 0$ with $\sum\limits_\alpha 1/p_\alpha = 1$,*

$$\int_\Omega \prod_\alpha |f^\alpha| \leq \frac{M}{2\sqrt{n}} \sum_\alpha \left(\int_\Omega |f^\alpha|^{2(p_\alpha - 1)} \right)^{1/2} \left(\int_\Omega |\nabla f^\alpha|^2 \right)^{1/2}.$$

Proof. This follows from Corollary 4.1 by putting $q_\alpha = 1$ for all α. Q.E.D.

Corollary 4.4 [2, 6]. *For any $f^\alpha \in C_0^1(\Omega)$ and any real numbers $q_\alpha \geq 0$ with $q := \sum q_\alpha \geq 1$,*

$$\int_\Omega \prod_\alpha |f^\alpha|^{q_\alpha} \leq \frac{M}{2\sqrt{n}} \sum_\alpha q_\alpha \left(\int_\Omega |f^\alpha|^{2(q-1)} \right)^{1/2} \left(\int_\Omega |\nabla f^\alpha|^2 \right)^{1/2}.$$

Proof. This follows from Corollary 4.1 by putting $p_\alpha = q$ for all α. Q.E.D.

Corollary 4.5. *For any $f^\alpha \in C_0^1(\Omega)$,*

$$\int_\Omega \prod_\alpha |f^\alpha| \leq \frac{M}{2\sqrt{n}} \sum_\alpha \left(\int_\Omega |f^\alpha|^{2(m-1)} \right)^{1/2} \left(\int_\Omega |\nabla f^\alpha|^2 \right)^{1/2}.$$

In particular, when $m = 2$ and $f, g \in C_0^1(\Omega)$,

$$\int_\Omega |fg| \leq \frac{M}{2\sqrt{n}} \left(\|f\| \|\nabla f\| + \|g\| \|\nabla g\| \right),$$

where, as usual, $\|\cdot\|$ is the L^2-norm on $C_0^1(\Omega)$.

Proof. This follows immediately from Corollary 4.3 by setting $p_\alpha = m$ for all α or from Corollary 4.4 by setting $q_\alpha = 1$ for all α. Q.E.D.

To establish Theorem 4.1, we need the following basic lemma.

Lemma 4.1. *If $f \in C_0^1(\Omega)$, then for any $t \in \Omega$ and $s \geq 1$,*

$$|f(t)|^s \leq \frac{s}{2n} \sum_i \int_{a_i}^{b_i} |f(t_1, \dots, u_i, \dots, t_n)|^{s-1} |f_i(t_1, \dots, u_i, \dots, t_n)| du_i.$$

Proof. Since f vanishes on $\partial\Omega$, for any $i = 1, \dots, n$, $s \geq 1$, and any fixed $t \in \Omega$, we have

$$f(t)^s = s \int_{a_i}^{t_i} f(t_1, \dots, u_i, \dots, t_n)^{s-1} f_i(t_1, \dots, u_i, \dots, t_n) du_i,$$

thus

$$nf(t)^s = s \sum_i \int_{a_i}^{t_i} f(t_1, \ldots, u_i, \ldots, t_n)^{s-1} f_i(t_1, \ldots, u_i, \ldots, t_n) du_i \ .$$

Similarly, we have

$$nf(t)^s = -s \sum_i \int_{t_i}^{b_i} f(t_1, \ldots, u_i, \ldots, t_n)^{s-1} f_i(t_1, \ldots, u_i, \ldots, t_n) du_i \ .$$

Taking absolute value of these equations and summing up, we have

$$|f(t)|^s \le \frac{s}{2n} \sum_i \int_{a_i}^{b_i} |f(t_1, \ldots, u_i, \ldots, t_n)|^{s-1} |f_i(t_1, \ldots, u_i, \ldots, t_n)| du_i \ . \quad \text{Q.E.D.}$$

Proof of Theorem 4.1. By Lemma A and Lemma 4.1, we have

$$\prod_\alpha |f^\alpha(t)|^{q_\alpha} = C \prod_\alpha |C_\alpha f^\alpha(t)|^{q_\alpha}$$

$$\le C \sum_\alpha \frac{q_\alpha}{p_\alpha} C_\alpha^{p_\alpha} |f^\alpha(t)|^{p_\alpha}$$

$$\le C \sum_\alpha \frac{q_\alpha}{2n} C_\alpha^{p_\alpha} \sum_i \int_{a_i}^{b_i} |f^\alpha(t_1, \ldots, u_i, \ldots, t_n)|^{p_\alpha - 1}$$
$$\times |f_i^\alpha(t_1, \ldots, u_i, \ldots, t_n)| du_i$$

$$= \frac{C}{2n} \sum_{\alpha,i} q_\alpha C_\alpha^{p_\alpha} \int_{a_i}^{b_i} |f^\alpha(t_1, \ldots, u_i, \ldots, t_n)|^{p_\alpha - 1}$$
$$\times |f_i^\alpha(t_1, \ldots, u_i, \ldots, t_n)| du_i$$

for all $t \in \Omega$. Thus

$$\int_\Omega \prod_\alpha |f^\alpha(t)|^{q_\alpha} dt \le \frac{C}{2n} \sum_{\alpha,i} q_\alpha C_\alpha^{p_\alpha} \int_\Omega \int_{a_i}^{b_i} |f^\alpha(t_1, \ldots, u_i, \ldots, t_n)|^{p_\alpha - 1}$$
$$\times |f_i^\alpha(t_1, \ldots, u_i, \ldots, t_n)| du_i \, dt$$

$$= \frac{C}{2n} \sum_{\alpha,i} q_\alpha C_\alpha^{p_\alpha} \int_{a_i}^{b_i} \int_\Omega |f^\alpha(t)|^{p_\alpha - 1} |f_i^\alpha(t)| dt \, du_i$$

$$\le \frac{CM}{2n} \sum_\alpha q_\alpha C_\alpha^{p_\alpha} \int_\Omega |f^\alpha(t)|^{p_\alpha - 1} \sum_i |f_i^\alpha(t)| dt$$

$$\le \frac{CM}{2n} \sum_\alpha q_\alpha C_\alpha^{p_\alpha} \left(\int_\Omega |f^\alpha(t)|^{2(p_\alpha - 1)} dt \right)^{1/2}$$
$$\times \left(\int_\Omega \left[\sum_i |f_i^\alpha(t)| \right]^2 dt \right)^{1/2}$$

$$\le \frac{CM}{2\sqrt{n}} \sum_\alpha q_\alpha C_\alpha^{p_\alpha} \left(\int_\Omega |f^\alpha(t)|^{2(p_\alpha - 1)} dt \right)^{1/2}$$
$$\times \left(\int_\Omega |\nabla f^\alpha(t)|^2 dt \right)^{1/2} ,$$

where the last two inequalities follow from Hölder's inequality and Lemma B, respectively.

Q.E.D.

Similar to the situation for Poincaré-type integral inequalities, our results here for Sobolev-type integral inequalities for the case of many functions ($m \geq 2$) can generate integral inequalities for the case of one function ($n = 1$). For instance, we have

Corollary 4.6. *For any $f \in C_0^1(\Omega)$ and any real number $q \geq 1$,*

$$\int_\Omega |f|^q \leq \frac{Mq}{2\sqrt{n}} \left(\int_\Omega |f|^{2(q-1)} \right)^{1/2} \left(\int_\Omega |\nabla f|^2 \right)^{1/2}.$$

Proof. This is obvious by letting $f^\alpha = f$ for all α in Corollary 4.4.　　　　Q.E.D.

5. Applications

With slight modification of the techniques used in the proofs of results in the previous sections, we can easily arrive at some new integral inequalities which are seen to be variations of Poincaré-type and Sobolev-type inequalities.

As above, let $M = \max\{b_i - a_i : i = 1, \dots, n\}$.

Theorem 5.1. *For any $f^\alpha \in C_0^1(\Omega)$, any real numbers $p_\alpha \geq 2$, $q_\alpha \geq 0$ with $\sum_\alpha q_\alpha/p_\alpha = 1$, and any $C_\alpha > 0$,*

$$\int_\Omega \sum_\beta \left(\prod_{\alpha \neq \beta} |f^\alpha|^{q_\alpha} \right) |\nabla f^\beta|^{q_\beta} \leq C\, K(p,q) \cdot \sum_\alpha \frac{q_\alpha}{p_\alpha} C_\alpha^{p_\alpha} \int_\Omega |\nabla f^\alpha|^{p_\alpha},$$

where

$$C := \prod_\beta C_\beta^{-q_\beta}$$

and

$$K(p,q) = K(p_\alpha, q_\alpha) := \sum_\beta \left(\frac{1}{n} \right)^{1 - \frac{q_\beta}{p_\beta}} \left(\frac{M}{2} \right)^{\sum_{\alpha \neq \beta} q_\alpha}.$$

Proof. By a generalization of Hölder's inequality for the case of many functions and by Corollary 3.6, we have

$$\int_\Omega \sum_\beta \Big(\prod_{\alpha \neq \beta} |f^\alpha|^{q_\alpha} \Big) |\nabla f^\beta|^{q_\beta}$$

$$= \sum_\beta \int_\Omega C \Big(\prod_{\alpha \neq \beta} |C_\alpha f^\alpha|^{q_\alpha} \Big) |C_\beta \nabla f^\beta|^{q_\beta}$$

$$\leq C \sum_\beta \Big\{ \Big[\prod_{\alpha \neq \beta} \Big(\int_\Omega |C_\alpha f^\alpha|^{p_\alpha} \Big)^{\frac{q_\alpha}{p_\alpha}} \Big] \Big(\int_\Omega |C_\beta \nabla f^\beta|^{p_\beta} \Big)^{\frac{q_\beta}{p_\beta}} \Big\}$$

$$\leq C \sum_\beta \Big(\frac{1}{n} \Big)^{\sum_{\alpha \neq \beta} \frac{q_\alpha}{p_\alpha}} \Big(\frac{M}{2} \Big)^{\sum_{\alpha \neq \beta} q_\alpha} \prod_\alpha \Big(\int_\Omega |C_\alpha \nabla f^\alpha|^{p_\alpha} \Big)^{\frac{q_\alpha}{p_\alpha}}$$

$$= C\,K(p,q) \cdot \prod_\alpha \Big(\int_\Omega |C_\alpha \nabla f^\alpha|^{p_\alpha} \Big)^{\frac{q_\alpha}{p_\alpha}}$$

and so by Lemma A, we conclude that

$$\int_\Omega \sum_\beta \Big(\prod_{\alpha \neq \beta} |f^\alpha|^{q_\alpha} \Big) |\nabla f^\beta|^{q_\beta}$$

$$\leq C\,K(p,q) \cdot \sum_\alpha \frac{q_\alpha}{p_\alpha} \int_\Omega |C_\alpha \nabla f^\alpha|^{p_\alpha}$$

$$= C\,K(p,q) \cdot \sum_\alpha \frac{q_\alpha}{p_\alpha} C_\alpha^{p_\alpha} \int_\Omega |\nabla f^\alpha|^{q_\alpha} . \qquad\qquad \text{Q.E.D.}$$

Theorem 5.1 generalizes and improves some existing results in the literature [3, 6, 7, 17, 21]. The following are some immediate consequences of the theorem.

Corollary 5.1 [7]. *For any $f^\alpha \in C_0^1(\Omega)$ and any real numbers $p_\alpha \geq 2$, $q_\alpha \geq 0$ with $\sum_\alpha q_\alpha/p_\alpha = 1$,*

$$\int_\Omega \sum_\beta \Big(\prod_{\alpha \neq \beta} |f^\alpha|^{q_\alpha} \Big) |\nabla f^\beta|^{q_\beta} \leq K(p,q) \sum_\alpha \frac{q_\alpha}{p_\alpha} \int_\Omega |\nabla f^\alpha|^{p_\alpha} ,$$

where $K(p,q)$ is defined as in Theorem 5.1.

Proof. This follows from Theorem 5.1 by letting $C_\alpha = 1$ for all α. Q.E.D.

Corollary 5.2. *For any $f^\alpha \in C_0^1(\Omega)$ and any real numbers $p_\alpha \geq 2$, $q_\alpha > 0$ with $\sum q_\alpha/p_\alpha = 1$,*

$$\int_\Omega \sum_\beta \Big(\prod_{\alpha \neq \beta} |f^\alpha|^{q_\alpha} \Big) |\nabla f^\beta|^{q_\beta} \leq C\,K(p,q) \sum_\alpha \int_\Omega |\nabla f^\alpha|^{p_\alpha} ,$$

where $K(p,q)$ is defined as in Theorem 5.1. and

$$C := \prod_\alpha \left(\frac{q_\alpha}{p_\alpha}\right)^{\frac{q_\alpha}{p_\alpha}} .$$

Proof. This follows from Theorem 5.1 by setting

$$C_\alpha = \left(\frac{p_\alpha}{q_\alpha}\right)^{\frac{1}{p_\alpha}}$$

for all α.   Q.E.D.

Corollary 5.3 [3, 6]. *For any $f^\alpha \in C_0^1(\Omega)$ and any real numbers $p_\alpha \geq 2$ satisfying $\sum_\alpha 1/p_\alpha = 1$,*

$$\int_\Omega \sum_\beta \left(\prod_{\alpha\neq\beta} |f^\alpha|\right) |\nabla f^\beta| \leq \left(\frac{M}{2}\right)^{m-1} \left[\sum_\alpha \left(\frac{1}{n}\right)^{1-\frac{1}{p_\alpha}}\right] \left[\sum_\alpha \frac{1}{p_\alpha} \int_\Omega |\nabla f^\alpha|^{p_\alpha}\right] .$$

Proof. This follows immediately from Corollary 5.1 by setting $q_\alpha = 1$ for all α.   Q.E.D.

Corollary 5.4. *For any $f^\alpha \in C_0^1(\Omega)$ and any real numbers $q_\alpha \geq 0$ with $q := \sum_\alpha q_\alpha \geq 2$,*

$$\int_\Omega \sum_\beta \left(\prod_{\alpha\neq\beta} |f^\alpha|^{q_\alpha}\right) |\nabla f^\beta|^{q_\beta} \leq \frac{K(q)}{q} \sum_\alpha q_\alpha \int_\Omega |\nabla f^\alpha|^q ,$$

where

$$K(q) = K(q_\alpha) := \sum_\beta \left(\frac{1}{n}\right)^{1-\frac{q_\beta}{q}} \left(\frac{M}{2}\right)^{q-q_\beta} .$$

Proof. It is immediate from Corollary 5.1 by letting $p_\alpha = q$ for all α.   Q.E.D.

Corollary 5.5. *For any $f^\alpha \in C_0^1(\Omega)$,*

$$\int_\Omega \sum_\beta \left(\prod_{\alpha\neq\beta} |f^\alpha|\right) |\nabla f^\beta| \leq \left(\frac{M}{2}\right)^{m-1} \left(\frac{1}{n}\right)^{1-\frac{1}{m}} \sum_\alpha \int_\Omega |\nabla f^\alpha|^m .$$

Proof. This follows from Corollary 5.3 by letting $p_\alpha = m$ for all α or from Corollary 5.4 by letting $q_\alpha = 1$ for all α.   Q.E.D.

Again Theorem 5.1 and its consequences also give new integral inequalities for the case $m = 1$ (i.e., with only one dependent function) for free. For instance, we have

Corollary 5.6. *For any $f \in C_0^1(\Omega)$ and any real numbers $q_\alpha \geq 0$ with $q := \sum_\alpha q_\alpha \geq 2$,*

$$\int_\Omega \sum_\alpha |f|^{q-q_\alpha} |\nabla f|^{q_\alpha} \leq K(q) \int_\Omega |\nabla f|^q \,,$$

where $K(q)$ is defined as in Corollary 5.4. In particular,

$$\int_\Omega |f|^{m-1} |\nabla f| \leq \left[\left(\frac{1}{n} \right)^{1-\frac{1}{m}} \left(\frac{M}{2} \right)^{m-1} \right] \cdot \int_\Omega |\nabla f|^m \,.$$

Proof. These follow from Corollary 5.4 by letting $f^\alpha = f$ for all α and then $q_\alpha = 1$ for all α.   Q.E.D.

Remark. Further interesting integral inequalities can easily be generated by the results in the preceding sections. For instance, by taking $m = 3$ in Corollary 5.5, we have

$$\int_\Omega |fg \nabla h + gh \nabla f + hf \nabla g| \leq \frac{M^2}{4n^{2/3}} \left[\int_\Omega |\nabla f|^3 + \int_\Omega |\nabla g|^3 + \int_\Omega |\nabla h|^3 \right] ;$$

taking $m = 2$ in Corollary 5.5, we have

$$\int_\Omega |f \nabla g + g \nabla f| \leq \frac{M}{2\sqrt{n}} \left[\int_\Omega |\nabla f|^2 + \int_\Omega |\nabla g|^2 \right] .$$

By putting $f = g = h$ in these inequalities (or by using Corollary 5, 6 directly), we obtain respectively

$$\int_\Omega f^2 |\nabla f| \leq \frac{M^2}{4n^{2/3}} \int_\Omega |\nabla f|^3$$

and

$$\int_\Omega |f| |\nabla f| \leq \frac{M}{2\sqrt{n}} \int_\Omega |\nabla f|^2 .$$

On the other hand, using Hölder's inequality, we have

$$\int_\Omega |f| |\nabla g| \leq \left(\int_\Omega |f|^2 \right)^{1/2} \left(\int_\Omega |\nabla g|^2 \right)^{1/2} ,$$

hence by Corollary 3.6,

$$\int_\Omega |f| |\nabla g| \leq \frac{M}{2\sqrt{n}} \left(\int_\Omega |\nabla f|^2 \right)^{1/2} \left(\int_\Omega |\nabla g|^2 \right)^{1/2}$$

and by Corollary 4.6,

$$\int_\Omega |f| |\nabla g| \le \left(\frac{M}{\sqrt{n}}\right)^{1/2} \left(\int_\Omega |f|^2\right)^{1/4} \left(\int_\Omega |\nabla f|^2\right)^{1/2} \left(\int_\Omega |\nabla g|^2\right)^{1/2}.$$

Similarly, other interesting integral inequalities involving the gradient of $C_0^1(\Omega)$ functions can easily be generated. Integral inequalities of such forms are in general of great interest in its own right and meanwhile are important in the study of both quantitative and qualitative properties of solutions of differential and integral equations. The importance of our results also lie in that by choosing different combinations of the parameters m, n, p_α, q_α, etc., we can obtain as many as we wish new integral inequalities involving the gradient of $C_0^1(\Omega)$ functions.

References

1. E.F. Beckenbach and R. Bellman, *Inequalities*, Springer-Verlag, New York, 1965.
2. W.S. Cheung, *On integral inequalities of the Sobolev type*, Aeq. Math. **49** (1995), 153-159.
3. _____, *On Poincaré-type integral inequalities*, Proc. Amer. Math. Soc. **119**, no. 3 (1993), 857-863.
4. _____, *On some new integrodifferential inequalities of the Gronwell and Wendroff type*, J. Math. Anal. Appl. **178**, no.2 (1993),438-449.
5. _____, *Opial-type inequalities with m functions in n variables*, Mathematika **39** (1992), 319-326.
6. _____, *Some multi-dimensional integral inequalities*, to appear in Nonlinear Math. Anal. Appl.
7. _____, *Some new Poincaré-type inequalities*, preprint.
8. W.S. Cheung, Z. Hanjš and J. Pečarić, *Some Hardy-type inequalities*, preprint.
9. W.S. Cheung and J. Pečarić, *Multi-dimensional integral inequalities of the Wirtinger-type*, to appear in Math. Ineq. Appl.
10. A. Friedman, *Partial differential equations of parabolic type*, Prentice-Hall, Englewood Cliffs, NY, 1964.
11. G.H. Hardy, J.E. Littlewood, and G. Pólya, *Inequalities*, Cambridge Univ. Press, Cambridge, 1952.
12. C.O. Horgan, *Eigenvalue estimates and the trace problem*, J. Math. Anal. Appl. **69** (1979), 231-242.
13. _____, *Integral bounds for solutions of nonlinear reaction-diffusion equations*, J. Appl. Math. Phys. (ZAMP) **28** (1977), 197-204.
14. C.O. Horgan and R.R. Nachlinger, *On the domain of attraction for steady states in heat conduction*, Internat. J. Engrg. Sci. **14** (1976), 143-148.
15. C.O. Horgan and L.T. Wheeler, *Spatial decay estimates for the Navier-Stokes equations with applications to the problem of entry flow*, SIAM J. Appl. Math. **35** (1978), 97-116.
16. G.V. Milovanović, D.S. Mitrinović and Th.M. Rassias, *Topics in Polynomials: Extremal Problems, Inequalities, Zeros*, World Scientific Publ. Co., 1994.

17. D.S. Mitrinović, *Analytic Inequalities*, Springer-Verlag, New York, 1970.
18. L. Nirenberg, *On elliptic partial differential equations*, Ann. Scuola Norm. Sup. Pisa. Cl. Sci. (4) **13** (1959), 116-162.
19. B.G. Pachpatte, *On Poincaré type integral inequalities*, J. Math. Anal. Appl. **114** (1986), 111-115.
20. ———, *On Sobolev type integral inequalities*, Proc. Royal Soc. Edinburgh **103A** (1986), 1-14.
21. ———, *On some new integral inequalities in several independent variables*, Chinese J. Math. **14**, no. 2 (1986), 69-79.
22. ———, *On two inequalities of the Sobolev type*, Chinese J. Math. **15**, no. 4 (1987), 247-252.
23. H. Poincaré, *Sur les équations de la physique mathématique*, Rend. Circ. Mat. Palermo **8** (1894) 57-156, or Oeuvres de Henri Poincaré, Paris, 1954.
24. Th.M. Rassias, *On certain properties of eigenvalues and the Poincaré Inequality*, Global Analysis - Analysis on Manifolds, Teubner-Texte zur Math, Teubner Leipzig, Band **57** (1983), 282-300.
25. Th.M. Rassias, *Un contre-exemple à l'inégalité de Poincaré*, C.R. Acad. Sciences Paris **284** (1977), 409-412.

Department of Mathematics
University of Hong Kong
Hong Kong
wscheung@ hkucc.hku.hk

Department of Mathematics
National Technical University of Athens
Zografou Campus
15780 Athens
Greece
trassias@math.ntua.gr

SOME INEQUALITIES FOR TRIANGLE:
OLD AND NEW RESULTS

RADOSAV Ž. DJORDJEVIĆ

Faculty of Electronic Engineering, Department of Mathematics, P.O. Box 73, 18000 Niš, Yugoslavia

Abstract. In this paper a brief review of known results from the area of inequalities for triangle is given, the classification of these results is done, and for ineqalities chosen in this way, their hierarchy, severity and some of their generalizations are pointed out. Also, a three new inequalities for elements of triangle are proved.

0. Introduction

Geometrical inequalities take very important place in the huge theory of inequalities. These geometrical inequalities are, first of all, those which relate to the elements of triangle, but also to the elements of other geometrical forms, for example, quadrangle, polygon, sphere, tetraedar.

The importance of geometrical inequalities is certainly proved by a great number of mathematicians who worked or are still working on the area of geometrical inequalities. That number is almost 800 mathematicians, and some of them are very famous ones.

It is known that geometrical inequalities are not a new mathematical area. The first geometrical inequalities date from Euclide, i.e. from the known inequalities for sides of a triangle: $a + b > c$ and $|a - b| < c$.

But it is really interesting that this area is up-to-date nowadays too. Naturally, at the beginning, geometrical inequalities appeared casually. They appeared almost spontaneously during the work on some other problems of geometry, especially during the analytical approach to their solution.

Later on, a good number of geometrical inequalities appeared due to their more organized study. A great number of scientific works dedicated to geometrical inequalities were published. Mathematical problems dedicated to this area were also presented in many scientific journals. Published solutions of these problems induced a lot of mathematicians to work more intensively on geometrical inequalities.

In fact, as in other areas of mathematics, a lot of generalizations, as well as a lot of intensifications of already known geometrical inequalities appeared.

1991 *Mathematics Subject Classification.* Primary 51M05, 51M16.
Key words and phrases. Geometric inequalities; Triangle; Inequality; Equality.
This work was supported in part by the Serbian Scientific Foundation, grant number 04M03.

T.M. Rassias and H.M. Srivastava (eds.), Analytic and Geometric Inequalities and Applications, 69–92.

However, neither the geometrical inequalities, nor any of their part, were gathered, systematized and published.

The first steps in that sense dated from the appearance of relatively modest books, sometimes a little book, which paid attention, in the whole or only with its one part, to the geometrical inequalities.

But nowadays, several important monographs on geometrical inequalities have appeared, and some of them can be mentioned as follows:

1. D. S. MITRINOVIĆ, P. M. VASIĆ, R. Ž. DJORDJEVIĆ and R. R. JANIĆ: *Geometrijske nejednakosti.* Zavod za izdavanje udžbenika. Beograd 1966.

2. O. BOTTEMA, R. Ž. DJORDJEVIĆ, R. R. JANIĆ, D. S. MITRINOVIĆ and P. M. VASIĆ: *Geometric Inequalities.* Wolters-Noordhoff Publishing. Groningen 1969.

3. V. P. SOLTAN AND S. I. MEJDMAN: *Toždestva i neravenstva v treugol'nike.* Stiinca. Kišinev 1982.

4. D. S. MITRINOVIĆ, J. E. PEČARIĆ AND V. VOLENEC: *Recent Advances in Geometric Inequalities.* Kluwer Academic Publishers. Dordrecht-Boston-London 1989.

In each of these monographs, the corresponding classification of treated geometrical inequalities is done.

And, naturally, it is shown that the number of inequalities for elements of triangle exceeds, to a great extent, the number of all other geometrical inequalities.

In contrast to previous classifications, the classification of geometrical inequalities will be done here according to the expressions, in symbol S, which exceptions are determined.

So, if it is not indicated in some different way, in all inequalities, equalities hold if and only if the triangle is equilateral.

Known inequalities are mentioned under certain number, and prooves of each inequalities, as well as possible comments related to them, can be found in texts of the corresponding references, cited in each inequality.

Our comments will be presented through the corresponding *Remarks*

We use the standard notation in the theory of geometric inequalities.

New results are also cited in the classification already done, and their prooves can be found in theorems formulated and proved at the end of this paper, in the section *Some New Results.*

1. The Oldest Geometric Inequalities

We start this section with the oldest geometric inequalities which are well-known in literature and which were discovered before 1930. Some of them were rediscovered several times.

Corresponding inequalities from this section are connected, with each other, by comments also presented through *Remarks*.

1.1. Inequality for $\underline{S = 2r}$

$$2r \leq R.$$

Some details about this inequality can be found in:

1. L. Euleri, Novi commentarii academiae scientarium Petropolitanae **11** (1765), 103–123.
2. L. Euleri, *Opera omnia*, I 26, 1953, pp. 139–157.
3. Ramus and É. Rouché, *Question* 233, Nouv. Ann. Math. **10** (1851), 353–355.
4. O. Bottema, R. Ž. Djordjević, R. R. Janić, D. S. Mitrinović and P. M. Vasić, *Geometric Inequalities*, Wolters-Noordhoff Publishing, Groningen, 1969. (Ineq. 5.1).

1.2. Inequalities for $\underline{S = F^2}$

$$r^2\left(2R^2 + 10Rr - r^2 - 2(R - 2r)\sqrt{R(R - 2r)}\right)$$
$$\leq F^2$$
$$\leq r^2\left(2R^2 + 10Rr - r^2 + 2(R - 2r)\sqrt{R(R - 2r)}\right).$$

If one of the signs $\leq$ in **1.2** means $=$, then the triangle is isosceles; and vice versa. If both signs $\leq$ in **1.2** stand for $=$, then the triangle is eqilateral; and vice versa.

Details about these inequalities one can find in:

1. Ramus and É. Rouché, *Question* 233, Nouv. Ann. Math. **10** (1851), 353–355.
2. A. Laisant, *Géométrie du triangle*, Paris, 1896.
3. O. Bottema, R. Ž. Djordjević, R. R. Janić, D. S. Mitrinović and P. M. Vasić, *Geometric Inequalities*, Wolters-Noordhoff Publishing, Groningen, 1969. (Ineq. 7.11).
4. D. S. Mitrinović, J. E. Pečarić and V. Volonec, *Recent Advances in Geometric Inequalities*, Kluwer Academic Publishers, Dordrecht-Boston-London, 1989 (Th. A)..

1.3. Inequality for $\underline{S = a^3 + b^3 + c^3 + 3abc}$

$$a^3 + b^3 + c^3 + 3abc \leq \frac{2}{3}(a + b + c)(a^2 + b^2 + c^2).$$

This inequality were treated in:

1. M. Collins and A. B. Evans, *Problem* 2837, Ed. Times **13** (1870), 30–31.
2. O. Bottema, R. Ž. Djordjević, R. R. Janić, D. S. Mitrinović and P. M. Vasić, *Geometric Inequalities*, Wolters-Noordhoff Publishing, Groningen, 1969. (Ineq. 1.6).

1.4. Inequalities for $\underline{S = s^2}$

$$3r(4R + r) \leq s^2 \leq \frac{1}{3}(4R + r)^2.$$

Some details about these inequalities can be found in:

1. G. Colombier and T. Doucet, *Problem* 1051, Nouv. Ann. Math. **31** (1872), 467.
2. O. Bottema, R. Ž. Djordjević, R. R. Janić, D. S. Mitrinović and P. M. Vasić, *Geometric Inequalities*, Wolters-Noordhoff Publishing, Groningen, 1969. (Ineq. 5.5).

Remark. See inequality **1.5**.

1.5. Inequality for $\underline{S = s^2}$

$$27r^2 \leq s^2.$$

Some details on this inequality one can find in:

1. N. A. Edwards, *Problem* 1273, Nouv. Ann. Math. **37** (1878), 475.
2. J. M. Child, Math. Gaz. **23** (1939), 138–143.
3. O. Bottema, R. Ž. Djordjević, R. R. Janić, D. S. Mitrinović and P. M. Vasić, *Geometric Inequalities*, Wolters-Noordhoff Publishing, Groningen, 1969. (Ineq. 5.11).

Remark. The first inequality in **1.4** is stronger than the inequality **1.5**.

1.6. Inequality for $\underline{S = 2s^2}$

$$6r(4R + r) \le 2s^2 \le 2(2R + r) + R^2.$$

Some details about these inequalities can be found in:

1. A. Emerich, *Problem* 10656, Math. Questions **54** (1891), 100.
2. F. Leuenberger and L. Carlitz, *Problem* E 1481, Amer. Math. Monthly **68** (1961), 803 and **69** (1962), 312.
3. O. Bottema, R. Ž. Djordjević, R. R. Janić, D. S. Mitrinović and P. M. Vasić, *Geometric Inequalities*, Wolters-Noordhoff Publishing, Groningen, 1969. (Ineq. 5.6).
4. D. S. Mitrinović, J. E. Pečarić and V. Volonec, *Recent Advances in Geometric Inequalities*, Kluwer Academic Publishers, Dordrecht-Boston-London, 1989. (p. 681, A. Bager).

1.7. Inequality for $\underline{S = (a^2 + b^2 + c^2)/(a^t + b^t + c^t)}$

$$\frac{a^2 + b^2 + c^2}{a^t + b^t + c^t} < \frac{2(a^{t-2} + b^{t-2} + c^{t-2})}{b^{t-2}c^{t-2} + c^{t-2}a^{t-2} + a^{t-2}b^{t-2}} \qquad (t \in \mathbb{R}).$$

For details see:

1. I. Bénézech, *Problem* 412, J. Math. Elém. Paris (4) **1** (1892), 234.
2. O. Bottema, R. Ž. Djordjević, R. R. Janić, D. S. Mitrinović and P. M. Vasić, *Geometric Inequalities*, Wolters-Noordhoff Publishing, Groningen, 1969. (Ineq. 1.22).

1.8. Inequality for $\underline{S = (s - a)\sqrt{3}}$

$$(s - a)\sqrt{3} < 4R - r_a.$$

Some details on this inequality one can find in:

1. E. Lemoine, *Problem* 578, J. Math. Elem. (4) **3** (1894), 263.
2. O. Bottema, R. Ž. Djordjević, R. R. Janić, D. S. Mitrinović and P. M. Vasić, *Geometric Inequalities*, Wolters-Noordhoff Publishing, Groningen, 1969. (Ineq. 5.31).

1.9. Inequalities for $\underline{S = a^2 + b^2 + c^2}$

$$36r^2 \le a^2 + b^2 + c^2 \le 9R^2.$$

Some details about these inequalities can be found in:

1. J. Neuberg, Ed. Times, News Ser. **9** (1906), 51–52.
2. T. Kubota, Tôhoku Math. J. **25** (1925), 122–126.
3. J. Steinig, Elem. Math. **18** (1963), 127–131.
4. O. Bottema, R. Ž. Djordjević, R. R. Janić, D. S. Mitrinović and P. M. Vasić, *Geometric Inequalities*, Wolters-Noordhoff Publishing, Groningen, 1969. (Ineq. 5.13).

Remark. See inequality **1.11.**
Remark. See inequality **1.18.**

1.10. Inequality for $\underline{S = (b^2 + c^2 - a^2)(c^2 + a^2 - b^2)(a^2 + b^2 - c^2)}$

$$(b^2 + c^2 - a^2)(c^2 + a^2 - b^2)(a^2 + b^2 - c^2) \leq \left(\frac{4F}{\sqrt{3}}\right)^3.$$

Some details on this inequality one can find in:

1. F. Balitrand, Interméd. Math. **23** (1916), 86–87.
2. O. Bottema, R. Ž. Djordjević, R. R. Janić, D. S. Mitrinović and P. M. Vasić, *Geometric Inequalities*, Wolters-Noordhoff Publishing, Groningen, 1969. (Ineq. 4.20).

1.11. Inequalities for $\underline{S = (a^2 + b^2 + c^2)/(a + b + c)^2}$

$$\frac{1}{3} \leq \frac{a^2 + b^2 + c^2}{(a + b + c)^2} < \frac{1}{2}.$$

Some details about these inequalities can be found in:

1. M. Petrović, *Sur quelques fonctions des côtés et des angles d'une triangle*, Ens. Math. **18** (1916), 153–163.
2. O. Bottema, R. Ž. Djordjević, R. R. Janić, D. S. Mitrinović and P. M. Vasić, *Geometric Inequalities*, Wolters-Noordhoff Publishing, Groningen, 1969. (Ineq. 1.19).

Remark. The first inequality in **1.11** is reduced to an inequality which is more stringet then the first inequality in **1.9**.

1.12. Inequality for $\underline{S = \cos\alpha + \cos\beta}$

If $\alpha < \pi/2$ and $\beta < \pi/2$, then

$$\cos\alpha + \cos\beta > \sin\gamma.$$

If $\alpha > \pi/2$ or $\beta > \pi/2$, then reversed inequality is valid.

Some details on that can be found in:

1. A. Pantazi, Gaz. Mat. **23** (1917/18), 144.
2. V. Cristescu, Gaz. Mat. **26** (1920/21), 88–89.
3. B. M. Barbalatt, Gaz. Mat. **30** (1924/25), 380–381.
4. O. Bottema, R. Ž. Djordjević, R. R. Janić, D. S. Mitrinović and P. M. Vasić, *Geometric Inequalities*, Wolters-Noordhoff Publishing, Groningen, 1969. (Ineq. 2.25).

1.13. Inequality for $\underline{S = a^2 + b^2 + c^2}$

$$a^2 + b^2 + c^2 \geq 4\sqrt{3}\,F.$$

Some details on this inequality one can find in:

1. R. Weitzenböck, Math. Z. **5** (1919), 137–146.
2. P. Finsler and H. Hadwigwr, Comment. Math. Helv. **10** (1937/38), 316–326.
3. L. A. Santalo, Math. Notae **3** (1943), 65–73.

4. F. Goldner, *Problem* 69, Elem. Math. **4** (1949), 120.
5. O. Bottema, R. Ž. Djordjević, R. R. Janić, D. S. Mitrinović and P. M. Vasić, *Geometric Inequalities*, Wolters-Noordhoff Publishing, Groningen, 1969. (Ineq. 4.4).

Remark. See inequality **1.22**.

1.14. Inequality for $S = (a^2 + b^2 + c^2)^2$

$$(a^2 + b^2 + c^2)^2 \leq 64R^4 + 48F^2.$$

This inequality were treated in:

1. T. Hayashi, Tôhoku Science Rep. **11** (1922), 115–121.
2. O. Bottema, R. Ž. Djordjević, R. R. Janić, D. S. Mitrinović and P. M. Vasić, *Geometric Inequalities*, Wolters-Noordhoff Publishing, Groningen, 1969. (Ineq. 7.6).

1.15. Inequality for $S = a + b + c$

$$a + b + c \leq 3\sqrt{3}\, R.$$

Details about this inequality one can find in:

1. S. Nakajima, *Some Inequalities between the Fundamental Quantities of the Triangle*, Tôhoku Math. J. **25** (1925), 115–121.
2. A. Padoa, Period. Mat. (4) **5** (1925), 80–85.
3. O. Bottema, R. Ž. Djordjević, R. R. Janić, D. S. Mitrinović and P. M. Vasić, *Geometric Inequalities*, Wolters-Noordhoff Publishing, Groningen, 1969. (Ineq. 5.3).

Remark. The inequality **1.15** is lighter than the inequality which comes from the first inequality in **1.11**.
Remark. See inequality **1.16**.
Remark. See inequality **1.21**.

1.16. Inequalities for $S = \sin\alpha + \sin\beta + \sin\gamma$

$$0 < \sin\alpha + \sin\beta + \sin\gamma \leq \frac{3\sqrt{3}}{2}.$$

Some details on these inequalities one can find in:

1. A. Padoa, Period. Mat. (4) **5** (1925), 80–85.
2. T. R. Curry, *Problem* E 1644, Amer. Math. Monthly **70** (1963), 1099 and **71** (1964), 915–916.
4. O. Bottema, R. Ž. Djordjević, R. R. Janić, D. S. Mitrinović and P. M. Vasić, *Geometric Inequalities*, Wolters-Noordhoff Publishing, Groningen, 1969. (Ineq. 2.1).

Remark. Inequalities **1.16** and **1.15** are equivalent.

1.17. Inequality for $S = s^2$

$$s^2 \leq 4R^2 + \frac{11}{3\sqrt{3}}\, F.$$

This inequality were treated in:

1. S. Nakajima, *Some Inequalities between the Fundamental Quantities of the Triangle*, Tôhoku Math. J. **25** (1925), 115–121.
2. O. Bottema, R. Ž. Djordjević, R. R. Janić, D. S. Mitrinović and P. M. Vasić, *Geometric Inequalities*, Wolters-Noordhoff Publishing, Groningen, 1969. (Ineq. 7.3).

1.18. Inequality for $S = a^2 + b^2 + c^2$

$$a^2 + b^2 + c^2 \leq 8R^2 + \frac{4}{3\sqrt{3}}\,F.$$

Details about this inequality one can find in:

1. S. Nakajima, *Some Inequalities between the Fundamental Quantities of the Triangle*, Tôhoku Math. J. **25** (1925), 115–121.
2. O. Bottema, R. Ž. Djordjević, R. R. Janić, D. S. Mitrinović and P. M. Vasić, *Geometric Inequalities*, Wolters-Noordhoff Publishing, Groningen, 1969. (Ineq. 7.5).

Remark. Inequality 1.18 is stronger than the second inequality in 1.9.

1.19. Inequality for $S = (s-a)(s-b)(s-c)$

$$(s-a)(s-b)(s-c) \leq \frac{1}{8}\,abc.$$

Some details on this inequality one can find in:

1. A. Padoa, Period. Mat. (4) **5** (1925), 80–85.
2. O. Bottema, R. Ž. Djordjević, R. R. Janić, D. S. Mitrinović and P. M. Vasić, *Geometric Inequalities*, Wolters-Noordhoff Publishing, Groningen, 1969. (Ineq. 1.3).

1.20. Inequality for $S = \cos\alpha\,\cos\beta\,\cos\gamma$

$$\cos\alpha\,\cos\beta\,\cos\gamma \leq \frac{1}{8}\,.$$

This inequality were treated in:

1. C. C. Popovici, Gaz. Mat. **31** (1925), 132.
2. J. M. Child, Math. Gaz. **23** (1939), 138–143.
2. O. Bottema, R. Ž. Djordjević, R. R. Janić, D. S. Mitrinović and P. M. Vasić, *Geometric Inequalities*, Wolters-Noordhoff Publishing, Groningen, 1969. (Ineq. 2.23).

1.21. Inequality for $S = s^3(s-a)(s-b)(s-c)$

$$s^3(s-a)(s-b)(s-c) \leq \frac{27}{64}\,a^2b^2c^2.$$

Some details on this inequality one can find in:

1. A. Padoa, Period. Mat. (4) **6** (1926), 38–40.
2. O. Bottema, R. Ž. Djordjević, R. R. Janić, D. S. Mitrinović and P. M. Vasić, *Geometric Inequalities*, Wolters-Noordhoff Publishing, Groningen, 1969. (Ineq. 1.12).

Remark. Inequalities 1.21 and 1.15 are equivalent.

1.22. Inequality for $S = a^\lambda + b^\lambda + c^\lambda$

$$a^\lambda + b^\lambda + c^\lambda \geq 3 \cdot \left(\frac{4F}{\sqrt{3}}\right)^{\lambda/2} \qquad (\lambda > 0).$$

Details about this inequality one can find in:

1. C. N. Mills and O. Dunkel, *Problem* 3207, Amer. Math. Monthly **34** (1927), 382-384.
1. A. Padoa, Period. Mat. (4) **5** (1925), 80–85.
2. O. Bottema, R. Ž. Djordjević, R. R. Janić, D. S. Mitrinović and P. M. Vasić, *Geometric Inequalities*, Wolters-Noordhoff Publishing, Groningen, 1969. (Ineq. 4.21).

Remark. The inequality **1.22** represents a generalization of **1.13** and it is reduced to it for $\lambda = 2$.

2. Inequalities for $S(a)$

2.1. Inequalities for $S = a + b + c$

$$6\sqrt{3}\,r \leq a + b + c$$
$$\leq \frac{2\sqrt{3}}{3}(4R + r) \leq 4R + 2(3\sqrt{3} - 4)r \leq 3\sqrt{3}\,R.$$

Some details about these inequalities can be found in: [64], [66], [32], [14], [52], [54], [58], [78], [8], [10], [21], [43], [75], [12, Ineq. 4.13, 7.9, 7.1, 7.2, 5.3, 5.4, 5.29, 5.24, 5.33], [69], [69], [63, Ineq. 3.3, 3.4].

2.2. Inequalities for $S = a + b + c$

$$\frac{abc}{R^2} \leq a + b + c \leq \frac{abc}{4r^2}.$$

Details about these inequalities one can find in: [52], [45], [12, Ineq. 5.24, 5.26].

2.3. Inequality for $S = a^2 + b^2 + c^2$

$$a^2 + b^2 + c^2 \leq 8R^2 + \frac{11}{3\sqrt{3}}\,F.$$

Some details about this inequality can be found in: [64], [12, Ineq 7.5].

2.4. Inequalities for $S = a^2 + b^2 + c^2$

$$4F\sqrt{3} \leq 4F\sqrt{3} + Q \leq a^2 + b^2 + c^2 \leq 4F\sqrt{3} + \frac{1}{2}\,Q.$$

where $Q = (b - c)^2 + (c - a)^2 + (a - b)^2$.
These inequalities were treated in: [81], [30], [41], [74], [44], [33], [46], [78], [83], [12, Ineq. 4.4, 4.7].

2.5. Inequalities for $S = a^2 + b^2 + c^2$

$$36r^2 \leq 18Rr \leq 12r(2R - r)$$
$$\leq 4R^2 + 16Rr - 3r^2 - 4(R - 2r)\sqrt{R(R - 2r)}$$
$$\leq a^2 + b^2 + c^2$$
$$\leq 4R^2 + 16Rr - 3r^2 + 4(R - 2r)\sqrt{R(R - 2r)}$$
$$\leq 8R^2 + 4r^2 \leq 9R^2.$$

Some details on that can be found in: [65], [48], [32], [78], [8], [12, Ineq. 5.13, 5.14, 5.15], [63, Ineq. 5.1].

2.6. Inequalities for $S = ab + bc + ca$

$$36r^2 \leq 4F\sqrt{3} \leq 18Rr \leq 4r(5R - r)$$
$$\leq 2R^2 + 14Rr - 2(R - 2r)\sqrt{R(R - 2r)}$$
$$\leq ab + bc + ca$$
$$\leq 2R^2 + 14Rr + 2(R - 2r)\sqrt{R(R - 2r)} \leq 9R^2.$$

These inequalities were treated in: [52], [78], [8], [12, Ineq. 5.16, 5.17, 5.18, 5.19], [63 Ineq. 5.2].

2.7. Inequalities for $S = (b - c)^2 + (c - a)^2 + (a - b)^2$

$$8r(R - 2r) \leq 4R^2 + 2Rr - 3r^2 - 4(R - 2r)\sqrt{R(R - 2r)}$$
$$\leq (b - c)^2 + (c - a)^2 + (a - b)^2$$
$$\leq 4R^2 + 2Rr - 3r^2 + 4(R - 2r)\sqrt{R(R - 2r)} \leq 8R(R - 2r).$$

Some details about these inequalities can be found in: [32], [17], [12, Ineq. 5.25], [63, Ineq. 53].

2.8. Inequalities for $S = a^3 + b^3 + c^3$

$$72\sqrt{3}\,r^3 \leq 36\sqrt{3}\,Rr^2 \leq 12\sqrt{3}\,r^2(5R - 4r) \leq 4sr(5R - 4r)$$
$$\leq a^3 + b^3 + c^3$$
$$\leq 4sR(2R - r) \leq 4R(2R + (3\sqrt{3} - 4)r)(2R - r)$$
$$\leq 6\sqrt{3}\,R^2(2R - r)$$

Details about these inequalities one can find in: [37], [13], [79], [63, Ineq. 5.5].

2.9. Inequality for $S = a^\lambda + b^\lambda + c^\lambda$

$$a^\lambda + b^\lambda + c^\lambda \geq 3 \cdot \left(\frac{4F}{\sqrt{3}}\right)^{\lambda/2} \qquad (\lambda > 0).$$

Some details on these inequalities one can find in: [62], [12, Ineq, 4.21].

Remark. The inequality **2.9** is inequality **1.21**.

2.10. Inequalities for $\underline{S = abc}$

$$24\sqrt{3}\,r^3 \le 12\sqrt{3}\,Rr^2$$
$$\le abc$$
$$\le 4Rr\left(2R + (3\sqrt{3} - 4)r\right) \le 6\sqrt{3}\,R^2r \le 3\sqrt{3}\,R^3.$$

Some details about these inequalities can be found in: [82], [14], [39], [12, Ineq. 2.7, 5.27]. [63, Ineq. 5.4].

2.11. Inequalities for $\underline{S = (b+c)(c+a)(a+b)}$

$$192\sqrt{3}\,r^3 \le 96\sqrt{3}\,Rr^2 \le 12\sqrt{3}\,r^2(9R - 2r) \le 4sr(9R - 2r)$$
$$\le (b+c)(c+a)(a+b)$$
$$\le 4s(2R^2 + 3Rr + 2r^2) \le 4\left(2R + (3\sqrt{3} - 4)r\right)(2R^2 + 3Rr + 2r^2)$$
$$\le 6\sqrt{3}\,R(2R^2 + 3Rr + 2r^2) \le 24\sqrt{3}\,R^3.$$

These inequalities were treated in: [63, Ineq. 5.6].

2.12. Inequalities for $\underline{S = 1/a + 1/b + 1/c}$

$$\frac{\sqrt{3}}{R} \le \frac{3\sqrt{3}}{2(R+r)} \le \frac{1}{a} + \frac{1}{b} + \frac{1}{c} \le \frac{\sqrt{3}}{2r}\,.$$

Details about these inequalities one can find in: [53], [78], [12, Ineq. 5.22, 5.23].

2.13. Inequalities for $\underline{S = 1/a + 1/b + 1/c}$

$$\frac{\sqrt{3}}{R} \le \frac{2(5R - r)}{3\sqrt{3}\,R^2} \le \frac{5R - r}{R(2R + (3\sqrt{3} - 4)r)} \le \frac{5R - r}{Rs}$$
$$\le \frac{1}{a} + \frac{1}{b} + \frac{1}{c}$$
$$\le \frac{(R+r)^2}{Rrs} \le \frac{(R+r)^2}{3\sqrt{3}\,Rr^2} \le \frac{\sqrt{3}\,R}{4r^2}\,.$$

These inequalities were treated in: [63, Ineq. 5.7].

2.14. Inequalities for $\underline{S = 1/a + 1/b + 1/c}$

$$\frac{R^2 + 7Rr - (R - 2r)\sqrt{R(R - 2r)}}{2Rrs}$$
$$\le \frac{1}{a} + \frac{1}{b} + \frac{1}{c}$$
$$\le \frac{R^2 + 7Rr + (R - 2r)\sqrt{R(R - 2r)}}{2Rrs}\,.$$

Some details on these inequalities one can find in: [23].

Remark. Inequalities **2.14** are stronger than inequalities **2.13**.

2.15. Inequalities for $\underline{S = s^2}$

$$\frac{r(4R+r)^2}{2R-r} \le 3r(4R+r) \le \frac{r(4R+r)^2}{R+r} \le r(16R-5r)$$

$$\le s^2$$

$$\le \frac{R(4R+r)^2}{2(2R-r)} \le 4R^2 + 4Rr + 3r^2 \le \frac{(4R+r)^2}{3} \le \frac{(R+r)^4}{r^2}.$$

These inequalities were treated in: [19], [64], [41], [32], [46], [14], [56], [78], [8], [38], [12, Ineq. 4.9, 5.5, 5.6, 5.7, 5.8, 5.9, 7.1, 7.2, 7.4], [3], [4], [63, Ineq. 3.5, 3.6].

2.16. Inequalities for $\underline{S = s^2}$

$$3\sqrt{3}\,F + Q \le s^2 \le 3\sqrt{3}\,F + 2Q.$$

where $Q = (b-a)^2 + (c-a)^2 + (a-b)^2$.

Details about these inequalities one can find in: [41], [12, Ineq. 4.9].

2.17. Inequalities for $\underline{S = s^2}$

$$8Rr + \frac{11}{3\sqrt{3}}\,F \le s^2 \le 4R^2 + \frac{11}{3\sqrt{3}}\,F.$$

Some details on these inequalities one can find in: [64], [12, Ineq. 7.3, 7.4].

2.18. Inequality for $\underline{S = s^2}$

$$3r(4R+r) \le s^2 \le (2R+r)^2 + \frac{1}{2}\,R^2.$$

Some details about these inequalities can be found in: [56], [12, Ineq. 5.6].

2.19. Inequalities for $\underline{S = s^2}$

$$2R^2 + 10Rr - r^2 - 2(R-2r)\sqrt{R^2 - 2Rr}$$

$$\le s^2$$

$$\le 2R^2 + 10Rr - r^2 + 2(R-2r)\sqrt{R^2 - 2Rr}.$$

Details about these inequalities one can find in: [8], [12, Ineq. 5.10].

Remark. Inequalities **2.19** are stronger than inequalities **2.15**.

Remark. Inequalities **2.19** and **1.2** are equivalent.

2.20. Inequalities for $\underline{S = (a^2 + b^2 + c^2)/(a + b + c)^2}$

$$\frac{1}{3} \leq \frac{a^2 + b^2 + c^2}{(a + b + c)^2} < \frac{1}{2}.$$

Some details on these inequalities one can find in: [70], [12, Ineq. 1.19].

2.21. Inequality for $\underline{S = (s - a)(s - b)(s - c)}$

$$(s - a)(s - b)(s - c) \leq \frac{1}{8} abc.$$

Some details about this inequality can be found in: [66], [12, Ineq. 1.3].

2.22. Inequality for $\underline{S = s^3(s - a)(s - b)(s - c)}$

$$s^3(s - a)(s - b)(s - c) \leq \frac{27}{64} a^2 b^2 c^2.$$

Some details on this inequality one can find in: [67], [12, Ineq. 1.12].

2.23. Inequalities for $\underline{S = \sqrt{s - a} + \sqrt{s - b} + \sqrt{s - c}}$

$$\sqrt{s} < \sqrt{s - a} + \sqrt{s - b} + \sqrt{s - c} \leq \sqrt{3s}.$$

Details about these inequalities one can find in: [74], [35], [12, Ineq. 1.20].

2.24. Inequality for $\underline{S = a(s - a) + b(s - b) + c(s - c)}$

$$a(s - a) + b(s - b) + c(s - c) \leq 9Rr.$$

This inequality were treated in: [38], [12, Ineq. 5.20].

2.25. Inequality for $\underline{S = a^2(s - a) + b^2(s - b) + c^2(s - c)}$

$$a^2(s - a) + b^2(s - b) + c^2(s - c) \leq \frac{3}{2} abc.$$

Some details about this inequality can be found in: [12, Ineq. 1.7].
Remark. See inequality **2.27.**

2.26. Inequality for $\underline{S = a^3(s - a) + b^3(s - b) + c^3(s - c)}$

$$a^3(s - a) + b^3(s - b) + c^3(s - c) \leq abcs.$$

Details about this inequality one can find in: [1], [12, Ineq. 1.9].
Remark. See inequality **2.27.**

2.27. Inequality for $S = a^\lambda(s-a) + b^\lambda(s-b) + c^\lambda(s-c)$

$$a^\lambda(s-a) + b^\lambda(s-b) + c^\lambda(s-c) \leq \frac{1}{2} abc(a^{\lambda-2} + b^{\lambda-2} + c^{\lambda-2}) \quad (\lambda \in \mathbb{R}).$$

Some details on this inequality one can find in: [23].

Remark. The inequality **1.2.27** represents a generalization of inequalities **2.25** and **2.26** and it is reduced to im for $\lambda = 2$ and $\lambda = 3$, rspectively.

2.28. Inequalities for $S = abc/(a+b+c)$

$$4r^2 \leq \frac{abc}{a+b+c} \leq R^2.$$

Details about these inequalities one can find in: [21], [45], [12, Ineq. 4.13, 5.26].

3. Inequalities for $S(\alpha)$

3.1. Inequalities for $S = \sin\alpha + \sin\beta + \sin\gamma$

$$0 < \sin\alpha + \sin\beta + \sin\gamma \leq 2 + (3\sqrt{3} - 4) \cdot \frac{r}{R} \leq \frac{3\sqrt{3}}{2}.$$

This inequality were treated in: [66], [11], [47], [22], [8], [12, Ineq. 2.1, 2.2, 3.15], [60], [63, Ineq. 6.1].

3.2. Inequalities for $S = \sin^2\alpha + \sin^2\beta + \sin^2\gamma$

$$\frac{15r}{2R} \leq \frac{3r}{R^2}(2R - r) \leq \sin^2\alpha + \sin^2\beta + \sin^2\gamma \leq \frac{1}{R^2}(2R^2 + r^2) \leq \frac{9}{4}.$$

Some details about these inequalities can be found in: [47], [12, Ineq. 2.3], [60], [63, Ineq. 6.5].

3.3. Inequalities for $S = \sin^3\alpha + \sin^3\beta + \sin^3\gamma$

$$\frac{9\sqrt{3}r^2}{2R^2} \leq \frac{3\sqrt{3}r^2}{2R^3}(5R - 4r) \leq \frac{sr}{2R^3}(5R - 4r)$$
$$\leq \sin^3\alpha + \sin^3\beta + \sin^3\gamma$$
$$\leq \frac{s}{2R^2}(2R - r) \leq \frac{2R + (3\sqrt{3} - 4)r}{2R^2}(2R - r)$$
$$\leq \frac{3\sqrt{3}}{4R}(2R - r) < \frac{3\sqrt{3}}{2}.$$

These inequalities were treated in: [63, Ineq. 6.6].

3.4. Inequality for $\underline{S = \sin^3 \alpha + \sin^3 \beta + \sin^3 \gamma}$

$$\sin^3 \alpha + \sin^3 \beta + \sin^3 \gamma \le (\sin \alpha + \sin \beta + \sin \gamma)\left(\sin \frac{\alpha}{2} + \sin \frac{\beta}{2} + \sin \frac{\gamma}{2}\right).$$

Details about this inequality one can find in: [2], [63, Ineq. 2.2.5].

3.5. Inequalities for $\underline{S = \sin \alpha \sin \beta \sin \gamma}$

$$\frac{3\sqrt{3}r^2}{2R^2} \le \sin \alpha \sin \beta \sin \gamma \le \frac{r}{2R^2}(2R + (3\sqrt{3} - 4)r) \le \frac{3\sqrt{3}}{8}.$$

Details about these inequalities one can find in: [82], [47], [12, 2.7, 2.8].

3.6. Inequalities for $\underline{S = \cos \alpha + \cos \beta + \cos \gamma}$

$$\frac{3r}{R} \le \cos \alpha + \cos \beta + \cos \gamma \le \frac{3}{2}.$$

Some details on these inequalities one can find in: [47], [17], [12, Ineq. 2.16], [63, Ineq. 6.9].

3.7. Inequalities for $\underline{S = \cos^2 \alpha + \cos^2 \beta + \cos^2 \gamma}$

$$\frac{3}{4} \le \cos^2 \alpha + \cos^2 \beta + \cos^2 \gamma < 3.$$

Some details about these inequalities can be found in: [47], [12, Ineq. 2.21].

3.8. Inequalities for $\underline{S = \cos \alpha \cos \beta \cos \gamma}$

$$-1 < \cos \alpha \cos \beta \cos \gamma \le \frac{1}{8}.$$

These inequalities were treated in: [71], [17], [47], [12, Ineq. 2.23, 2.24].
Remark. See inequality **3.7.**

3.9. Inequalities for $\underline{S = \cos \alpha \cos \beta \cos \gamma}$

$$-1 - 3k - \frac{3}{2}k^2 \le \cos \alpha \cos \beta \cos \gamma \le \frac{1}{2}k^2 \qquad (k = r/R).$$

Some details on these inequalities one can find in: [9], [12, Ineq. 3.11].
Remark. Second inequality of **3.9** reduced to inequality **3.8** for $k = 1/2$, i.e., if and only if the triangle is equilateral.

3.10. Inequality for $\underline{S = \cos \alpha \cos \beta \cos \gamma}$

$$\cos \alpha \cos \beta \cos \gamma \ge \sqrt{3} \sin \alpha \sin \beta \sin \gamma - 1.$$

Details about this inequality one can find in: [40], [12, Ineq. 2.59].

3.11. Inequalities for $S = \cos^2(\alpha/2) + \cos^2(\beta/2) + \cos^2(\gamma/2)$

$$2 < \cos^2 \frac{\alpha}{2} + \cos^2 \frac{\beta}{2} + \cos^2 \frac{\gamma}{2} \leq \frac{9}{4}.$$

Some details about these inequalities can be found in: [47], [12, Ineq. 2.29].

3.12. Inequalities for $S = \cos \beta \cos \gamma + \cos \gamma \cos \alpha + \cos \alpha \cos \beta$

$$\frac{7r - 2R}{2R} \leq \frac{4Rr - r^2 - R^2}{R^2}$$
$$\leq \cos \beta \cos \gamma + \cos \gamma \cos \alpha + \cos \alpha \cos \beta$$
$$\leq \frac{r(R+r)}{R^2} \leq \frac{3r}{2R} \leq \frac{3}{4}.$$

These inequalities were treated in: [17], [12, Ineq. 2.22], [60], [63, Ineq. 6.10].
Remark. See inequality **3.13**.

3.13. Inequalities for $S = \cos \beta \cos \gamma + \cos \gamma \cos \alpha + \cos \alpha \cos \beta$

$$-1 + 4k - k^2 \leq \cos \beta \cos \gamma + \cos \gamma \cos \alpha + \cos \alpha \cos \beta \leq k + k^2 \quad (k = r/R).$$

Some details on these inequalities one can find in: [9], [12, 3.11].
Remark. Second inequality of **3.13** reduced to inequality **3.12** for $k = 1/2$, i.e., if and only if the triangle is equilateral.

3.14. Inequality for $S = \tan(\alpha/2) + \tan(\beta/2) + \tan(\gamma/2)$

$$\tan \frac{\alpha}{2} + \tan \frac{\beta}{2} + \tan \frac{\gamma}{2} \geq \sqrt{3}.$$

Some details about this inequality can be found in: [44], [47], [12, Ineq. 2.33].

3.15. Inequality for $S = \tan^2(\alpha/2) + \tan^2(\beta/2) + \tan^2(\gamma/2)$

$$\tan^2 \frac{\alpha}{2} + \tan^2 \frac{\beta}{2} + \tan^2 \frac{\gamma}{2} \geq 1.$$

Some details on this inequality one can find in: [47], [12, Ineq. 2.35].

3.16. Inequality for $S = \tan(\alpha/2) \tan(\beta/2) \tan(\gamma/2)$

$$\tan \frac{\alpha}{2} \tan \frac{\beta}{2} \tan \frac{\gamma}{2} \leq \frac{\sqrt{3}}{9}.$$

This inequality were treated in: [47], [12, Ineq. 2.34].

Remark. See inequality **3.19**.

3.17. Inequalities for $S = \cot\alpha + \cot\beta + \cot\gamma$

$$\cot\alpha + \cot\beta + \cot\gamma \geq \frac{\sqrt{3}}{9}\frac{(a^2 + b^2 + c^2)(a + b + c)}{abc} \geq \sqrt{3}.$$

Details about these inequalities one can find in: [80], [47], [21], [12, Ineq. 2.38, 3.8].

3.18. Inequality for $S = \cot(\alpha/2) + \cot(\beta/2) + \cot(\gamma/2)$

$$\cot\frac{\alpha}{2} + \cot\frac{\beta}{2} + \cot\frac{\gamma}{2} \geq 3\sqrt{3}.$$

Some details about this inequality can be found in: [47], [12, Ineq. 2.41].

3.19. Inequality for $S = \cot^2(\alpha/2) + \cot^2(\beta/2) + \cot^2(\gamma/2)$

$$\cot^2\frac{\alpha}{2} + \cot^2\frac{\beta}{2} + \cot^2\frac{\gamma}{2} \geq 9.$$

Some details about these inequalities can be found in: [47], [31], [12, Ineq. 2.43].

3.20. Inequality for $S = \cot^n(\alpha/2) + \cot^n(\beta/2) + \cot^n(\gamma/2)$

$$\cot^n\frac{\alpha}{2} + \cot^n\frac{\beta}{2} + \cot^n\frac{\gamma}{2} \geq 3^{(n+2)/2}.$$

This inequality were treated in: [84].

Remark. For $n = 1$ and $n = 2$ inequalities **3.18** and **3.19** follow.

3.21. Inequality for $S = \cot(\alpha/2)\cot(\beta/2)\cot(\gamma/2)$

$$\cot\frac{\alpha}{2}\cot\frac{\beta}{2}\cot\frac{\gamma}{2} \geq 3\sqrt{3}.$$

Some details on this inequality one can find in: [47], [12, Ineq. 2.42].

Remark. Inequalities **3.21** and **3.16** are ekvivalent.

3.22. Inequality for $S = \sin 2\alpha + \sin 2\beta + \sin 2\gamma$

$$\sin 2\alpha + \sin 2\beta + \sin 2\gamma \leq \sin\alpha + \sin\beta + \sin\gamma.$$

This inequality were treated in: [12, Ineq. 2.4].

3.23. Inequality for $S = \cot^n\alpha + \cot^n\beta + \cot^n\gamma$

$$\cot^n\alpha + \cot^n\beta + \cot^n\gamma \geq 3\cdot 3^{-n/2}.$$

Details about this inequality one can find in: [49], [12, Ineq. 2.65].

Remark. The inequality **3.22** represents a generalization of inequality **3.17** and it is reduced to it for $n = 1$.

4. Inequalities for $S(a, \alpha)$, $S(a, w_a)$

4.1. Inequalities for $\underline{S = (\alpha a + \beta b + \gamma c)/(a + b + c)}$

$$\frac{\pi}{3} \leq \frac{\alpha a + \beta b + \gamma c}{a + b + c} \leq \frac{\pi}{2} - \frac{1}{2} \min\{\alpha, \beta, \gamma\}.$$

Details about these inequalities one can find in: [61], [12, Ineq. 3.4].

4.2. Inequalities for $\underline{S = a \sin \alpha + b \sin \beta + c \sin \gamma}$

$$2\sqrt{3} \cdot \frac{F}{R} \leq a \sin \alpha + b \sin \beta + c \sin \gamma \leq \frac{9}{2} R.$$

Details about this inequality one can find in: [12, Ineq. 3.13, 3.14].

4.3. Inequalities for $\underline{S = a^p \sin^q \alpha + b^p \sin^q \beta + c^p \sin^q \gamma}$

$$a^p \sin^q \alpha + b^p \sin^q \beta + c^p \sin^q \gamma \geq 2^p \cdot \frac{3}{3^{(p+q)/4}} \cdot \frac{F^{(p+q)/2}}{R^q} \qquad (p, q \in \mathbb{R}, \ p + q > 0)$$

Remark. For $p = q = 1$ the inequality **4.2** follows.

4.4. Inequalities for $\underline{S = a^n \cos \alpha + b^n \cos \beta + c^n \cos \gamma}$

$$a^n \cos \alpha + b^n \cos \beta + c^n \cos \gamma \leq \frac{1}{3} (a^n + b^n + c^n)(\cos \alpha + \cos \beta + \cos \gamma)$$

$$\leq \frac{1}{2} (a^n + b^n + c^n).$$

Details about these inequalities one can find in: [63, Ineq. 4.6(1)].

4.5. Inequalities for $\underline{S = \sum a^2 \cot(\alpha/2)/\sum a^2 \tan(\alpha/2)}$

$$3 \cdot \frac{r}{R} < \frac{\sum a^2 \cot(\alpha/2)}{\sum a^2 \tan(\alpha/2)} \leq 3 \leq \frac{3}{2} \cdot \frac{R}{r}.$$

Details about these inequalities one can find in: [25].

4.6. Inequality for $\underline{S = a^\lambda w_a + b^\lambda w_b + c^\lambda w_c}$

$$a^\lambda w_a + b^\lambda w_b + c^\lambda w_c \leq \sqrt{\frac{3}{2} abcs(a^{2(\lambda-2)} + b^{2(\lambda-2)} + c^{2(\lambda-2)})} \qquad (\lambda \in \mathbb{R}).$$

Details about this inequality one can find in: [23], [12, Ineq. 8.12].

4.7. Inequalities for $\underline{S = a^\lambda w_a^2 + b^\lambda w_b^2 + c^\lambda w_c^2}$

$$a^\lambda w_a^2 + b^\lambda w_b^2 + c^\lambda w_c^2 \leq \frac{1}{2} abcs(a^{\lambda-2} + b^{\lambda-2} + c^{\lambda-2}) \qquad (\lambda \in \mathbb{R}).$$

Details about this inequality one can find in: [23], [12, Ineq. 8.13].

5. Inequalities for $S(m_a)$, $S(w_a)$, $S(r_a)$

5.1. Inequalities for $S = m_a + m_b + m_c$

$$s < m_a + m_b + m_c < 2s.$$

Some details about these inequalities can be found in: [76], [74], [12, Ineq. 8.1].

5.2. Inequalities for $S = m_a + m_b + m_c$

$$9r \le m_a + m_b + m_c \le 4R + r \le \frac{9}{2} R.$$

Some details about these inequalities can be found in: [52], [54], [36], [12, Ineq. 8.2, 8.3].

5.3. Inequalities for $S = m_a + m_b + m_c$

$$w_a + w_b + w_c \le m_a + m_b + m_c \le r_a + r_b + r_c.$$

These inequalities were treated in: [54], [12, Ineq. 8.20].

5.4. Inequalities for $S = m_a^2 + m_b^2 + m_c^2$

$$3\sqrt{3}\, F \le w_a^2 + w_b^2 + w_c^2 \le s^2 \le m_a^2 + m_b^2 + m_c^2$$

Details about this inequality one can find in: [74], [55], [77], [15], [73], [34], [12, Ineq. 8.6].

5.5. Inequalities for $S = w_b w_c + w_c w_a + w_a w_b$

$$w_b w_c + w_c w_a + w_a w_b \le s^2.$$

Some details on these inequalities one can find in: [59], [12, Ineq. 8.16].

5.6. Inequalities for $S = (w_b w_c)^2 + (w_b w_a)^2 + (w_a w_b)^2$

$$(w_b w_c)^2 + (w_b w_a)^2 + (w_a w_b)^2 \le rs^2(4R + r).$$

Some details about this inequality can be found in: [16], [12, Ineq. 8.15].

5.7. Inequalities for $S = r_a r_b r_c$

$$w_a w_b w_c \le r_a r_b r_c \le m_a m_b m_c.$$

Details about this inequality one can find in: [54], [12, Ineq. 8.21].

6. Some New Results

In this section we prove three theorems for geometric inequalities:

Theorem 6.1. *For each triangle and $n \in \mathbb{N}$ the inequality*

$$a^n \sin 2\alpha + b^n \sin 2\beta + c^n \sin 2\gamma \leq \frac{1}{2R}(a^{n+1} + b^{n+1} + c^{n+1})$$

holds. Equality is attained if and only if the triangle is equilateral.

Proof. Since

$$\sin 2\alpha = 2 \sin \alpha \cos \alpha, \qquad \text{i.e,} \qquad \sin 2\alpha = \frac{a}{R} \cos \alpha$$

and

$$\sin 2\beta = \frac{b}{R} \cos \beta, \qquad \sin 2\gamma = \frac{c}{R} \cos \gamma,$$

we conclude immediately that

$$a^n \sin 2\alpha + b^n \sin 2\beta + c^n \sin 2\gamma = \frac{1}{R}(a^{n+1} \cos \alpha + b^{n+1} \cos \beta + c^{n+1} \cos \gamma),$$

wherefrom, applying the inequality **4.4**, we obtain

$$a^n \sin 2\alpha + b^n \sin 2\beta + c^n \sin 2\gamma$$
$$\leq \frac{1}{R} \cdot \frac{1}{3}(a^{n+1} + b^{n+1} + c^{n+1})(\cos \alpha + \cos \beta + \cos \gamma)$$
$$\leq \frac{1}{2R}(a^{n+1} + b^{n+1} + c^{n+1}).$$

Here, as in **4.4**, equality holds if and only if the triangle is equilateral. $\square$

Theorem 6.2. *If n is a natural number, we have for each triangle*

$$a^n \cos(\beta - \gamma) + b^n \cos(\gamma - \alpha) + c^n \cos(\alpha - \beta)$$
$$\geq \frac{abc}{2R^2}(a^{n-1} + b^{n-1} + c^{n-1}) - \frac{1}{2}(a^n + b^n + c^n). \tag{6.1}$$

Equality holds if and only if the triangle is equilateral.

Proof. Using the sine law

$$\frac{a}{\sin \alpha} = \frac{b}{\sin \beta} = \frac{c}{\sin \gamma} = 2R,$$

we have

$$a \sin \beta = b \sin \alpha, \qquad b \sin \gamma = c \sin \beta, \qquad a \sin \gamma = c \sin \alpha. \tag{6.2}$$

Also, using the cosine theorem

$$a^2 = b^2 + c^2 - 2bc \cos \alpha, \quad b^2 = c^2 + a^2 - 2ca \cos \beta, \quad c^2 = a^2 + b^2 - 2ab \cos \gamma,$$

we get

$$b \cos \gamma + c \cos \beta = a, \quad c \cos \alpha + a \cos \gamma = b, \quad a \cos \beta + b \cos \alpha = c, \qquad (6.3)$$

and

$$\begin{aligned}
(b^2 + c^2) \cos \alpha &= a^2 \cos \alpha + 2bc \cos^2 \alpha, \\
(c^2 + a^2) \cos \beta &= b^2 \cos \beta + 2ca \cos^2 \beta, \\
(a^2 + b^2) \cos \gamma &= c^2 \cos \gamma + 2ab \cos^2 \gamma.
\end{aligned} \qquad (6.4)$$

However, since

$$a^2 \cos(\beta - \gamma) = a \cos \beta \cdot a \cos \gamma + a \sin \beta \cdot a \sin \gamma,$$

according to (6.2) and (6.3), we find

$$\begin{aligned}
a^2 \cos(\beta - \gamma) &= \left(a \cos \beta \cdot a \cos \gamma + bc \sin^2 \alpha \right) \qquad (6.5) \\
&= \left((c - b \cos \alpha)(b - c \cos \alpha) + bc \sin^2 \alpha \right) \\
&= 2bc - (b^2 + c^2) \cos \alpha,
\end{aligned}$$

and then, basing on (6.2), we obtain

$$a^2 \cos(\beta - \gamma) = 2bc - a^2 \cos \alpha - 2bc \cos^2 \alpha = 2bc \sin^2 \alpha - a^2 \cos \alpha.$$

Thus, the following equalities hold:

$$\begin{aligned}
a^2 \cos(\beta - \gamma) &= 2bc \sin^2 \alpha - a^2 \cos \alpha, \\
b^2 \cos(\gamma - \alpha) &= 2ca \sin^2 \beta - b^2 \cos \beta, \\
c^2 \cos(\alpha - \beta) &= 2ab \sin^2 \gamma - c^2 \cos \gamma.
\end{aligned} \qquad (6.6)$$

Since

$$\begin{aligned}
a^n &\cos(\beta - \gamma) + b^n \cos(\gamma - \alpha) + c^n \cos(\alpha - \beta) \\
&= a^{n-2} a^2 \cos(\beta - \gamma) + b^{n-2} b^2 \cos(\gamma - \alpha) + c^{n-2} c^2 \cos(\alpha - \beta),
\end{aligned}$$

using (6.6), we obtain

$$\begin{aligned}
a^n &\cos(\beta - \gamma) + b^n \cos(\gamma - \alpha) + c^n \cos(\alpha - \beta) \\
&= a^{n-2}(2bc \sin^2 \alpha - a^2 \cos \alpha) + b^{n-2}(2ca \sin^2 \beta - b^2 \cos \beta) \\
&\qquad + c^{n-2}(2ab \sin^2 \gamma - c^2 \cos \gamma),
\end{aligned}$$

i.e.,

$$a^n \cos(\beta - \gamma) + b^n \cos(\gamma - \alpha) + c^n \cos(\alpha - \beta)$$
$$= a^{n-2} 2bc \sin^2 \alpha + b^{n-2} 2ca \sin^2 \beta + c^{n-2} 2ab \sin^2 \gamma$$
$$- (a^n \cos \alpha + b^n \cos \beta + c^n \cos \gamma),$$

from which, because of

$$2bc \sin \alpha = 2ca \sin \beta = 2ab \sin \gamma = 4F = \frac{abc}{R}$$

and

$$\sin \alpha = \frac{a}{R}, \qquad \sin \beta = \frac{b}{R}, \qquad \sin \gamma = \frac{c}{R},$$

we conclude that the following equality

$$a^n \cos(\beta - \gamma) + b^n \cos(\gamma - \alpha) + c^n \cos(\alpha - \beta)$$
$$= \frac{abc}{2R^2} \left(a^{n-1} + b^{n-1} + c^{n-1}\right) - (a^n \cos \alpha + b^n \cos \beta + c^n \cos \gamma). \quad (6.7)$$

holds. Since (see **4.4**)

$$a^n \cos \alpha + b^n \cos \beta + c^n \cos \gamma \le \frac{1}{2} \left(a^n + b^n + c^n\right), \qquad (6.8)$$

from (6.7) we get the inequality

$$a^n \cos(\beta - \gamma) + b^n \cos(\gamma - \alpha) + c^n \cos(\alpha - \beta)$$
$$\ge \frac{abc}{2R^2} \left(a^{n-1} + b^{n-1} + c^{n-1}\right) - \frac{1}{2} \left(a^n + b^n + c^n\right),$$

i.e., (6.1). Since in (6.8) equality holds if and only if the triangle is equilateral, our theorem is proved. $\square$

Theorem 6.3. *In each triangle the following inequality*

$$(a^2 + b^2 + c^2 - 6R^2) abc \le (a^3 + b^3 + c^3) R^2. \qquad (6.9)$$

holds. Equality holds if and only if the triangle is equilateral.

Proof. Since

$$a^3 \cos(\beta - \gamma) + b^3 \cos(\gamma - \alpha) + c^3 \cos(\alpha - \beta)$$
$$= a \cdot a^2 \cos(\beta - \gamma) + b \cdot b^2 \cos(\gamma - \alpha) + c \cdot c^2 \cos(\alpha - \beta),$$

90

according to (6.6), we conclude that

$$a^3 \cos(\beta - \gamma) + b^3 \cos(\gamma - \alpha) + c^3 \cos(\alpha - \beta)$$
$$= a\big(2bc - (b^2 + c^2)\cos\alpha\big) + b\big(2ca - (c^2 + a^2)\cos\beta\big)$$
$$+ c\big(2ab - (a^2 + b^2)\cos\gamma\big)$$
$$= 6abc - bc(b\cos\gamma + c\cos\beta) - ca(c\cos\alpha + a\cos\gamma)$$
$$- ab(a\cos\beta + b\cos\alpha)$$
$$= 6abc - 3abc = 3abc.$$

On the other side, basing on Theorem 6.2, the inequality

$$a^3 \cos(\beta - \gamma) + b^3 \cos(\gamma - \alpha) + c^3 \cos(\alpha - \beta)$$
$$\geq \frac{abc}{2R^2}\,(a^2 + b^2 + c^2) - \frac{1}{2}\,(a^3 + b^3 + c^3),$$

holds, with an equality if and only if the triangle is equlateral. This means that

$$3abc \geq \frac{abc}{2R^2}\,(a^2 + b^2 + c^2) - \frac{1}{2}\,(a^3 + b^3 + c^3),$$

which is evidently equivalent to the inequality (6.9). □

References

1. J. Andersson, *Problem* E 1779, Amer. Math. Monthly **59** (1952), 41.
2. A. Bager and H. Frischknecht, *Aufgabe* 716, Elem. Math. **27** (1972), 68 and **28** (1973), 75.
3. A. Bager and O. Reuter, *Aufgabe* 688, Elem. Math. **29** (1974), 18–19..
4. A. Bager and O. Reuter, *Aufgabe* 690, Elem. Math. **29** (1974), 46–47.
5. F. Balitrand, Interméd. Math. **23** (1916), 86–87.
6. B. M. Barbalatt, Gaz. Mat. **30** (1924/25), 380–381.
7. I. Bénézech, *Problem* 412, J. Math. Elém. Paris (4) **1** (1892), 234.
8. W. J. Blundon, *Inequalities Associated with the Triangle*, Canad. Math. Bull. **8** (1965), 615–626.
9. W. J. Blundon, *Problem* E 1925, Amer. Math. Monthly **73** (1966), 1016.
10. W. J. Blundon, *Problem* E 1935, Amer. Math. Monthly **73** (1966), 1122.
11. O. Bottema, Euclides **30** (1954/55), 114–116.
12. O. Bottema, R. Ž. Djordjević, R. R. Janić, D. S. Mitrinović and P. M. Vasić, *Geometric Inequalities*, Wolters-Noordhoff Publishing, Groningen, 1969.
13. O. Bottema and G. R. Veldkamp, *Problem* 364, Nieuw Arch. Wisk. **22** (1974), 79 and **22** (1974), 266-267.
14. L. Carlitz and F. Leuenberger, *Problem* E 1454, Amer. Math. Monthly **68** (1961), 177 and **68** (1961), 805–806.
15. L. Carlitz and S. Philipp, *Problem* E 1828, Amer. Math. Monthly **70** (1963), 891 and **71** (1964), 681.
16. L. Carlitz, *Problem* E 1628, Amer. Math. Monthly **70** (1963), 891 and **71** (1964), 687.
17. J. M. Child, Math. Gaz. **23** (1939), 138–143.
18. M. Collins and A. B. Evans, *Problem* 2837, Ed. Times **13** (1870), 30–31.
19. G. Colombier and T. Doucet, *Problem* 1051, Nouv. Ann. Math. **31** (1872), 467.

20. V. Cristescu, Gaz. Mat. **26** (1920/21), 88–89.
21. T. R. Curry, *Problem* E 1861, Amer. Math. Monthly **73** (1966), 199.
22. T. R. Curry, *Problem* E 1644, Amer. Math. Monthly **70** (1963), 1099 and **71** (1964), 915–916.
23. R. Ž. Djordjević, *Some Inequalities for Triangle*, Univ. Beograd. Publ. Elektrotehn. Fak. Ser. Mat. Fiz. № **247** – № **273** (1969), 35–39.
24. R. Ž. Djordjević, *Some Inequalities for the Elements of a Triangle*, Univ. Beograd. Publ. Elektrotehn. Fak. Ser. Mat. Fiz. № **274** – № **301** (1969), 124–126.
25. R. Ž. Djordjević, *Some Inequalities for Elements of a Triangle*, Facta Universitatis (Niš) Ser. Math. Inform. **8** (1993), 73–75.
26. N. A. Edwards, *Problem* 1273, Nouv. Ann. Math. **37** (1878), 475.
27. A. Emerich, *Problem* 10656, Math. Questions **54** (1891), 100.
28. L. Euleri, Novi commentarii academiae scientarium Petropolitanae **11** (1765), 103–123.
29. L. Euleri, *Opera omnia*, I 26, 1953, pp. 139–157.
30. P. Finsler and H. Hadwigwr, Comment. Math. Helv. **10** (1937/38), 316–326.
31. Ju. I. Gerasimov, Mat. v škole № **3** (1964), 75.
32. J. C. H. Gerretsen, *Ongelijkheden in de Driehoek*, Nieuw Tijdschr. Wisk. **41** (1953), 1–7.
33. F. Goldner, *Problem* 69, Elem. Math. **4** (1949), 120.
34. V. O. Gordon, Mat. v škole № **5** (1966), 80.
35. E. G. Gotman, Mat. v škole № **1** (1965), 76.
36. E. G. Gotman, Mat. v škole № **5** (1966), 76.
37. S. G. Guba, *Problem* Mat. v škole № **5** (1965), 69.
38. S. Guba, *Problem* 317, Mat. v škole № **6** (1966), 67.
39. H. W. Guggenheimer, *Problem* E 1724, Amer. Math. Monthly **71yr 1964**, 911 and **72** (1965), 791-793.
40. H. W. Guggenheimer, *Plane Geometry and its Groups*, San Francisco-Cambridge-London-Amsterdam 1967.
41. H. Hadwiger, Jber. Deutsch. Math.-Verein. **49** (1939), *35–39*.
42. T. Hayashi, Tôhoku Science Rep. **11** (1922), 115–121.
43. R. R. Janić, *On a Geometric Inequality of D. F. Barrow*, Univ. Beograd. Publ. Elektrotehn. Fak. Ser. Mat. Fiz. № **181** – № **196** (1967), 75–76.
44. J. Karamata, *Problem* 119, Glasnik Mat.-Fiz. Astronom. **3** (1948), 223.
45. M. S. Klamkin, Math. Teacher **60** (1967), 323–328.
46. O. Kooi, Simon Stevin **32** (1958), 97–101.
47. R. Kooistra, Nieuw Tijdschr. Wisk. **45** (1957/58), 108–115.
48. T. Kubota, Tôhoku Math. J. **25** (1925), 122–126.
49. M. N. Kritikos, Actes du Congrès interbalkanique de mathématiciens, Athènes, 1934, pp. 157–158.
50. A. Laisant, *Géométrie du triangle*, Paris, 1896.
51. E. Lemoine, *Problem* 578, J. Math. Elem. (4) **3** (1894), 263.
52. F. Leuenberger, *Einige Dreiecksungleichungen*, Elem. Math. **13** (1958), 121–126.
53. F. Leuenberger, Elem. Math. **15** (1960), 77-79.
54. F. Leuenberger, *Gegensätzliches Verhalten der arithmetischen und geometrischen Mittel*, Elem. Math. **16** (1961), 127–129.
55. F. Leuenberger, Elem. Math. **18** (1963), 35–36.
56. F. Leuenberger and L. Carlitz, *Problem* E 1481, Amer. Math. Monthly **68** (1961), 803 and **69** (1962), 312.
57. A. Lupaş, *Problem* 441, Mat. Vesnik **2** (**15**)(**30**) (1978), 293.
58. A. Makowski, Elem. Math. **16** (1961), 134.
59. A. Makowski and J. Schopp, *Problem* E 1675, Amer. Math. Monthly **72** (1965), 187–188.
60. M. Marčev, *Neravenstva meždu perimetr'a i radiusite na vpisanata i opisanata okr'užnost na tri'g'lnika i njakoi sledstvija ot tjah'*, Ob. po matematika (Sofija), № **6** (1976), 3–7.
61. D. Marković, Bull. Soc. Math. Phys. Serbie (3–4) **4** (1952), 71.
62. C. N. Mills and O. Dunkel, *Problem* 3207, Amer. Math. Monthly **34** (1927), 382-384.

92

63. D. S. Mitrinović, J. E. Pečarić and V. Volonec, *Recent Advances in Geometric Inequalities*, Kluwer Academic Publishers, Dordrecht-Boston-London, 1989.
64. S. Nakajima, *Some Inequalities between the Fundamental Quantities of the Triangle*, Tôhoku Math. J. **25** (1925), 115–121.
65. J. Neuberg, Ed. Times, News Ser. **9** (1906), 51–52.
66. A. Padoa, Period. Mat. (4) **5** (1925), 80–85.
67. A. Padoa, Period. Mat. (4) **6** (1926), 38–40..
68. A. Pantazi, Gaz. Mat. **23** (1917/18), 144.
69. I. Pasche, *Problem* 165, Mat. Vesnik **10 (25)** (1973), 97.
70. M. Petrović, *Sur quelques fonctions des côtés et des angles d'une triangle*, Ens. Math. **18** (1916), 153–163.
71. C. C. Popovici, Gaz. Mat. **31** (1925), 132.
72. Ramus and É. Rouché, *Question* 233, Nouv. Ann. Math. **10** (1851), 353–355.
73. F. Ryžkov, Mat. v škole $N^{\underline{o}}$ 1 (1964), 80 and $N^{\underline{o}}$ 4 (1964), 73.
74. L. A. Santaló, Math. Notae (2) **3** (1943), 65–73.
75. K. Schuler, Praxis der Math. **9** (1967), 344.
76. C. Sebastiano, Boll. Math. (Firenze) (1938), 59.
77. Z. A. Skopec, Mat. v škole $N^{\underline{o}}$ 3 (1963), 89.
78. J. Steinig, Elem. Math. **18** (1963), 127–131.
79. G. Tsintsifas and M. S. Klamkin, *Problem* 816, Crux Math. **2** (1983), 46 and 10 (1984), 157.
80. T. Varopoulos, Bull. Soc. Grèce 15_1 (1934), 17.
81. R. Weitzenböck, Math. Z. **5** (1919), 137–146.
82. K. P. Wilkins, Amer. Math. Monthly **44** (1937), 579–583.
83. S. Zetel', Mat. v škole $N^{\underline{o}}$ 3 (1965), 66–69..
84. V. G. Zvezdin, Mat. v škole (1968), 65.

A TREATISE ON GRÜSS' INEQUALITY

A.M. FINK

Department of Mathematics, Iowa State University, Ames, IA 50011 USA

Abstract. We explore the possibility of proving Grüss' type inequality for measures other than Lebesgue measure including signed measures. In addition, we give new proofs to some old results.

1. Introduction

Chebyshev created quite a stir in 1882 when he submitted a paper [1, 1882] in which he proved a sequence of inequalities, the first of which is sometimes called "Chebyshev's Other Inequality". To wit,

$$(1) \qquad \int_a^b f(x)g(x)p(x)dx \int_a^b p(x)dx \geq \int_a^b f(x)p(x)dx \int_a^b g(x)p(x)dx$$

which is to hold for functions f, g which are both increasing or both decreasing and $p \geq 0$. It was soon realized that the proof was a lot easier than the one Chebeshev gave. Namely (1) is equivalent to

$$(2) \qquad \int_a^b \int_a^b [f(x) - f(y)][g(x) - g(y)]p(x)p(y)dx\,dy \geq 0$$

and so holds when $[f(x) - f(y)][g(x) - g(y)] \geq 0$. This condition is called 'similarly ordered'. Various authors purport to generalize this inequality and some actually do. See the book Mitrinović, Pečarić, and Fink [2, 1993] for an entire chapter devoted to the history of this inequality. Fink and Jodeit, [3, 1976], showed that when f and g are both increasing or both decreasing, then the 'measure' p need not be non-negative. This paper was circulated privately and finally published in [4, 1982]. In the mean time, Pečarić

1991 *Mathematics Subject Classification.* 26D10, 26D15.
Key words and phrases. Grüss, inequality, signed measures.

T.M. Rassias and H.M. Srivastava (eds.), Analytic and Geometric Inequalities and Applications, 93–113.
© 1999 *Kluwer Academic Publishers.*

[5, 1980] observed the same thing. The condition turns out to be "end positive" whose definition is given below. This condition is closely associated with inequalities in which functions are monotone. Indeed, it is the same condition for which Jensen's inequality can be proved with monotone functions in the integrand. For some results in this direction see Fink and Jodeit [6, 1990].

Chebyshev's usual Inequality is the familiar one from probability and statistics. Chebyshev's other Inequality above is also used in Statistics because it is a condition about the expectation of correlated variables.

For purposes of this paper we will write Chebyshev's Inequality as

$$
(3) \qquad \int_0^1 fg \, d\mu \int_0^1 d\mu - \int_0^1 f \, d\mu \int_0^1 g \, d\mu \geq 0.
$$

To simplify the typography, we will replace this by

$$
(4) \qquad T = T(f, g, \mu) = \int_0^1 fg \, d\mu - \int_0^1 f \, d\mu \int_0^1 g \, d\mu \geq 0
$$

where we assume that μ is a normalized measure, i.e. $\int_0^1 d\mu = 1$.

If we translate T to other real intervals, and use Lebesgue Measure, then T is modified by the multiplying the first term by the length of the interval.

Now comes Grüss [7, 1935] who asks the question, Chebyshev gives the lower bound for T, what is an upper bound? Now this question has many answers since one may ask what form the upper bound should take. With the exception of Pečarić, [8, 1984], all the writers on Grüss' Inequality have used either Lebesgue measure or other positive measures related to certain L^p spaces. Pečarić has one result for end positive measures.

In this paper we explore the derivation of upper bounds of several types in which we at first allow the measures to be normalized signed measures. This does not allow us to use the basic step in the previous discussions of Grüss' Inequality. That is, one usually uses the Cauchy-Schwarz-Buniakovsky Inequality to reduce the problem to bounds for $T(f, f)$. We call this the Cauchy-Schwarz argument for positive measures. What we mean for example. Suppose one can show that for all admissible P

$$
T(f, f, \mu) \leq Ah(f)^2
$$

where h is some given functional. Then

$$T(f,g,\mu) \le (T(f,f,\mu)^{\frac{1}{2}}(T(g,g,\mu))^{\frac{1}{2}} \le (Ah^2(f))^{\frac{1}{2}}(Ah^2(g))^{\frac{1}{2}} = Ah(f)h(g).$$

Equality holds here when $f = g$. So we get the best constant by doing the special case for $T(f,f,\mu)$. The monograph [2, 1993] has a chapter on the history of this inequality.

A comment about extremals. Since $T(\alpha f + \beta, g) = T(f,g)$ for any constants α, β, if f is an extremal so is $\alpha f + \beta$. So the statement that f_0 is the only extremal means that the affine class $\alpha f_0 + \beta$ are the only functions for which equality holds.

2. Absolutely and Completely Monotonic Functions

A very interesting class of functions are those whose derivatives are constant sign. We first look at absolutely monotonic functions,

$$AC = \{f | f^{(k)}(t) \ge 0 \text{ for } k = 0, 1, 2, \ldots, t \in [0, \infty]\}.$$

Grüss has shown the estimate for such functions

$$(5) \qquad\qquad T(f, g, dx) \le (4/45)[f(1) - f(0)][g(1) - g(0)]$$

using Bernstein polynomials. Here and in subsequent places, dx refers to Lebesgue measure. Pečarić has shown the same thing by using the fact that the functions in AC are the restrictions of entire functions to the real axis. We will give such a proof, noting what changes are to be made for general measures and specifically give the proof for Lebesgue measure since it impacts our further discussions.

Let $f, g \in AC$ so that

$$(5) \qquad\qquad f(x) = \sum_{k=0}^{\infty} a_k x^k, \quad g(x) = \sum_{k=0}^{\infty} b_k x^k$$

with $a_k \ge 0$, $b_k \ge 0$ for $k = 0, 1, \ldots$.

Then by the linearity of T we have

$$(6) \qquad\qquad T(f, g, \mu) = \sum_{k,j=1}^{\infty} a_k b_j T(x^k, x^j, \mu).$$

$(T(a, f, \mu) = 0$ when a is a constant.) We have

Theorem 1. *Let $f, g \in AC$ and μ any normalized (signed) measure. Then*

(7)
$$T(f, g, \mu) \le C[f(1) - f(0)][g(1) - g(0)]$$

where

(8)
$$C = \sup_{k,j \ge 1} T(x^k, x^j, \mu).$$

At this point we compute C for Lebesgue measure. T is just a combination of moments of μ.

(9)
$$T(x^k, x^j, dx) = \int_0^1 x^{k+j} dx - \int_0^1 x^k dx \int_0^1 x^j dx$$
$$= \frac{kj}{(k+1)(j+1)(k+j+1)}.$$

For $k + j + 1 = a$ this quantity $\frac{kj}{(kj+1)a}$ is an increasing function of kj and so is dominated by

$$h(a) = \frac{\left(\frac{a-1}{2}\right)^2}{\left[\left(\frac{a-1}{2}\right)^2 + a\right] a} = \frac{(a-1)^2}{a(a+1)^2}.$$

It is easily verified that $h'(a) \le 0$ for $a \ge 5$. Consequently for $k + j \ge 4$, the expression in (9) is dominated by $h(5) = \frac{4}{45}$. For $k + j = 2, 3$, $k, j \ge 1$, the expression is $\frac{1}{12} < \frac{4}{45}$. We have $T(x^k, x^j, dx) = \frac{4}{45}$ only for $k = j = 2$.

Thus we have

Corollary 1. *(Grüss) for $f, g \in AC$*

$$T(f, g, dx) \le \frac{4}{45}[f(1) - f(0)][g(1) - g(0)].$$

Equality holds if and only if $f(x) = g(x) = x^2$.

The original proof of Grüss used approximation by Bernstein polynomials. A different estimate for the expression in (9) is also possible. We have

(10)
$$\frac{kj}{(k+1)(j+1)(k+j+1)} \le \frac{1}{3} \frac{kj}{(k+1)(j+1)}$$

with equality only for $k = j = 1$.

Theorem 2. *Let $f, g \in AC$ then*

$$T(f, g, dx) \leq \frac{1}{3} \left[f(1) - \int\limits_0^1 f \right] \left[g(1) - \int\limits_0^1 g \right]$$

with equality if and only if $f(x) = g(x) = x$.

Proof.

$$T(f, g, dx) = \sum_{k_j=1}^{\infty} a_k b_j T(x^k, x^j, dx)$$

$$\leq \frac{1}{3} \sum_{k,j=1}^{\infty} \left(\frac{k}{k+1} a_k \right) \left(\frac{j}{j+1} b_j \right) = \frac{1}{3} \left(\sum_{k=1}^{\infty} \frac{k}{k+1} a_k \right) \left(\sum_{j=1}^{\infty} \frac{j}{j+1} b_j \right)$$

$$= \frac{1}{3} \left[f(1) - \int\limits_0^1 f \right] \left[g(1) - \int\limits_0^1 g \right].$$

Equality holds here if and only if $a_k = b_k = 0$ unless $k = 1$.

We can give a second example:

Example 1. Let $d\mu = 2x \, dx$ then

$$T(x^k, x^j, d\mu) = 2 \int\limits_0^1 x^{k+j+1} dx - 4 \int\limits_0^1 x^{k+1} dx \int\limits_0^1 x^{j+1} dx$$

$$= \frac{2kj}{[kj + 2(k+j+2)](k+j+2)}.$$

Letting $k + j + 2 = a$ this is $2\frac{kj}{(kj+2a)a}$ which as above is increasing in kj so is dominated by

$$k(a) = 2\frac{\left(\frac{a-2}{2}\right)^2}{\left(\left(\frac{a-2}{2}\right)^2 + 2a\right)a} = 2\frac{(a-2)^2}{a(a+2)^2}.$$

This function is increasing on $(4, 8)$ and deceasing on $(9, \infty)$. The values of $k(a)$ are $k(4) = \frac{1}{18}$, $k(5) = \frac{18}{245}$, $k(6) = \frac{1}{12}$, $k(7) = \frac{25}{294}$, $k(8) = \frac{9}{100}$ and $k(9) = \frac{98}{1089} = .08999\ldots$. Now $k(8) = \frac{9}{100}$ is only attained at $k = j = 3$. So we have

$$T(f, g, 2x dx) \leq \frac{9}{100}(f(1) - f(0)(g(1) - f(0))$$

with equality only for $f(x) = g(x) = x^3$.

To get a bound for an estimate of the form $c \left[f(1) - \int\limits_0^1 f \right] \left[g(1) - \int\limits_0^1 g \right]$ one finds the

maximum of $2\frac{kj}{(k+j+2)(kj+2(k+j+2))} \left(\frac{k+1}{k} \right) \left(\frac{j+1}{j} \right)$. Letting $k + j + 1 = a$ this is easily done and we find that for $f, g \in AC$

$$T(f, g, 2x\ dx) \leq \frac{2}{9} \left(f(1) - \int\limits_0^1 f \right) \left(g(1) - \int\limits_0^1 g \right)$$

with equality only for $f(x) = g(x) = x$.

For general positive measures we have

Theorem 3. *If $\mu \geq 0$ and $\mu\{1\} = 0$, then C in (8) is given by*

$$C = \sup_{k \geq 1} T(x^k, x^k, \mu).$$

Proof. Let $h(t) = \int\limits_0^1 x^t d\mu(x)$ then $T(x^k, x^j, d\mu) = h(k + j) - h(k)h(j)$. So consider

$h(t + u) - h(t)h(u) = k(t, u)$. Then $k(0, u) = k(t, 0) = 0$; $h(\infty, u) = h(t, \infty) = 0$ unless

μ has an atom at 1. So the maximum is in the interior.

Now $k_1(t, u) = h'(t + u) - h'(t)h(u)$ and $k_2(t, u) = h'(t + u) - h(t)h'(u)$. If $k_1 = k_2 = 0$ then $\frac{h'(t)}{h(t)} = \frac{h'(u)}{h(u)}$. (Note $h(0) = 1$) We shall show that $\frac{h'}{h}$ is increasing so this

implies that $t = u$ which is what we want to show. We show $h(t)h''(t) - h'(t)^2 \geq 0$.

Now $h(t)h''(t) = \int\limits_0^1 x^t d\mu(x) \int\limits_0^1 (\ln y)^2 y^t d\mu(y) = \frac{1}{2} \int\limits_0^1 \int\limits_0^1 x^t y^t [(\ln x)^2 + (\ln x)^2] d\mu(x) d\mu(y)$ while

$h'(t)^2 = \int\limits_0^1 \int\limits_0^1 x^t y^t (\ln x)(\ln y) d\mu(x) d\mu(y)$ so

$$h(t)h''(t) - h'(t)^2 = \frac{1}{2} \int\limits_0^1 \int\limits_0^1 x^t y^t [(\ln x)^2 + (\ln y)^2 - 2(\ln x)(\ln y)] d\mu(x) d\mu(y) \geq 0.$$

Now this is 0 only if the measure $d\mu(x)d\mu(y)$ lives on the diagonal $x = y$. But for such to

happen μ has to be a single atom and $h(t) = x_0^t$ in which case $T(x^k, x^t, d\mu) = x_0^{k+j} - x_0^k x_0^j \equiv$

0. Discarding this case, the theorem is proved.

We use this theorem to compute C for Lebesgue measure.

$$T(x^k, x^k, dx) = \frac{1}{2k+1} - \left(\frac{1}{k+1}\right)^2 = \frac{k^2}{(k+1)^2(2k+1)}.$$

The function $\frac{x^2}{(x+1)^2(2x+1)}$ is decreasing for $x \geq 2$ so the maximum is either at $k = 1$ or 2.

These numbers are $\frac{1}{2}$ and $\frac{4}{45}$. The answer is $\frac{4}{45}$, and equality holds only for $f = g = x^2$.

Remark. Theorem 3 can also be proved by using the Cauchy-Schwarz inequality but it is not so easy to retain exact cases of equality.

We now turn our attention to completely monotonic function, CC on $[0, \infty]$. By Bernstein's Theorem [9, 1941, p.160] such functions are precisely those with the representation

$$\tag{11} f(x) = \int_0^\infty e^{-sx} d\sigma(s)$$

for σ a non-negative measure with $f(0) = \int_0^\infty d\sigma$.

Theorem 4. Let $f, g \in CC$ and μ any signed measure, then

$$\tag{12} T(f, g, \mu) \leq K(f(1) - f(0))[g(1) - g(0)]$$

where

$$K = \sup_{s,t \geq 0} \frac{T(e^{-sx}, e^{-tx}, \mu(x))}{(1 - e^{-s})(1 - e^{-t})}$$

Proof. The inequality (12) is with $g(x) = \int_0^\infty e^{-sx} d\lambda(s)$

$$\int_0^\infty \int_0^\infty \left[\frac{1}{2} \int_0^1 \int_0^1 (e^{-sx} - e^{-sy})(e^{-tx} - e^{-ty}) d\mu(x) d\mu(y) - K(1 - e^{-s})(1 - e^{-t}) \right] d\sigma(s) d\lambda(t) \leq 0$$

the [ ] is $T(e^{-s}, e^{-t}, d\mu) - K(1 - e^{-s})(1 - e^{-t})$ which must be ≤ 0 if (12) is to hold for all $f, g \in CC$. This gives the result.

Again we want to compute this for Lebesgue measure. So we want

(13)
$$\int_0^1 e^{-(s+t)x}\,dx - \int_0^1 e^{-sx}\,dx \int_0^1 e^{-tu}\,du \le K(1-e^{-s})(1-e^{-t}).$$

Writing $1 - e^{-s} = -s\int_0^1 e^{-sx}\,dx$, this inequality becomes after symmetrization

$$\int_0^1 \int_0^1 \left[e^{-(s+t)x} + e^{-(s+t)y} - (e^{-sx}e^{-ty} + e^{-sy}e^{-tx}) - Kst(e^{-sx}e^{-ty} + e^{-sy}e^{-tx}) \right] dx\,dy \le 0.$$

This is to hold for $s, t \ge 0$. Solving this for minimal K seems difficult to do. By the Cauchy-Schwarz argument we may take $s = t$ and look back to (13), which becomes

$$\frac{(1-e^{2s})}{2s} - \left(\frac{1-e^{-s}}{s} \right)^2 \le K(1-e^{-s})^2.$$

Cancelling the factor $(1-e^{-s})$ this amounts to finding the smallest K for which

(14)
$$g(s) = s(1+e^{-s}) - 2(1-e^{-s}) - 2Ks^2(1-e^{-s}) \le 0$$

on $[0,\infty)$. It may be verified that $g(0) = g'(0) = g''(0) = 0$ for any K and $g'''(s) = e^{-s}[1 - s - 12K + 12Ks - 2Ks^2]$. If g is to be negative it is necessary that $g'''(0) \le 0$, or $K \ge \frac{1}{12}$. But if $K = \frac{1}{12}, g'''(s) = -\frac{1}{6}e^{-s}s^2 \le 0$ so $g(s) \le 0$ on $[0,\infty)$, and the least $K = \frac{1}{12}$.

Corollary 2. *For $f, g \in CC$,*

$$T(f,g,dx) \le \frac{1}{12}[f(1) - f(0)][g(1) - g(0)].$$

The constant $\frac{1}{12}$ is the best constant. There are no extremals, but the functions $\{e^{-sx}\}_{s\downarrow 0}$ are a minimizing sequence.

Proof. We find that $\frac{1}{12}$ works by the preceding argument. By the argument above $g(s) < 0$ on $(0,\infty)$ so no extremal can exist. But it is easily verified that

$$\lim_{s\to 0+} \frac{T(e^{-sx}, e^{-sx}, dx)}{(1-e^{-s})^2} = \lim_{s\to 0+} \frac{\left(\frac{1-e^{-2s}}{2s} \right)^2 - \left(\frac{1-e^{-s}}{s} \right)^2}{(1-e^{-s})^2} = \frac{1}{12}.$$

Although we offer the above proof, we might observe that if $f \in CC$ then $f(1-x)$ has all of its derivatives ≥ 0. So the proof for AC would work. But of course $f(1-x)$ may not be in AC. For example, the $(1-x)^2$ has opposite signed derivatives on $(0,1)$ but is not in CC. But this should be the extremal if we reduce the argument to AC. In any case our proof suggests the minimizing sequence.

As a different type of estimate, suppose we look for

$$(15) \qquad T(f,g,d\mu) \leq D \left(f(1) - \int_0^1 f \right) \left(g(1) - \int_0^1 g \right)$$

or

$$(16) \qquad T(f,g,d\mu) \leq E \left(\int_0^1 f - f(0) \right) \left(\int_0^1 f - g(0) \right).$$

As above these numbers D and E may be found by for positive measures by

$$(17) \qquad D = \sup_{t \geq 0} \frac{\frac{1-e^{-2t}}{2t} - \left(\frac{1-e^{-t}}{t} \right)^2}{\left(e^{-t} - \frac{1-e^{-t}}{t} \right)^2}$$

and

$$(18) \qquad E = \sup_{t \geq 0} \frac{\frac{1-e^{-2t}}{2t} - \left(\frac{1-e^{-t}}{t} \right)^2}{\left(\frac{1-e^{-t}}{t} - 1 \right)^2}$$

But

$$D = \sup_{t \geq 0} \frac{\frac{t}{2}(1-e^{-t}) - (1-e^{-t})^2}{(te^{-t} - 1 + e^{-t})^2} = +\infty$$

since the lim as $t + \infty$ is infinite, so no inequality of the form (15) exists for completely monotonic functions. However, for E we have to find a minimum E for which

$$k(t) = (2E + t + 2) + e^t(4Et - 4E - 4) + e^{2t}(2Et^2 - 4Et - t + 2 + 2t)$$

is non-negative on $(0, \infty)$. It may be verified that regardless of E, k has four zeros at $t = 0$ and $k^{iv}(t) = e^t[12E - 4 + 4Et] + e^{2t}[32Et^2 + (64E - 16)t]$. But $h^{iv}(0) \geq 0$ is a necessary condition so that $E \geq \frac{1}{3}$. With $E = \frac{1}{3}, k^{iv}(t) \geq 0$ for on $(0, \infty)$. Again there are no extremals but

$$\lim_{a \to 0} \frac{T(e^{-at}, e^{-at}, dt)}{\left[e^{-a} - \left(\frac{1-e^{-a}}{a} \right) \right]^2} = \frac{1}{3}.$$

Theorem 5. *Let* $f, g \in CC$ *then*

$$T(f, g, dx) \leq \frac{1}{3} \left[\int\limits_0^1 f - f(0) \right] \left[\int\limits_0^1 g - g(0) \right]$$

and $\frac{1}{3}$ *is the best constant, and* $\{e^{st}\}_{s \downarrow 0}$ *is a minimizing sequence.*

3. Anchored Monotone Functions

Suppose we define

$M_n = \{f | f$ has n zeros at 0 and there is a measure $\sigma \geq 0$ so that $f(x) = \frac{1}{(n-1)!} \int\limits_0^x (x-t)^{n-1} d\sigma(t)\}$. Those functions $f \in C^n[0,1]$ with the zeros at 0 satisfy this if $f^{(n)} \geq 0$ and $d\sigma = f^{(n)}(t)dt$. This formulation allows f with $f^{(n-1)}$ being absolutely continuous and $f^{(n)}$ existing a.e. and in L_1. We propose to find k_n so that

$$(15) \qquad T(f, g, dx) \leq k_n[f(1) - f(0)][g(1) - g(0)]$$

for $f, g, \in M_n$.

Let $x_+ = \max(0, x)$ with x_+^n meaning $(x_+)^n$ and $0^0 \equiv 1$.

Theorem 6. *With the above definitions*

$$k_1 = \frac{1}{4} \quad \text{with extremal} \quad \left(x - \frac{1}{2}\right)_+^0 ;$$

$$k_2 = \frac{1}{9} \quad \text{with extremal} \quad \left(x - \frac{1}{3}\right)_+ ;$$

$$k_3 = \frac{9}{100} \quad \text{with extremal} \quad \left(x - \frac{1}{10}\right)_+^2 ;$$

$$k_n = \frac{(n-1)^2}{n^2(2n-1)} \quad \text{with extremal} \quad x^n \quad \text{for} \quad n \geq 4.$$

Proof. We write

$$f(x) = \frac{1}{(n-1)!} \int\limits_0^1 (x-t)_+^{n-1} d\sigma(t)$$

and use the Cauchy-Schwarz argument so that the inequality (15) is equivalent to

$$(16) \qquad \int\limits_0^1 \int\limits_0^1 \int\limits_0^1 \int\limits_0^1 \left[\frac{(x-t)^{2n-2}}{(n-1)!^2} - \frac{(x-t)_+^{n-1}(y-s)_+^{n-1}}{(n-1)!^2} \right] dx \, dy d\sigma(s) d\sigma(t)$$

$$\leq k_n \int\limits_0^1 \int\limits_0^1 \frac{(1-s)^{n-1}(1-t)^{n-1}}{(n-1)!^2} d\sigma(s) d\sigma(t).$$

If this is to hold for all measures σ we want

(17) $T((\cdot - t)_+^{n-1}, (\cdot - s)_+^{n-1}, dx) \le k_n(1-s)^{n-1}(1-t)^{n-1}$ for $0 \le s, t \le 1$.

Invoking the Cauchy-Schwarz argument again we may take $s = t$ in (17); i.e.

(18) $$k_n = \sup_{0 \le t \le 1} \frac{\int\limits_0^1 (x-t)_+^{2n-2} dx - \left(\int\limits_0^1 (x-t)_+^{n-1} dx\right)^2}{(1-t)^{2n-2}}.$$

So

$$k_n = \max_{0 \le t \le 1} \left[\frac{1-t}{2n-1} - \frac{(1-t)^2}{n^2}\right].$$

For $n = 1, 2, 3$ the maximum is attained at $\frac{1}{2}, \frac{1}{3}, \frac{1}{10}$ giving the extremals as noted. For $n \ge 4$, the maximum is at $t = 0$. This ends the proof.

The general measure case follows from (16) and (17).

Theorem 7. *For any normalized measure μ and $f, g \in M_n$,*

$$T(f, g, \mu) \le \sup_{0 \le s \le t \le 1} \frac{T((\cdot - t)_+^{n-1}, (\cdot - s)_+^{n-1}, \mu)}{(1-s)^{n-1}(1-t)^{n-1}}[f(1) - f(0)][g(1) - g(0)].$$

If in addition $\mu \ge 0$ then

$$T(f, g, \mu) \le \sup_{0 \le t \le 1} \frac{T((\cdot - t)_+^{n-1}, (\cdot - t)_+^{n-1}, \mu)}{(1-t)^{2n-2}}[f(1) - f(0)][g(1) - g(0)].$$

There is no inequality of the form

$$T(f, g, d\mu) \le B \int\limits_0^1 f \int\limits_0^1 g \text{ for } f, g \in M_n$$

since for $f(x) = (x - t)_+^{n-1} = g(x)$ this would require

$$\frac{\frac{(1-t)^{2n-1}}{2n-1} - \frac{(1-t)^{2n}}{n^2}}{\frac{(1-t)^{2n}}{n^2}} \le B \text{ on } (0, 1).$$

This cannot hold for t near 1. On the other hand there is an inequality of the form

$$T(f, q, d\mu) \le B \left(f(1) - \int\limits_0^1 f\right)\left(g(1) - \int\limits_0^1 g\right)$$

Theorem 8. *If $f, g \in M_n$ then*

$$T(f, g, dx) \le \frac{1}{2n-1} \left[f(1) - \int_0^1 f \right] \left[g(1) - \int_0^1 g \right].$$

Equality holds for $f = g = x^{n-1}$.

Proof. This inequality is of the form

$$\frac{(1-t)^{2n-1}}{(2n-1)(n-1)!^2} - \frac{(1-t)^{2n}}{(n!)^2} \le B \left[\frac{(1-t)^{n-1}}{(n-1)!} - \frac{(1-t)^n}{n!} \right]^2$$

or $\frac{n^2}{2n-1} - (1-t) \le B[n - (1-t)]^2$.

It is necessary that this holds at $t = 0$ and $t = 1$. From $t = 0$ we get $B \ge \frac{1}{2n-1}$. But if we replace B by $\frac{1}{2n-1}$ one verifies that the inequality collapses to $-\frac{t^2}{2n-1} \le 0$. This means that $(x - 0)_+^{n-1}$ is the only extremal.

4. Landau's Problem

Landau [10, 1935] considered the class $\widehat{M}_n = \{f | f, f', \ldots, f^{(n)} \ge 0\}$. Note that $\widehat{M}_n \supset M_n$.

Theorem 9. *(Landau) For $f, g, \in \widehat{M}_n$*

$$T(f, g, dx) \le d_n(f(1) - f(0))[g(1) - g(0)]$$

where $d_1 = \frac{1}{4}$, $d_2 = \frac{1}{9}$, $d_3 = \frac{9}{10}$, $d_n = \frac{4}{45}$ for $n \ge 4$.
   The only extremals are $\left(x - \frac{1}{2} \right)_+^0$, $\left(x - \frac{1}{3} \right)_+$, $\left(x - \frac{1}{10} \right)_+^2$, and x^n for $n \ge 4$.

Comment: Landau did not write down the extremals. This proof used Bernstein polynomials and discrete inequalities. We will give a proof that does not do this so we can give the cases of equality.

Proof. Again it is sufficient to prove the inequality for $f = g$. We write

$$f(x) = \sum_{k=0}^{n-1} a_k x^k + \int_0^1 \frac{(x-t)_+^{n-1}}{(n-1)!} f^{(n)}(t) dt = p(x) + R(x)$$

where $p(x) = \sum\limits_{k=0}^{n-1} a_k x^k$ and $R(x) = \int\limits_0^1 \frac{(x-t)_+^{n-1}}{(n-1)!} f^{(n)}(t)dt$ and note $a_k \geq 0$ and $R \in M_n$.

Using the linearity of T we have

(20) $$T(f, f, dx) = T(p, p, dx) + 2T(p, R, dx) + T(R, R, dx).$$

and

$$T(p, p, dx) = \sum_{k,j=1}^{n-1} a_k a_j T(x^k, x^j, dx) = \sum_{k,j=1}^{n-1} a_k a_j \frac{kj}{(k+1)(j+1)(k+j+1)}.$$

When $n = 1$, this sum is empty. In fact for $n = 1, p(x) = a_0$ and $T(p, p, dx) = T(p, R, dx) = 0$, so

$$T(f, f, dx) = T(R, R, dx) \leq k_1 R(1)^2 = k_1[f(1) - f(0)]^2 = \frac{1}{4}[f(1) - f(0)]^2$$

by Theorem 6. Recall the proof of Theorem 1.

(For $n = 2, 3$), $\quad T(p, p, dx) \leq \frac{1}{12} \sum\limits_{k,j=1}^{n-1} a_k a_j = \frac{1}{12}[p(1) - p(0)]^2.$

For $n \geq 4, T(p, p, dx) \leq \frac{4}{45}[p(1) - p(0)]^2.$

So let $\beta_n = \frac{1}{4}$ for $n = 1, \beta_2 = \frac{1}{12}$ and $\beta_n = \frac{4}{45}$ for $n \geq 3$. Then $T(p, p, dx) \leq \beta_n[p(1) - p(0)]^2.$

Moreover, $T(R, R, dx) \leq k_n[R(1) - R(0)]^2 = k_n R(1)^2$ where k_n is defined in Theorem 6.

Finally

$$T(p, R, dx) \leq T(p, p, dx)^{\frac{1}{2}} T(p, R, dx)^{\frac{1}{2}} \leq \max(\beta_n, k_n)(p(1) - p(0))[R(1)].$$

From (20) we find

$$T(f, f, dx) \leq \max(\beta_n, k_n)[(p(1) - p(0))^2 + 2[p(1) - p(0)]R(1) + R(1)^2]$$
$$= \max(\beta_n, k_n)[p(1) - p(0) + R(1)]^2 = \max(\beta_n, k_n)[f(1) - f(0)]^2.$$

To get equality, note that for $n = 1, 2, 3$ $k_n > \beta_n$ so we must have $p = 0$ and R the extremal for M_n as in Theorem 6. If $n \geq 4, \beta_n > k_n$ so for equality we need $R = 0$ and $p(x) = x^2$.

5. The General Monotone Case

For general signed measures we have no general universal upper bound. For non-negative measure, the identity for $\underline{f} \leq f \leq \overline{f}$

$$(21) \qquad T(f, f, \mu) = \left(\overline{f} - \int_0^1 f d\mu\right)\left(\int_0^1 f d\mu - \underline{f}\right) - \int_0^1 [\overline{f} - f(x)][f(x) - \underline{f}]d\mu$$

is valid. Hence T is dominated by the first term which in turn is dominated by $\frac{1}{4}(\overline{f} - \underline{f})^2$.

Theorem 10. *If $\mu \geq 0$, then*

$$T(f, g, \mu) \leq \frac{1}{4}(\max f - \min f)(\max g - \min g)$$

and $\frac{1}{4}$ is the best constant.

Extremals are given by $f_0(t) = g_0(t) = (t - t_0)_+^0$ where $\int_{t_0}^1 d\mu = \frac{1}{2}$.

Proof. The estimate is proved above. But

$$T(f_0, f_0, \mu) = \int_{t_0}^1 d\mu - \left(\int_{t_0}^1 d\mu\right)^2 = \frac{1}{4}$$

If μ is not a non-negative measure, we introduce the nation of 'end-positive measure'. This notion is common to inequalities for monotone functions.

Let μ be a signed measure. Define

$$(22) \qquad L(x) = \int_0^x d\mu \quad \text{and} \quad \int_x^1 d\mu = R(x)$$

(L for left and R for right).

The measure μ is end-positive if $0 \leq L(x), R(x)$ for $x \in [0, 1]$ with $L(1) = R(0) = 1$.

Remark. If f is montone and μ is endpositive, then $\int_0^1 f d\mu$ lies between the extremes of f. To see this, consider

$$\int_0^1 f d\mu = f(0) + \int_0^1 f'R = f(1) - \int_0^g f'L.$$

Theorem 11. *Let μ be an endpositive measure and $TV(f)$ be the total variation. Then for differentiable f, g*

$$(23) \qquad\qquad T(f, g, \mu) \leq F\ TV(f)TV(g),$$

where $F = \max\limits_{0 \leq t \leq x \leq 1} L(t)R(x)$.
If f and g have derivatives of the same sign then

$$(24) \qquad\qquad T(f, g, \mu) \leq F[f(1) - f(0)][g(1) - g(0)]$$

Both (23) and (24) are best possible if $F = L(t_0)R(x_0)$, $t_0 \leq x_0$ and the extremals are $f(x) = (x - t)^0_+$ and $g(x) = (x - x_0)^0_+$.

Proof. We use an identity of Pečarić [5, 1984]

$$(25) \quad T(f, g, \mu) = \int_0^1 R(x)f'(x) \left(\int_0^x L(t)g'(t)dt \right) dx + \int_0^1 R(x)g'(x) \left(\int_0^x L(t)f'(t)dt \right) dx.$$

We write this as

$$T(f, g, \mu) = \int_0^1 \int_0^1 L(t)R(x) \left[f'(t)g'(x) + g'(t)f'(x) \right] (x - t)^0_+ dt\ dx.$$

If $f'g' \geq 0$ we immediately have

$$T(f, g, \mu) \leq F \int_0^1 \int_0^1 \left[f'(t)g'(x) + g'(t)f'(x) \right] (x - t)^0_+ dt\ dx$$

$$(26)$$
$$= F \left[\int_0^1 f'(x) \int_0^x g'(t)dt\ dx + \int_0^1 g'(x) \int_0^x f'(t)dt\ dx \right]$$

$$= F \left[\int_0^1 f'(x) \int_0^x g'(t)dt\ dx + \int_0^1 f'(t) \int_t^1 g'(x)dx\ dt \right]$$

$$= F \int_0^1 f' \int_0^1 g'$$

yielding (24).

To do the general case (23) replace f' and g' in this argument by $|f'|$ and $|g'|$.

If $F = L(t_0)R(x_0)$ with $t_0 \leq x_0$ then

$$T((x - t_0)^0_+, (x - x_0)^0_+, \mu) = R(x_0) - R(x_0)R(t_0) = R(x_0)[1 - R(t_0)] = R(x_0)L(t_0)$$

so these give equality. Since they are monotone, these serve as extremals for both inequalities.

Note that if $\mu > 0$ the above computations lead to

$$F = \max_{0 \leq t \leq x \leq 1} L(t)R(x) = \max_{0 \leq x \leq 1} L(x)R(x) = \max_{0 \leq y \leq 1} y(1 - y) = \frac{1}{4}.$$

With $t_0 = x_0 = \frac{1}{2}$ we get $f(x) = g(x) = \left(x - \frac{1}{2}\right)^0_+$ as the (usual) extremals.

Example 2. Let $d\mu = \frac{3\pi}{2}\sin 3\pi x \, dx$. Then $L(x) = \frac{1 - \cos 3\pi x}{2}$ and $R(x) = \frac{1}{2}[1 + \cos 3\pi x]$. By drawing the graphs of these functions one sees that

$$F = \max_{0 \leq t \leq x \leq 1} L(t)R(x) = L(t_0)R(x_0) = 1$$

with $t_0 = \frac{1}{3}$ and $x_0 = \frac{2}{3}$. So that $f(x) = \left(x - \frac{1}{3}\right)^0_+$ and $g(x) = \left(x - \frac{2}{3}\right)^0_+$ should be the extremal and we find $T(f, g, \mu) = 1$.

Example 3. Let $d\mu = (1 + 2\cos 2\pi t)dt$, with $L(x) = x + \frac{\sin 2\pi x}{2\pi}$ and $R(x) = 1 - x - \frac{\sin 2\pi x}{\pi}$. As above, we find that $F = L\left(\frac{1}{3}\right)R\left(\frac{2}{3}\right) = \left(\frac{1}{3} + \frac{\sqrt{3}}{2\pi}\right)^2 = .609\ldots$.

Remark. If we look at the constant C of Theorem 1, we need to compute T on powers of x. Note that if μ is endpositive, then

$$\int_0^1 x^k d\mu = \int_0^1 \int_0^x kt^{k-1} dt \, d\mu(x) = \int_0^1 kt^{k-1} \int_t^1 d\mu(t)dt = k\int_0^1 t^{k-1}R(t)dt$$

so that

$$T(x^k, x^j, \mu) = (k + j)\int_0^1 x^{k+j-1}R(x)dx - kj\int_0^1 x^{k-1}R(x)dx \int_0^1 x^{j-1}R(x)dx$$

is an expression with a positive measure Rdx. Usual inequalities may be used.

6. Concave Functions

Grüss [7, 1935] quotes a result of M. Krafft to the effect that

(27)
$$|T(f, g, dx)| \leq \frac{1}{3}\int_0^1 f \int_0^1 g$$

if $f, g \geq 0$ and concave.

Theorem 12. *Suppose f, g are monotone $C^1[0,1]$ concave functions and $\mu \geq 0$. Then*

(27)
$$T(f, g, \mu) \leq M[f(1) - f(0)][g(1) - g(0)]$$

where $M = \max\limits_{0 \leq x \leq 1} \frac{1}{x} R_1(x) L_2(x)$ and

$$R_1(x) = R(x) = \int_x^1 d\mu \quad \text{and} \quad L_2(x) = \int_0^x L_1(t)dt, \ L_1(t) = \int_0^t d\mu.$$

If $d\mu = dx$ then $M = \frac{1}{8}$.

Proof. Suppose first that f, g are decreasing and concave. Using Pečarić's identity we have

$$T(f, g, \mu) = \int_0^1 Rf'(x) \int_0^x Lg'(t)dt \ dx + \int_0^1 Rg'(x) \int_0^x Lf'(t)dt \ dx.$$

Now L is increasing and f' is decreasing, so by Chebyshev's inequality $\int_0^x Lg'(t)dt \leq$

$\frac{1}{x} \int_0^x L \int_0^x g' = \frac{L_2(x)}{x} \int_0^x g'$. Thus

$$T(f, g, \mu) \leq \int_0^1 \frac{R(x)L_2(x)}{x} f'(x) \left(\int_0^x g'dt \right) du + \int_0^1 \frac{R(x)L_2(x)}{x} g'(x) \left(\int_0^x f'dt \right) d\mu$$

$$\leq M \left[\int_0^1 f'(x) \int_0^x g'(t)dt \ dx + \int_0^1 g'(x) \int_0^x f'(t)dt \ dx \right]$$

$$= M[f(1) - f(0)][g(1) - g(0)].$$

If f, g are decreasing, we apply the above argument to $f(1 - x)$ and $g(1 - x)$. For Lebesgue measure $L_2(x) = \frac{x^2}{2}$ and $R(x) = R_1(x) = 1 - x$.

The estimate for Lebesgue measure is not best possible.

Theorem 13. *If f, g are monotone and concave and $\mu \geq 0$, then*

$$T(f, g, dx) \leq \frac{1}{12}[f(1) - f(0)][g(1) - g(0)].$$

Proof. Use the Cauchy-Schwarz argument. If $f' \leq 0$ and f is concave, then $h = -f + f(0) \in M_2$ and $T(f, f, dx) = T(h, h, dx) \leq \frac{1}{12}[h(1) - h(0)]^2 = \frac{1}{12}[f(1) - f(0)]^2$. If f, g are increasing, again graph backwards.

7. Other bounds

We have mostly looked for bounds in terms of the oscillation of f or its deviation from the average. One other natural bound is $\|f'\|_\infty$. For example, using Pečarić's identity for end positive measures we have.

Theorem 14. *Let μ be an end positive measure and $f, g \in C^1(0,1)$, then*

$$T(f,g,\mu) \leq N\|f'\|_\infty\|g'\|_\infty, \text{ where}$$

$$N = 2\int_0^1 R_1 L_2(x)dx, \ L_2(x) = \int_0^x L(t)dt.$$

This constant is best possible.

Proof.

$$T(f,g,u) = \int_0^1 Rf' \int_0^x Lg' + \int_0^1 Rg' \int_0^x Lf'$$

$$\leq \|f'\|_\infty\|g'\|_\infty \left[\int_0^1 R(x)\left(\int_0^x L(t)dt\right)dx + \int_0^1 R(x)\left(\int_0^x L(t)dt\right)dx\right]$$

$$= 2\int_0^1 R_1 L_2(x)dx\|f'\|_\infty\|g'\|_\infty.$$

Note that $\int_0^1 R(x)\int_0^x L(x)dt\ dx + \int_0^1 R(x)\int_0^x L(t)dt\ dx = T(x,x,\mu)$ by Pečarić's identity, so equality holds if $f' = g' = 1$.

Corollary 3. *(Pečarić) For Lebesgue measure*

$$T(f,g,dx) \leq \frac{1}{12}\|f'\|_\infty\|g'\|_\infty.$$

Theorem 15. *For absolutely monotonic functions f, g we have*

$$T(f,g,dx) \leq \frac{1}{12}\|f'\|_\infty\|g'\|_\infty = \frac{1}{12}f'(1)g'(1).$$

Equality holds for $f = g = x$.

Proof. See (6) and (9)

$$T(f,f,dx) = \sum_{k,j=1}^\infty a_k a_j \frac{kj}{(k+1)(j+1)(k+j+1)}$$

Noting that $f'(1) = \sum\limits_{1}^{\infty} ka_k$, we find that

$$T(f, f, dx) \leq \frac{1}{12} \sum_{k,j=1}^{\infty} kj a_k a_j = \frac{1}{12} f'(1)^2$$

Theorem 16. *Let f, g be in CC then*

(28) $$T(f, g, dx) \leq \frac{1}{2}\|f'\|_\infty \|g'\|_\infty = \frac{1}{12} f'(0)g'(0)$$

This is best possible; $\{e^{-sx}\}_{s\downarrow 0}$ is a minimizing sequence.

Proof. Here one applies the previous result to $f(1-x)$ and $g(1-x)$. Again one cannot see an extremal. The inequality (28) is equivalent (see section on CC) to

$$\frac{1 - e^{-(s+t)}}{s+t} - \frac{(1-e^{-s})1 - e^t)}{st} \leq \frac{1}{12} st.$$

It can be verified that for $s = t$.
$$\lim_{s \to 0} \frac{\frac{1-e^{-s}}{2s} + \frac{(1-e^{-s})^2}{s^2}}{s^2} = \frac{1}{12}$$ so this gives a minimizing sequence.

Lemma.

$$T(f, g, \mu) = \int_0^1 f'g'(R_1 L_2 + L_1 R_2)dx - \int_0^1 R_2 f''(t)\left(\int_0^x g''(t)L_2(t)dt\right)dx$$
$$- \int_0^1 R_2 g''(x)\left(\int_0^x f''(t)L_2(t)dt\right)dx$$

where $L_2(t) = \int_0^t L_1(t)dt = \int_0^t L(s)ds$ and $R_2(t) = \int_t^1 R_1(s)ds = \int_t^1 R(s)ds.$

Proof. Starting with Pečarić's identity and parts

$$\int_0^x Lg' = g'(x)L_2(x) - \int_0^x g''L_2$$

so

$$\int_0^1 R_1 f' \int_0^x L_1 g' = \int_0^1 R_1 L_2 f'g' - \int_0^1 R_1 f' \int_0^x g''L_2$$

and parts again

$$= \int_0^1 R_1 L_2 f'g' - \int_0^1 f'' R_2(x) \int_0^x g'' L_2(t)dt \ dx - \int_0^1 R_2 L_2 f'g''$$

so

$$T(f,g,\mu) = 2\int_0^1 R_1 L_2 f'g' - \int_0^1 f'' R_2(x) \int_0^x g'' L_2(t)dt \ dx - \int_0^1 g'' R_2(x) \int_0^x f'' L_2(t)dt \ dx$$

$$- \int_0^1 R_2 L_2 (f'g'' + g'f'').$$

Integrating the last term by parts $\int_0^1 R_2 L_2 (f'g')' = \int_0^1 R_1 L_2 f'g' - \int_0^1 R_2 L_1 f'g'$ gives the identity.

Theorem 17. *If f and g are both concave or both convex and μ endpositive, then*

$$T(f,g,\mu) \le \int_0^1 f'g'(R_1 L_2 + L_1 R_2)$$

with equality for $f(x) = g(x) = x$.

$$T(f,g,\mu) \le \left(2\int_0^1 R_1 L_2\right) \|f'\|_\infty \|g'\|_\infty$$

with equality for $f(x) = g(x) = x$.

$$T(f,g,dx) \le \frac{1}{12}\|f'\|_\infty \|g'\|_\infty.$$

$$T(f,g,\mu) \le \max_{0 \le x \le 1}(R_1 L_2 + L_1 R_2)\|f'\|_2 \|g'\|_2.$$

$$T(f,g,dx) \le \frac{1}{8}\|f'\|_2 \|g'\|_2$$

but this is not best possible.

If μ is end positive and $d\sigma = (R_1 L_2 + L_1 R_2)dx$, then

$$T(f,g,\mu) \le \left(\int_0^1 (f')^2 d\sigma\right)^{\frac{1}{2}} \left(\int_0^1 (g')^2 d\sigma\right)^{\frac{1}{2}}.$$

Proof. Each of these follows from the lemma and the preceding inequality of the theorm.

References

1. Čebyšev, P.L., *O priblizennyh vyraženijah odnih integralov čerez drugie*, Soobščenija i Protokoly Zasedamiï Matematišeskogo Obšestva pri Imperatorskom Har'kovskom Universite No. 2 (1882), 93–98, Polnoe Sobranie Sočinenii P.L. Čebyševa. Moskva, Leningrad 1948, pp. 128–131.

2. Mitrinović, D.S., J.E. Pečarić, and A.M. Fink, *Classical and New Inequalities in Analysis*, Kluwer Academic Publishers, Dordrecht. Boston. London, 1993.

3. Fink, A.M. and Jodeit, Max, Jr., *Chebyshev Inequalities and Functions with Higher Monotonicities*, Technical Report, University of Minnesota and Iowa State University 1976.

4. _____, *On Čhebyšhev's other Inequality*, Symposium on Stat. Ineq., Institute of Mathematical Statistics Lecture Notes Edited by Y.L. Tong **5** (1984).

5. Pečarić, J.E., *On the Čebyšev inequality*, Bul. Inst. Politehn. Timisoara. Ser Math. Fiz. **26** (39) (1980), 5–9.

6. Fink, A.M. and Jodeit, Max, Jr., *Jensen Inequalities for functions with Higher Monotonicities*, Aeq. Math. **40** (1990), 26–43.

7. Grüss, G, *Über das Maximum des absoluten Betrages von* $\frac{1}{b-a} \int\limits_a^b f(x)g(x)dx - \frac{1}{(b-a)^2} \int\limits_a^b f(x)dx \int\limits_a^b g(x)dx$, Math. Z. **39** (1935), 215–226.

8. Pečarić, J.E., *On the Ostrowski generalization of Čebyšev's inequality*, J. Math. Anal. Appl. **102** (1984), 479–487.

9. Widder, David Vernon, *The Laplace Transform*, Princeton University Press, 1941.

10. Landau, E., *Über mehrfach monotone Folgen*, Prace Mat. Fiz. **44** (1936), 337–351.

$A \geq B \geq 0$ ENSURES $(B^{\frac{r}{2}} A^p B^{\frac{r}{2}})^{\frac{1}{q}} \geq (B^{\frac{r}{2}} B^p B^{\frac{r}{2}})^{\frac{1}{q}}$ FOR $r \geq 0$, $p \geq 0$, $q \geq 1$ WITH $(1+r)q \geq p+r$ AND ITS RECENT APPICATIONS

TAKAYUKI FURUTA

Department of Applied Mathematics, Faculty of Science, Science University of Tokyo, 1-3 Kagurazaka, Shinjukuku, Tokyo, 162-8601, Japan

Abstract. Inequality made in its title is Furuta inequality and a brief survey of some results selected from among our own recent applications of this inequality is given.

1. Introduction

A capital letter means a bounded linear operator on a Hilbert space H. An operator T is said to be positive (denoted by $T \geq 0$) if $(Tx, x) \geq 0$ for all $x \in H$ and also an operator T is said to be strictly positive (denoted by $T > 0$) if T is positive and invertible. It is well known that $A \geq B \geq 0$ does not always ensure $A^2 \geq B^2$ in general. Our main object is to establish "order preserving operator inequalities" on A and B in the case $A \geq B \geq 0$. We established the following order preserving operator inequalities as an extension of Löwner-Heinz inequality [L] and [H].

Theorem F (Furuta inequality) [F-1].

If $A \geq B \geq 0$, then for each $r \geq 0$,

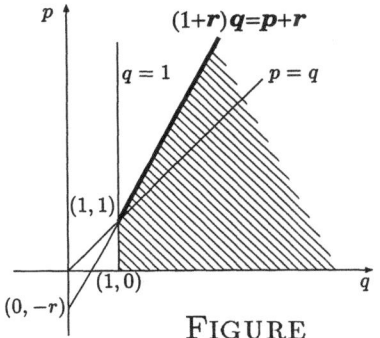

(i) $\quad (B^{\frac{r}{2}} A^p B^{\frac{r}{2}})^{\frac{1}{q}} \geq (B^{\frac{r}{2}} B^p B^{\frac{r}{2}})^{\frac{1}{q}}$

and

(ii) $\quad (A^{\frac{r}{2}} A^p A^{\frac{r}{2}})^{\frac{1}{q}} \geq (A^{\frac{r}{2}} B^p A^{\frac{r}{2}})^{\frac{1}{q}}$

hold for $p \geq 0$ and $q \geq 1$ with $(1+r)q \geq p+r$.

FIGURE

1991 *Mathematics Subject Classification.* Primary 47A63, 47A05, 47C10

Key word and phrases. Löwner-Heinz inequality, Furuta inequality

T.M. Rassias and H.M. Srivastava (eds.), Analytic and Geometric Inequalities and Applications, 115–134.

Theorem F yields the following famous Löwner-Heinz inequality when we put $r = 0$ in (i) or (ii) of Theorem F.

Theorem L-H ([L],[H]): $A \geq B \geq 0$ *ensures* $A^{\alpha} \geq B^{\alpha}$ *for any* $\alpha \in [0, 1]$.

We would like to explain the background of Theorem F. First of all, (i) and (ii) of Theorem F shoud be understood as follows: We first make the magic boxes

$$f(\Box) = (B^{\frac{r}{2}} \Box B^{\frac{r}{2}})^{\frac{1}{q}} \text{ and } g(\Box) = (A^{\frac{r}{2}} \Box A^{\frac{r}{2}})^{\frac{1}{q}}.$$

Although $A \geq B \geq 0$ does not always ensure $A^p \geq B^p$ for $p > 1$ in general, Theorem F asserts the following "two order preserving operator inequalities"

$$f(A^p) \geq f(B^p) \text{ and } g(A^p) \geq g(B^p)$$

hold whenever $A \geq B \geq 0$ under the condition p , q and r in Figure.

According to remarkable chievements of many mathematicians who have interested with operator inequalities, at present we have been finding a lot of applications of Theorem F in the following three branches (A) operator inequalities, (B) norm inequalities, and (C) operator equations.

As the area of applications is vast, we have to concentrate ourselves to some results selected from among our own recent applications.

(A) OPERATOR INEQUALITIES

(A-1) Several characterizations of operators $log A \geq log B$ and its applications;

(A-2) Generalizations of Ando's theorem;

(A-3) Applications to the relative operator entropy;

(A-4) Applications to Ando-Hiai log majorization;

(A-5) Generalized Aluthge transformation;

(A-6) Order preserving operator inequalities.

(A-7) Operator functions implying order preserving inequalities.

(B) NORM INEQUALITIES

(B-1) Several generalizations of Heinz-Kato theorem;

(B-2) Generalizations of some theorem on norms;

(B-3) An extension of Kosaki trace inequality and parallel results

(C) OPERATOR EQUATIONS

(C-1) Generalizations of Pedersen-Takesaki theorem and related results.

$$A \geq B \geq 0 \text{ ENSURES } (B^{\frac{r}{2}} A^p B^{\frac{r}{2}})^{\frac{1}{q}} \geq (B^{\frac{r}{2}} B^p B^{\frac{r}{2}})^{\frac{1}{q}}$$

2. An elementary and simplified proof of Theorem F in (A-6)

We shall give an elementary and simplified proof by refining [F-4].

Lemma A [F-13]. *Let X be a positive invertible operator and Y be an invertible operator. For any real number λ,*

$$(YXY^*)^\lambda = YX^{\frac{1}{2}}(X^{\frac{1}{2}}Y^*YX^{\frac{1}{2}})^{\lambda-1}X^{\frac{1}{2}}Y^*.$$

Proof. Let $YX^{\frac{1}{2}} = UH$ be the polar decomposition of $YX^{\frac{1}{2}}$, where U is unitary and $H = |YX^{\frac{1}{2}}|$. Then we have

$$(YXY^*)^\lambda = (UH^2U^*)^\lambda = YX^{\frac{1}{2}}H^{-1}H^{2\lambda}H^{-1}X^{\frac{1}{2}}Y^* = YX^{\frac{1}{2}}(X^{\frac{1}{2}}Y^*YX^{\frac{1}{2}})^{\lambda-1}X^{\frac{1}{2}}Y^*.$$

Proof of Theorem F. In the case $1 \geq p \geq 0$, the result is obvious by Theorem L-H. We have only to consider $p \geq 1$ and $q = \frac{p+r}{1+r}$ since (i) of Theorem F for values q larger than $\frac{p+r}{1+r}$ follows by Theorem L-H. We may assume that A and B are invertible without loss of generality. In the case $1 \geq r \geq 0$, $A^r \geq B^r$ holds Theorem L-H and $q = \frac{p+r}{1+r} \geq 1$ holds for $p \geq 1$. Then we have

$$
\begin{aligned}
(B^{\frac{r}{2}} A^p B^{\frac{r}{2}})^{\frac{1}{q}} &= B^{\frac{r}{2}} A^{\frac{p}{2}} (A^{\frac{-p}{2}} B^{-r} A^{\frac{-p}{2}})^{1-\frac{1}{q}} A^{\frac{p}{2}} B^{\frac{r}{2}} \quad \text{by Lemma A} \\
&\geq B^{\frac{r}{2}} A^{\frac{p}{2}} (A^{\frac{-p}{2}} A^{-r} A^{\frac{-p}{2}})^{\frac{p-1}{p+r}} A^{\frac{p}{2}} B^{\frac{r}{2}} \\
&= B^{\frac{r}{2}} A B^{\frac{r}{2}} \geq B^{1+r},
\end{aligned}
$$

and the first inequality follows by $B^{-r} \geq A^{-r}$ and Theorem L-H since $\frac{p-1}{p+r} \in [0,1]$ holds, and the last inequality follows by $A \geq B \geq 0$, so we have the following (2.1)

$$(2.1) \qquad (B^{\frac{r}{2}} A^p B^{\frac{r}{2}})^{\frac{1}{q}} \geq B^{1+r} \quad \text{for } p \geq 1, \ q = \frac{p+r}{1+r} \text{ and for } r \in [0,1].$$

Put $A_1 = (B^{\frac{r}{2}} A^p B^{\frac{r}{2}})^{\frac{1}{q}}$ and $B_1 = B^{1+r}$ in (2.1). Repeating (2.1) again for $A_1 \geq B_1 \geq 0$, $1 \geq r_1 \geq 0$ and $p_1 \geq 1$,

$$(B_1^{\frac{r_1}{2}} A_1^{p_1} B_1^{\frac{r_1}{2}})^{\frac{1}{q_1}} \geq B_1^{1+r_1} \qquad \text{for } q_1 = \frac{p_1+r_1}{1+r_1}.$$

Put $p_1 = q \geq 1$ and $r_1 = 1$, then

$$(2.2) \qquad (B^{r+\frac{1}{2}} A^p B^{r+\frac{1}{2}})^{\frac{1}{q_1}} \geq B^{2(1+r)} \quad \text{for } p \geq 1, \ q_1 = \frac{p_1+r_1}{1+r_1} \text{ and for } r \in [0,1]. \ .$$

Put $\frac{s}{2} = r + \frac{1}{2}$ in (2.2). Then $q_1 = \frac{p_1+r_1}{1+r_1} = \frac{p+s}{1+s}$ since $p_1 = q$ and $2(1+r) = 1+s$, so that (2.2) can be rewritten as follows;

$$(2.3) \qquad (B^{\frac{s}{2}} A^p B^{\frac{s}{2}})^{\frac{1}{q_1}} \geq B^{1+s} \quad \text{for } p \geq 1, \ q_1 = \frac{p+s}{1+s} \text{ and for } s \in [1,3].$$

Consequently (2.1) and (2.3) ensure that (2.1) holds for each $r \in [0,3]$ since $r \in [0,1]$ and $s = 2r + 1 \in [1,3]$ and repeating this method, (i) holds for each $r \geq 0$, (i) is shown. By hypothesis, $B^{-1} \geq A^{-1} \geq 0$. Then by (i), for each $r \geq 0$, $(A^{\frac{-r}{2}} B^{-p} A^{\frac{-r}{2}})^{\frac{1}{q}} \geq A^{\frac{-(p+r)}{q}}$ holds for each p and q such that $p \geq 0$, $q \geq 1$ and $(1+r)q \geq p+r$. Taking inverses gives (ii), so the proof of Theorem F is complete.

Remark 2.1. The original proof of Theorem F is given in [F-1] and alternative proofs have been given in [MF], [F-3] and [K] . Theorem F gives an affirmative answer to a conjecture in [C-K] and also Theorem F is a further extension of this conjecture.

Remark 2.2. The domain drawn for p,q and r in Figure is the best possible one for Theorem F and this result is shown in [T-1].

3. An extension of Theorem F in (A-6)

Here we state the following extension of Theorem F.

Theorem 3.1. *If $A \geq B \geq 0$ with $A > 0$, then for $1 \geq q \geq t \geq 0$ and $p \geq q$*

$$(3.1) \qquad A^{q-t+r} \geq \{A^{\frac{r}{2}}(A^{\frac{-t}{2}}B^{p}A^{\frac{-t}{2}})^{s}A^{\frac{r}{2}}\}^{\frac{q-t+r}{(p-t)s+r}} \qquad for\ s \geq 1\ and\ r \geq t.$$

Proof. First of all, we prove that if $A \geq B \geq 0$ with $A > 0$, then

$$(3.2) \qquad A^{q} \geq \{A^{\frac{t}{2}}(A^{\frac{-t}{2}}B^{p}A^{\frac{-t}{2}})^{s}A^{\frac{t}{2}}\}^{\frac{q}{(p-t)s+t}} \qquad for\ 1 \geq q \geq t \geq 0,\ p \geq q\ and\ s \geq 1.$$

In case $2 \geq s \geq 1$, as $s-1, \frac{q}{(p-t)s+t} \in [0,1]$ and $A^{t} \geq B^{t}$ by Löwner-Heinz inequality, so by Lemma A and Löwner-Heinz inequality we have

$$(3.3)\ B_{1} = \{A^{\frac{t}{2}}(A^{\frac{-t}{2}}B^{p}A^{\frac{-t}{2}})^{s}A^{\frac{t}{2}}\}^{\frac{q}{(p-t)s+t}} = \{B^{\frac{p}{2}}(B^{\frac{p}{2}}A^{-t}B^{\frac{p}{2}})^{s-1}B^{\frac{p}{2}}\}^{\frac{q}{(p-t)s+t}}$$

$$\leq \{B^{\frac{p}{2}}(B^{\frac{p}{2}}B^{-t}B^{\frac{p}{2}})^{s-1}B^{\frac{p}{2}}\}^{\frac{q}{(p-t)s+t}} = B^{q} \leq A^{q} = A_{1}$$

for $1 \geq q \geq t \geq 0$, $p \geq q$ and $2 \geq s \geq 1$. Repeating (3.3) for $A_{1} \geq B_{1} \geq 0$, then we have

$$(3.4)\ A_{1}^{q_{1}} \geq \{A_{1}^{\frac{t_{1}}{2}}(A_{1}^{\frac{-t_{1}}{2}}B_{1}^{p_{1}}A_{1}^{\frac{-t_{1}}{2}})^{s_{1}}A_{1}^{\frac{t_{1}}{2}}\}^{\frac{q_{1}}{(p_{1}-t_{1})s_{1}+t_{1}}} \text{ for } 1 \geq q_{1} \geq t_{1} \geq 0,\ p_{1} \geq q_{1}\ and$$

$2 \geq s_{1} \geq 1$. Put $1 = q_{1} \geq t_{1} = \frac{t}{q} \geq 0$ and $p_{1} = \frac{(p-t)s+t}{q} \geq q_{1} = 1$ in (3.4). Then we obtain

$$(3.5)\quad A^{q} \geq \{A^{\frac{t}{2}}[A^{\frac{-t}{2}}A^{\frac{t}{2}}(A^{\frac{-t}{2}}B^{p}A^{\frac{-t}{2}})^{s}A^{\frac{t}{2}}A^{\frac{-t}{2}}]^{s_{1}}A^{\frac{t}{2}}\}^{\frac{q}{(p-t)ss_{1}+t}}$$

$$= \{A^{\frac{t}{2}}(A^{\frac{-t}{2}}B^{p}A^{\frac{-t}{2}})^{ss_{1}}A^{\frac{t}{2}}\}^{\frac{q}{(p-t)ss_{1}+t}} \quad for\ 1 \geq q \geq t \geq 0,\ p \geq q\ and\ 4 \geq ss_{1} \geq 1.$$

Repeating this process from (3.3) to (3.5), we obtain (3.2) *for $1 \geq q \geq t \geq 0$, $p \geq q$ and any $s \geq 1$.* Put $A_{2} = A^{q}$ and $B_{2} = \{A^{\frac{t}{2}}(A^{\frac{-t}{2}}B^{p}A^{\frac{-t}{2}})^{s}A^{\frac{t}{2}}\}^{\frac{q}{(p-t)s+t}}$ in (3.2). Applying (ii) of Theorem F for $A_{2} \geq B_{2} \geq 0$ by (3.2) for $1 \geq q \geq t \geq 0$, $p \geq q$ and $s \geq 1$, so we have

$$(3.6) \qquad A_{2}^{1+r_{2}} \geq (A_{2}^{\frac{r_{2}}{2}}B_{2}^{p_{2}}A_{2}^{\frac{r_{2}}{2}})^{\frac{1+r_{2}}{p_{2}+r_{2}}} \text{ holds} \quad for\ p_{2} \geq 1\ and\ r_{2} \geq 0.$$

We have only to put $r_{2} = \frac{r-t}{q} \geq 0$ and $p_{2} = \frac{(p-t)s+t}{q} \geq 1$ in (3.6) to obtain the desired inequality (3.1) in Theorem 3.1, so the proof of Theorem 3.1 is complete.

Remark 3.1. The proof of Theorem 3.1 cited above is in [F-18] and we remark that the best possibility of Theorem 3.1 for $q = 1$ is obtained in [T-2] (see Remark 4.1).

4. Equivalence relation between generalized Furuta inequality and related operator functions in (A-7)

We shall show the following equivalence relation between Theorem 3.1 and related operator functions.

Theorem 4.1. *The following* (i), (ii), (iii) *and* (iv) *hold and follow from each other.*
(i) *If* $A \geq B \geq 0$ *with* $A > 0$, *then for each* $t \in [0,1]$ *and* $p \geq 1$,
$$A^{1-t+r} \geq \{A^{\frac{r}{2}}(A^{\frac{-t}{2}}B^pA^{\frac{-t}{2}})^sA^{\frac{r}{2}}\}^{\frac{1-t+r}{(p-t)s+r}} \quad \text{holds for } s \geq 1 \text{ and } r \geq t.$$

(ii) *If* $A \geq B \geq 0$ *with* $A > 0$, *then for each* $1 \geq q \geq t \geq 0$ *and* $p \geq q$,
$$A^{q-t+r} \geq \{A^{\frac{r}{2}}(A^{\frac{-t}{2}}B^pA^{\frac{-t}{2}})^sA^{\frac{r}{2}}\}^{\frac{q-t+r}{(p-t)s+r}} \quad \text{holds for } s \geq 1 \text{ and } r \geq t.$$

(iii) *If* $A \geq B \geq 0$ *with* $A > 0$, *then for each* $t \in [0,1]$ *and* $p \geq 1$,
$$F_{p,t}(A,B,r,s) = A^{\frac{-r}{2}}\{A^{\frac{r}{2}}(A^{\frac{-t}{2}}B^pA^{\frac{-t}{2}})^sA^{\frac{r}{2}}\}^{\frac{1-t+r}{(p-t)s+r}}A^{\frac{-r}{2}}$$

is decreasing for $r \geq t$ *and* $s \geq 1$.

(iv) *If* $A \geq B \geq 0$ *with* $A > 0$, *then for each* $t \in [0,1]$, $q \geq 0$ *and* $p \geq t$,
$$G_{p,q,t}(A,B,r,s) = A^{\frac{-r}{2}}\{A^{\frac{r}{2}}(A^{\frac{-t}{2}}B^pA^{\frac{-t}{2}})^sA^{\frac{r}{2}}\}^{\frac{q-t+r}{(p-t)s+r}}A^{\frac{-r}{2}}$$

is decreasing for $r \geq t$ *and* $s \geq 1$ *such that* $(p-t)s \geq q - t$.

Remark 4.1. Theorem 4.1 itself is shown in [FHI]. (i) and (iii) in Theorem 4.1 have been obtained in [F-13] and an excellent mean theoretic proof is shown in [MF-K-2]. (i) interpolates the inequality equivalent to the main result of the log majorization [A-H] and Theorem F itself. Recently (ii) is shown in [FYY-2] and (iv) is also shown in [FYY-1] as an extension of (iii) and [FW]. Related paper to Theorem 4.1 is in [JFK].

Proof of Theorem 4.1. We may assume that A and B are both invertible.
(iv) $\Longrightarrow$ (iii). We have only to put $q = 1$ in (iv).
(iii) $\Longrightarrow$ (i). $A \geq B \geq 0$ and monotonicity of $F_{p,t}(A,B,r,s)$ ensure
$$A^{1-t} \geq A^{\frac{-t}{2}}BA^{\frac{-t}{2}} = F_{p,t}(A,B,t,1) \geq F_{p,t}(A,B,r,s)$$
so that we have (i).
(i) $\Longrightarrow$ (ii). Put $A_1 = A^q$ and $B_1 = B^q$ for $q \in [0,1]$. Then $A_1 \geq B_1 \geq 0$ holds by Löwner-Heinz inequality. Put $p_1 = \frac{p}{q} \geq 1$, $t_1 = \frac{t}{q}$ and $r_1 = \frac{r}{q}$. Then we have only to apply (i) on $A_1 \geq B_1$.
(ii) $\Longrightarrow$ (iv). Put $q = t$ in (ii). Then if $A \geq B \geq 0$, then for each $t \in [0,1]$ and $p \geq t$

$$(4.1) \qquad A^r \geq \{A^{\frac{t}{2}}(A^{\frac{-t}{2}}B^p A^{\frac{-t}{2}})^s A^{\frac{t}{2}}\}^{\frac{r}{(p-t)s+r}} \qquad \text{for } s \geq 1 \text{ and } r \geq t.$$

(a) *Decreasing of* $G_{p,q,t}(A,B,r,s)$ *for* s. Put $D = A^{\frac{-t}{2}}B^p A^{\frac{-t}{2}}$. Applying Lemma A to (4.1) and Löwner-Heinz inequality, we obtain for each $t \in [0,1]$, $p \geq t$, $s \geq 1$ and $r \geq t$

$$(4.2) \qquad (D^{\frac{t}{2}}A^r D^{\frac{t}{2}})^{\frac{(p-t)w}{(p-t)s+r}} \geq D^w \qquad \text{for } s \geq w \geq 0.$$

Then we have

$$
\begin{aligned}
f(s) &= \{A^{\frac{t}{2}}(A^{\frac{-t}{2}}B^p A^{\frac{-t}{2}})^s A^{\frac{t}{2}}\}^{\frac{q-t+r}{(p-t)s+r}} \\
&= (A^{\frac{t}{2}}D^s A^{\frac{t}{2}})^{\frac{q-t+r}{(p-t)s+r}} \\
&= \{(A^{\frac{t}{2}}D^s A^{\frac{t}{2}})^{\frac{(p-t)(s+w)+r}{(p-t)s+r}}\}^{\frac{q-t+r}{(p-t)(s+w)+r}} \\
&= \{A^{\frac{t}{2}}D^{\frac{t}{2}}(D^{\frac{t}{2}}A^r D^{\frac{t}{2}})^{\frac{(p-t)w}{(p-t)s+r}} D^{\frac{t}{2}}A^{\frac{t}{2}}\}^{\frac{q-t+r}{(p-t)(s+w)+r}} \qquad \text{by Lemma A} \\
&\geq (A^{\frac{t}{2}}D^{s+w}A^{\frac{t}{2}})^{\frac{q-t+r}{(p-t)(s+w)+r}} \\
&= f(s+w)
\end{aligned}
$$

and the last inequality holds by (4.2) and Löwner-Heinz inequality since $\frac{q-t+r}{(p-t)(s+w)+r} \in [0,1]$ holds, so the proof of (a) is complete since $G_{p,q,t}(A,B,r,s) = A^{\frac{-r}{2}}f(s)A^{\frac{-r}{2}}$.

(b) *Decreasing of* $G_{p,q,t}(A,B,r,s)$ *for* r. Applying Löwner-Heinz inequality to (4.1), if $A \geq B \geq 0$, then for each $t \in [0,1]$, $p \geq t$, $s \geq 1$ and $r \geq t$

$$(4.3) \qquad A^u \geq (A^{\frac{t}{2}}D^s A^{\frac{t}{2}})^{\frac{u}{(p-t)s+r}} \qquad \text{for } r \geq u \geq 0.$$

Then we have

$$
\begin{aligned}
G_{p,q,t}(A,B,r,s) &= A^{\frac{-r}{2}}\{A^{\frac{t}{2}}(A^{\frac{-t}{2}}B^p A^{\frac{-t}{2}})^s A^{\frac{t}{2}}\}^{\frac{q-t+r}{(p-t)s+r}} A^{\frac{-r}{2}} \\
&= D^{\frac{t}{2}}(D^{\frac{t}{2}}A^r D^{\frac{t}{2}})^{\frac{q-t-(p-t)s}{(p-t)s+r}} D^{\frac{t}{2}} \qquad \text{by Lemma A} \\
&= D^{\frac{t}{2}}\{(D^{\frac{t}{2}}A^r D^{\frac{t}{2}})^{\frac{(p-t)s+r+u}{(p-t)s+r}}\}^{\frac{q-t-(p-t)s}{(p-t)s+r+u}} D^{\frac{t}{2}} \\
&= D^{\frac{t}{2}}\{D^{\frac{t}{2}}A^{\frac{t}{2}}(A^{\frac{t}{2}}D^s A^{\frac{t}{2}})^{\frac{u}{(p-t)s+r}} A^{\frac{t}{2}}D^{\frac{t}{2}}\}^{\frac{q-t-(p-t)s}{(p-t)s+r+u}} D^{\frac{t}{2}} \qquad \text{by Lemma A} \\
&\geq D^{\frac{t}{2}}(D^{\frac{t}{2}}A^{r+u}D^{\frac{t}{2}})^{\frac{q-t-(p-t)s}{(p-t)s+r+u}} D^{\frac{t}{2}} \\
&= G_{p,q,t}(A,B,r+u,s)
\end{aligned}
$$

and the last inequality holds by (4.3) and Löwner-Heinz inequality since $\frac{q-t-(p-t)s}{(p-t)s+r+u} \in [-1,0]$. Consequently we obtain (iv) by (a) and (b), so the proof of Theorem 4.1 is complete.

5. Application to Ando-Hiai log majorization in (A-4)

We are really impressed with beautiful and useful results on log-majorization in [A-H]. In what follows, a capital letter means $n \times n$ matrix. Following after [A-H], let us write $A \underset{(\log)}{\prec} B$ for positive semidefinite matrices $A, B \geq 0$ and call the *log majorization* if

$$\prod_{i=1}^{k}\lambda_i(A) \le \prod_{i=1}^{k}\lambda_i(B), \qquad\qquad k=1,2,...,n-1,$$

and

$$\prod_{i=1}^{n}\lambda_i(A) = \prod_{i=1}^{n}\lambda_i(B), \quad \text{i.e. } \det A{=}\det B,$$

where $\lambda_1(A) \ge \lambda_2(A) \ge ... \ge \lambda_n(A)$ and $\lambda_1(B) \ge \lambda_2(B) \ge ... \ge \lambda_n(B)$ are the eigenvalues of A and B respectively arranged in decreasing order. Note that when $A, B > 0$ (strictly positive) the log majorization $A \underset{(\log)}{\prec} B$ is equivalent to $\log A \prec \log B$. Also $A \underset{(\log)}{\prec} B$ ensures $\|A\| \le \|B\|$ holds for any unitarily invariant norm. When $0 \le \alpha \le 1$, the α-power mean of $A, B > 0$ is defined by $A \#_\alpha B = A^{\frac{1}{2}}(A^{\frac{-1}{2}}BA^{\frac{-1}{2}})^\alpha A^{\frac{1}{2}}$.

Further $A\#_\alpha B$ for $A, B \ge 0$ is defined by $A\#_\alpha B = \{lim_{\epsilon\downarrow 0}(A+\epsilon I)\#_\alpha(B+\epsilon I)$.

This α-power mean is the operator mean corresponding to the operator monotone function t^α. We can see [K-A] for general theory of operator means.

For the sake of convenience for symbolic expression , we define $A \,\natural_s B$ for any $s \ge 0$ and for $A > 0$ and $B \ge 0$ by the following $A \,\natural_s B = A^{\frac{1}{2}}(A^{\frac{-1}{2}}BA^{\frac{-1}{2}})^s A^{\frac{1}{2}}$.

$A \,\natural_\alpha B$ in the case $0 \le \alpha \le 1$ just coincides with the usual α-power mean denoted by $A\#_\alpha B$.

We can transform (3.1) of Theorem 3.1 into the following log majorization inequality by using the method by [A-H].

Theorem 5.1. *For every $A > 0$, $B \ge 0$, $0 \le \alpha \le 1$ and each $t \in [0,1]$*

$$(5.1) \qquad (A\#_\alpha B)^h \underset{(\log)}{\succ} A^{1-t+r}\#_\beta (A^{1-t} \,\natural_s B)$$

holds for $s \ge 1$, and $r \ge t \ge 0$, where $\beta = \dfrac{\alpha(1-t+r)}{(1-\alpha t)s + \alpha r}$ and $h = \dfrac{(1-t+r)s}{(1-\alpha t)s + \alpha r}$.

Proof. We may assume that $0 < \alpha \le 1$. In the same way as in the proof of [A-H, Theorem 2.1], by homogeneity of order in (5.1), to prove (5.1) we have only to show that $A^{-1} \ge (A^{\frac{-1}{2}}BA^{\frac{-1}{2}})^\alpha$ ensures the following inequality (5.2) :

$$(5.2) \qquad A^{-1-r+t} \ge \{A^{\frac{-1-r+t}{2}}(A^{1-t}\natural_s B)A^{\frac{-1-r+t}{2}}\}^\beta$$

for $s \ge 1$ and $r \ge t \ge 0$. Put $A_1 = A^{-1}$ and $B_1 = (A^{\frac{-1}{2}}BA^{\frac{-1}{2}})^\alpha$. By (3.1) in Theorem 3.1, $A_1 \ge B_1 \ge 0$ with $A_1 > 0$ ensures the following inequality for each $t \in [0,1]$,

$$A_1^{1-t+r} \ge \{A_1^{\frac{r}{2}}(A_1^{\frac{-t}{2}}B_1^p A_1^{\frac{-t}{2}})^s A_1^{\frac{r}{2}}\}^{\frac{1-t+r}{(p-t)s+r}}$$

for any $s \ge 1, p \ge 1$ and $r \ge t \ge 0$. Put $p = \frac{1}{\alpha} \ge 1$ in the above inequality, then we have

$$A^{-1-r+t} \geq \{A^{\frac{-r}{2}}(A^{\frac{t-1}{2}}BA^{\frac{t-1}{2}})^s A^{\frac{-r}{2}}\}^\beta$$

$$= \{A^{\frac{-1-r+t}{2}}(A^{1-t}\natural_s B)A^{\frac{-1-r+t}{2}}\}^\beta$$

that is, we have (5.2).

When $t = 1$ Theorem 5.1 becomes the following result.

Corollary 5.2. *For every $A, B \geq 0$ and $0 \leq \alpha \leq 1$,*

$$(A\#_\alpha B)^h \underset{(\log)}{\succ} A^r \#_{\frac{h\alpha}{s}} B^s \qquad for\ r \geq 1\ and\ s \geq 1$$

where $h = [\alpha s^{-1} + (1-\alpha)r^{-1}]^{-1}$.

Corollary 5.2 yields the following result [A-H, Theorem 2.1] when we put $r = s$.

Theorem A-H-1 [A-H]. *For every $A, B \geq 0$ and $0 \leq \alpha \leq 1$,*
$$(A\#_\alpha B)^r \underset{(\log)}{\succ} A^r \#_\alpha B^r \qquad for\ r \geq 1$$
or equivalently
$$(A^q \#_\alpha B^q)^{1/q} \underset{(\log)}{\succ} (A^p \#_\alpha B^p)^{1/p} \qquad for\ 0 < q \leq p.$$

Remark 2.1. We remark that $h = [\alpha s^{-1} + (1-\alpha)r^{-1}]^{-1}$ in Corollary 5.2 is a generalized harmonic mean of r and s and when $\alpha = \frac{1}{2}$, h is the usual harmonic mean of r and s.

Also we obtain the following log majorization by the same way as one in Theorem 5.1.

Theorem 5.3. *If $A > 0$ and $B \geq 0$, then for each $t \in [0,1]$ and $0 \leq \alpha \leq 1$*

$$(A^{\frac{1}{2}}BA^{\frac{1}{2}})^{\alpha ps} \underset{(\log)}{\succ} A^{\frac{1}{2}\alpha((p-t)s+r)}(A^{\frac{-(r-t)}{2}}(A^t\,\natural_s B^p)A^{\frac{-(r-t)}{2}})^\alpha A^{\frac{1}{2}\alpha((p-t)s+r)}$$

$$= A^{\frac{q}{2}}[A^{r-t}\#_\alpha(A^{-t}\natural_s B^p)]A^{\frac{q}{2}}$$

holds for any nonnegative numbers s, p and r such that $r \geq t$ and $(s-1)(p-1) \geq 0$ with $1 - t + r \geq ((p-t)s + r)\alpha$ where $q = \alpha(p-t)s + \alpha r - r + t$.

Theorem 5.4. *If $A > 0$ and $B \geq 0$, then for each $t \in [0,1]$ and $0 \leq \alpha \leq 1$*

$$A^{\frac{1}{2}}(A^p\#_\alpha B^p)^{\frac{q}{p}}A^{\frac{1}{2}}$$

$$\underset{(\log)}{\succ} A^{\frac{1}{2}(1-\frac{qt}{p}+\frac{rq}{ps})}\{A^{\frac{-r}{2}}[A^{\frac{t}{2}}(A^p\#_\alpha B^p)A^{\frac{t}{2}}]^s A^{\frac{-r}{2}}\}^{\frac{q}{sp}}A^{\frac{1}{2}(1-\frac{qt}{p}+\frac{rq}{ps})}$$

holds for every $p \geq q > 0$, $r \geq t$ and $s \geq 1$.

Corollary 5.5 . *If $A > 0$ and $B \geq 0$, then for every $0 \leq \alpha \leq 1$*

$$A^{\frac{1}{2}}(A^p \#_\alpha B^p)^{\frac{q}{p}} A^{\frac{1}{2}}$$

$$\underset{(\log)}{\succ} A^{\frac{1}{2}(1+\frac{rq}{ps})} \{A^{\frac{-r}{2}}(A^p \#_\alpha B^p)^s A^{\frac{-r}{2}}\}^{\frac{q}{sp}} A^{\frac{1}{2}(1+\frac{rq}{ps})}$$

holds for every $p \geq q > 0$, $r \geq 0$ and $s \geq 1$.

When $s = 1$ and $r = p$, Corollary 5.5 yields the following result.

Theorem A-H-2 [A-H] . *If $A > 0$ and $B \geq 0$, then*

$$A^{\frac{1}{2}}(A^p \#_\alpha B^p)^{\frac{q}{p}} A^{\frac{1}{2}}$$

$$\underset{(\log)}{\succ} A^{\frac{1+q}{2}}(A^{\frac{-p}{2}} B^p A^{\frac{-p}{2}})^{\frac{\alpha q}{p}} A^{\frac{1+q}{2}}$$

for every $0 \leq \alpha \leq 1$ and $0 < q \leq p$.

Taking $s = 2$ and $r = p$ in Corollary 5.5, we have

Corollary 5.6 . *If $A > 0$ and $B \geq 0$, then for every $0 \leq \alpha \leq 1$*

$$A^{\frac{1}{2}}(A^p \#_\alpha B^p)^{\frac{q}{p}} A^{\frac{1}{2}}$$

$$\underset{(\log)}{\succ} A^{\frac{1}{2}(1+\frac{q}{2})} \{(A^{\frac{-p}{2}} B^p A^{\frac{-p}{2}})^\alpha A^p (A^{\frac{-p}{2}} B^p A^{\frac{-p}{2}})^\alpha\}^{\frac{q}{2p}} A^{\frac{1}{2}(1+\frac{q}{2})}$$

holds for any $0 < q \leq p$.

Corollary 5.7. *If $A > 0$ and $B \geq 0$, then for every $0 \leq r \leq 1$*

$$A^{\frac{r}{2}} B^r A^{\frac{r}{2}} \underset{(\log)}{\succ} A^{\frac{r(1+\alpha)}{2}} (A^{\frac{-1}{2}} B^{\frac{1}{\alpha}} A^{\frac{-1}{2}})^{\alpha r} A^{\frac{r(1+\alpha)}{2}}$$

holds for every $0 < \alpha \leq 1$.

Corollary 5.8. *If $A > 0$ and $B \geq 0$, then for every $0 \leq r \leq 1$*

$$(A^{\frac{1}{2}} B A^{\frac{1}{2}})^r \underset{(\log)}{\succ} A^{\frac{\alpha u + r}{2}} (A^{\frac{-u}{2}} B^{r/\alpha} A^{\frac{-u}{2}})^\alpha A^{\frac{\alpha u + r}{2}}$$

holds for every $0 < \alpha \leq 1$ and $u \geq 0$.

Corollary 5.7 and Corollary 5.8 imply the following result.

Theorem A-H-3 [A-H]. *If $A > 0$ and $B \geq 0$, then for every $0 \leq r \leq 1$*

$$(A^{\frac{1}{2}} B A^{\frac{1}{2}})^r \underset{(\log)}{\succ} A^{\frac{r}{2}} B^r A^{\frac{r}{2}} \underset{(\log)}{\succ} A^r (A^{\frac{-1}{2}} B A^{\frac{-1}{2}})^r A^r.$$

Theorem 5.9. *If $A > 0$ and $B \geq 0$, then for every $0 \leq \alpha \leq 1$ and $t \in [0,1]$*

$$s \mathrm{Tr} A \log(A^p \#_\alpha B^p) - \mathrm{Tr} A \log\{A^{\frac{-r}{2}} [A^{\frac{1}{2}}(A^p \#_\alpha B^p) A^{\frac{1}{2}}]^s A^{\frac{-r}{2}}\}$$

$$\geq (r - st) \mathrm{Tr} \ A \log A$$

holds for any $s \geq 1$, $r \geq t$ and $p \geq 0$.

When $t = 0$ Theorem 5.9 yields the following result.

Corollary 5.10. *If $A > 0$ and $B \geq 0$, then for every $0 \leq \alpha \leq 1$*
$$s\mathrm{Tr}Alog(A^p \#_\alpha B^p) - \mathrm{Tr}Alog\{A^{\frac{-r}{2}}[A^p\#_\alpha B^p]^s A^{\frac{-r}{2}}\}$$
$$\geq r\mathrm{Tr}AlogA$$
holds for any $s \geq 1$, $r \geq 0$ and $p \geq 0$.

Taking $s = 1$ and $r = p > 0$ in Corollary 5.10 we have the following result [A-H,Theorem 5.3].

Theorem A-H-4 [A-H] . *If $A \geq 0$ and $B > 0$, then for every $0 \leq \alpha \leq 1$ and $p > 0$*
$$\frac{1}{p}\mathrm{Tr}Alog(A^p\#_\alpha B^p) + \frac{\alpha}{p}\mathrm{Tr}Alog(A^{\frac{p}{2}}B^{-p}A^{\frac{p}{2}}) \geq \mathrm{Tr}AlogA.$$

Put $s = 1$ in Theorem 5.9 and then replace $r - t$ by q, we have the following result.

Corollary 5.11 . *If $A > 0$ and $B > 0$, then for every $0 \leq \alpha \leq 1$*
$$\mathrm{Tr}Alog(A^p\#_\alpha B^p) + \mathrm{Tr}Alog\{A^{\frac{q}{2}}[A^{-p}\#_\alpha B^{-p}]A^{\frac{q}{2}}\} \geq q\mathrm{Tr}AlogA$$
holds for any $p \geq 0$ and $q \geq 0$.

We remark that Corollary 5.11 yields Theorem A-H-4 stated above taking $q = p$.

Also taking $s = 2$, $t = 0$ and $r = p \geq 0$ in Theroem 5.9 we have :

Corollary 5.12. *If $A > 0$ and $B > 0$, then for every $0 \leq \alpha \leq 1$*
$$\mathrm{Tr}Alog(A^p\#_\alpha B^p)^2 + \mathrm{Tr}Alog\{(A^{\frac{p}{2}}B^{-p}A^{\frac{p}{2}})^\alpha A^{-p}(A^{\frac{p}{2}}B^{-p}A^{\frac{p}{2}})^\alpha\}$$
$$\geq p\mathrm{Tr}AlogA$$
holds for any $p \geq 0$.

Other applications in detail are given in [F-13].

Remark 5.1. The inequality (5.1) of Theorem 5.1 interpolates Theorem F and Theorem F-H-1 on log majorization.

6. Application to Kosaki trace inequality and related results in (B-3)

It is well known that $A \geq B \geq 0$ ensures $Tr(f(A)) \geq Tr(f(B))$, where Tr denotes the usual trace and f is a continuous increasing function on $\mathbf{R}_+$ with $f(0) = 0$. Kosaki

[KO] shows the following very interesting trace inequality as a generalization of the above mentioned trace inequality.

Theorem KO [KO]. *Assume $A \geq B \geq 0$ and $p > 1$, $\alpha \geq Max\{-1, \frac{-p}{2}\}$.*

(i) *Then there exists a partial isometry operator U satisfying*
$$A^{\frac{\alpha}{2}} B^p A^{\frac{\alpha}{2}} \leq U^* A^{p+\alpha} U.$$

(ii) *For a continuous increasing function f on $\mathbf{R}_+$ with $f(0) = 0$, we have*
$$Tr(f(A^{\frac{\alpha}{2}} B^p A^{\frac{\alpha}{2}})) \leq Tr(f(A^{p+\alpha}))$$
In the above statements the invertibility of A is assumed when $\alpha < 0$.

By using Theorem 3.1, Theorem KO can be generalized to the following form.

Theorem 6.1. *Let A and B be positive operators such that $A \geq B \geq 0$ with $A > 0$. Assume that $p \geq 1$, $s \geq 1$, $t \in [0,1]$ and $\beta \geq Max[t-1, \frac{1}{2}\{t(s+1)-ps\}]$. Then the following inequalities hold.*

(I) *There exists the a partial isometry operator U satisfying*
$$A^{\frac{\beta}{2}}(A^{\frac{-t}{2}} B^p A^{\frac{-t}{2}})^s A^{\frac{\beta}{2}} \leq U^* A^{(p-t)s+\beta} U.$$

(II) *For a continuous increasing function f on $\mathbf{R}_+$ with $f(0) = 0$,*
$$Tr\{f(A^{\frac{\beta}{2}}(A^{\frac{-t}{2}} B^p A^{\frac{-t}{2}})^s A^{\frac{\beta}{2}})\} \leq Tr\{f(A^{(p-t)s+\beta})\}.$$

When we replace $A \geq B \geq 0$ by $log A \geq log B$ in Theorem 6.1, we have the following parallel result to Theorem 6.1.

Theorem 6.2. *Let A and B be positive invertible operators such that $A \gg B$ (i.e., $log A \geq log B$). Assume that $p \geq u > 0$, $s \geq 1$, $\alpha \in [0,1]$ and $\beta \geq -u\alpha$. Then the following inequalities hold.*

(I) *There exists the a partial isometry operator U satisfying*
$$A^{\frac{\beta}{2}}(A^{\frac{u\alpha}{2}} B^p A^{\frac{u\alpha}{2}})^s A^{\frac{\beta}{2}} \leq U^* A^{(u\alpha+p)s+\beta} U.$$

(II) *For a continuous increasing function f on $\mathbf{R}_+$ with $f(0) = 0$,*
$$Tr\{f(A^{\frac{\beta}{2}}(A^{\frac{u\alpha}{2}} B^p A^{\frac{u\alpha}{2}})^s A^{\frac{\beta}{2}})\} \leq Tr\{f(A^{(u\alpha+p)s+\beta})\}.$$

Corollary 6.3. *Let A and B be positive invertible operators such that $A \gg B$ (i.e., $log A \geq log B$). Assume that $p > 0$, and $\beta \geq 0$. Then the following inequalities hold.*

(I) *There exists the a partial isometry operator U satisfying*
$$A^{\frac{\alpha}{2}} B^p A^{\frac{\beta}{2}} \leq U^* A^{p+\beta} U.$$

(II) *For a continuous increasing function f on $\mathbf{R}_+$ with $f(0) = 0$,*

$$Tr\{f(A^{\frac{\beta}{2}}B^pA^{\frac{\beta}{2}})\} \le Tr\{f(A^{p+\beta})\}.$$

Results in this section are given in [F-15].

7. Extensions of Aluthge transformation on p-hyponormal operators in (A-5)

An operator T is said to be *p-hyponormal* if $(T^*T)^p \ge (TT^*)^p$ for positive number p.

The class of p-hyponormal has been defined as an extension of hyponormal and also it has been studied by many authors, for example [Al],[C-I],[D], [Hu] and [Xi].

For a p-hyponormal operator $T = U|T|$, Aluthge [Al] introduced the operator $\widetilde{T} = |T|^{\frac{1}{2}}U|T|^{\frac{1}{2}}$ which is called Aluthge transformation and Aluthge [Al] showed very interesting results on $\widetilde{T}$ as follows.

Theorem A [Al]. *Let $T = U|T|$ be p-hyponormal for $0 < p < \frac{1}{2}$ and U be unitary. Then $\widetilde{T} = |T|^{\frac{1}{2}}U|T|^{\frac{1}{2}}$ is $(p+\frac{1}{2})$-hyponormal .*

As an extension of Aluthge's result, we showed the following result.

Theorem B [F-17]. *Let $T = U|T|$ be the polar decomposition of p-hyponormal for $1 \ge p > 0$ with $N(T) = N(T^*)$. Then $\widetilde{T} = |T|^qU|T|^q$ is $\frac{1}{2}(1+\frac{p}{q})$-hyponormal for any q such that $q \ge p$.*

In this section, by introducing two notions of $\widetilde{T} = |T|^sU|T|^t$ and $\widehat{T} = |T|^tU|T|^s$ for some fixed positive number s and t, we shall extend Theorem B.

Theorem 7.1. *Let $T = U|T|$ be the polar decomposition of p-hyponormal for $1 \ge p > 0$ with $N(T) = N(T^*)$. Then $\widetilde{T} = |T|^sU|T|^t$ and $\widehat{T} = |T|^tU|T|^s$ are both $(\frac{p+s}{s+t})$-hyponormal operators for any $s \ge 0$ and $t \ge Max\{p,s\}$.*

Proof of Theorem 7.1. Firstly We recall that if T is p-hyponormal for $p > 0$, the following (7.1) holds obviously

(7.1) $$U^*|T|^{2p}U \ge |T|^{2p} \ge U|T|^{2p}U^* \text{ for any } p > 0,$$

and also we obtained the following (7.2) for any $r > 0$ in [F-17],

(7.2) $$(U^*|T|U)^r = U^*|T|^rU \quad \text{holds under the hypothesis } N(T) = N(T^*).$$

Let $A = U^*|T|^{2p}U$, $B = |T|^{2p}$ and $C = U|T|^{2p}U^*$. Then for any $p > 0$ and $q > 0$

$$A^{\frac{2q}{2p}} = (U^*|T|^{2p}U)^{\frac{2q}{2p}} = (U^*|T|^{2q}U) \text{ holds by (7.2)}$$

and also $C^{\frac{2q}{2p}} = (U|T|^{2p}U^*)^{\frac{2q}{2p}} = U|T|^{2q}U^*$ holds in general. $A \ge B \ge C$ holds by (7.1), and as $(1 + 2\frac{t}{2p})\frac{s+t}{p+s} \ge \frac{2s}{2p} + 2\frac{t}{2p}$ and $\frac{s+t}{p+s} \ge 1$ since $t \ge Max\{p,s\}$ holds , by using Theorem F , we obtain

(7.3) $$(\widetilde{T}^*\widetilde{T})^{\frac{p+s}{s+t}} = (|T|^tU^*|T|^{2s}U|T|^t)^{\frac{p+s}{s+t}}$$
$$= (B^{\frac{t}{2p}}A^{\frac{2s}{2p}}B^{\frac{t}{2p}})^{\frac{p+s}{s+t}} \quad \text{by (7.2)}$$
$$\ge (B^{\frac{t}{2p}}B^{\frac{2s}{2p}}B^{\frac{t}{2p}})^{\frac{p+s}{s+t}} \quad \text{by (i) of Theorem F}$$

$$= B^{\frac{p+s}{p}}$$
$$\geq (B^{\frac{s}{2p}}C^{\frac{2t}{2p}}B^{\frac{s}{2p}})^{\frac{p+s}{s+t}} \quad \text{by (ii) of Theorem F}$$
$$= (|T|^t U |T|^{2s} U^* |T|^t)^{\frac{p+s}{s+t}}$$
$$= (\widehat{T}\widehat{T}^*)^{\frac{p+s}{s+t}}$$

and as $(1+2\dfrac{s}{2p})\dfrac{s+t}{p+s} = \dfrac{2t}{2p}+2\dfrac{s}{2p}$ and $\dfrac{s+t}{p+s} \geq 1$ since $t \geq Max\{p,s\}$ holds, again by using Theorem F , we obtain

(7.4)
$$(\widehat{T}^*\widehat{T})^{\frac{p+s}{s+t}} = (|T|^s U^* |T|^{2t} U |T|^s)^{\frac{p+s}{s+t}}$$
$$= (B^{\frac{s}{2p}}A^{\frac{2t}{2p}}B^{\frac{s}{2p}})^{\frac{p+s}{s+t}} \quad \text{by (7.2)}$$
$$\geq B^{\frac{p+s}{p}} \quad \text{by (i) of Theorem F}$$
$$= (B^{\frac{s}{2p}}B^{\frac{2t}{2p}}B^{\frac{s}{2p}})^{\frac{p+s}{s+t}}$$
$$\geq (B^{\frac{s}{2p}}C^{\frac{2t}{2p}}B^{\frac{s}{2p}})^{\frac{p+s}{s+t}} \quad \text{by (ii) of Theorem F}$$
$$= (|T|^s U |T|^{2t} U^* |T|^s)^{\frac{p+s}{s+t}}$$
$$= (\widetilde{T}\widetilde{T}^*)^{\frac{p+s}{s+t}}.$$

Hence (7.3) and (7.4) ensure $(\widetilde{T}^*\widetilde{T})^{\frac{p+s}{s+t}} \geq B^{\frac{p+s}{p}} \geq (\widetilde{T}\widetilde{T}^*)^{\frac{p+s}{s+t}}$ and $(\widehat{T}^*\widehat{T})^{\frac{p+s}{s+t}} \geq B^{\frac{p+s}{p}} \geq (\widehat{T}\widehat{T}^*)^{\frac{p+s}{s+t}}$ by interpolating by $B^{\frac{p+s}{p}}$, that is, $\widetilde{T}$ and $\widehat{T}$ are both $\frac{p+s}{s+t}$ -hyponormal.

Remark 7.1. It is interesting to remark that $\widetilde{T} = |T|^s U |T|^t$ and $\widehat{T} = |T|^t U |T|^s$ are both $(\frac{p+s}{s+t})$-hyponormal operators for any $s \geq 0$ and $t \geq Max\{p,s\}$ by interpolating by $B^{\frac{p+s}{p}}$ in (7.3) and (7.4). When $s = t = q$ Theorem 7.1 becomes Theorem B. Results in this section are in [FY].

8. Application to the relative operator entropy in (A-3) and (A-1)

In [JIF-K-1] and [JIF-K-2], the relative operator entropy $S(A|B)$ is defined by

$$S(A|B) = A^{\frac{1}{2}}(log A^{\frac{-1}{2}}BA^{\frac{-1}{2}})A^{\frac{1}{2}}$$

for invertible positive operators A and B and this relative operator entropy can be considered as an extension of the entropy by Nakamura and Umegaki [N-U] and the relative entropy by Umegaki [U]. We remark that $S(A|I) = -AlogA$ is the usual well known operator entropy.

In this section, we introduce a generalized relative operator entropy as an extension of $S(A|B)$ as follows. For positive invertible operators A and B, $(s\text{-}\alpha)$-generalized relative operator entropy $W_{s,\alpha}(A|B)$ for each $\alpha \in [0,1]$ is defined by;

(*)
$$W_{s,\alpha}(A|B) = A^{\frac{1}{2}}\{log A^{\frac{-1}{2}}(A^{\frac{-\alpha}{2}}BA^{\frac{-\alpha}{2}})^s A^{\frac{-1}{2}}\}A^{\frac{1}{2}}$$

for any $s \geq 1$. As easily seen by definition (*), for each $\alpha \in [0,1]$ and any $s \geq 1$, $W_{s,\alpha}(A|A^{\alpha/2}BA^{\alpha/2}) = S(A|B^s)$. Especially $W_{s,\alpha}(A|B) = S(A|B)$ in the case $s = 1$ and $\alpha = 0$, that is, $W_{1,0}(A|B)$ coincides with the usual relative operator entropy $S(A|B)$. We write $A \gg B$ if $\log A \geq \log B$ for invertible operator A and B, which is called the chaotic order [MF-K-1]. It is shown in [A] and well known as a useful result that $A \gg B$ if and only if $A^p \geq (A^{\frac{p}{2}} B^p A^{\frac{p}{2}})^{\frac{1}{2}}$ for all $p \geq 0$. As an application of Theorem 3.1, we shall discuss monotonicity property of a generalized relative entropy $W_{s,\alpha}(A|B)$ and also we shall show order preserving property of $W_{s,\alpha}(A|B)$ as a continuation of the results in [F-6] and [F-10]. Finally we shall attempt to extend the differential geometric view in [CPR] between the geodestic from A to B and the relative operator entropy $S(A|B)$, that is, we shall show a relation between a generalized geodestic from A to B and a generalized relative entropy $W_{s,\alpha}(A|B)$. Results in this section are in [F-14] and [F-16].

Theorem 8.1. *Let A and B be positive invertible operators. If $A \gg B$ (i.e., $\log A \geq \log B$) , then the following (I) and (II) hold.*
(I) *For each $\alpha \in [0,1]$, and all $p \geq 0$ and $u \geq 0$,*
$$G_{p,u,\alpha}(A,B,s) = A^{\frac{-u}{2}} \{A^{\frac{u}{2}}(A^{\frac{u\alpha}{2}} B^p A^{\frac{u\alpha}{2}})^s A^{\frac{u}{2}}\}^{\frac{u(\alpha+1)}{(u\alpha+p)s+u}} A^{\frac{-u}{2}}$$
is a decreasing function of s such that $s \geq 1$.
(II) *For each $\alpha \in [0,1]$, and all $p \geq 0$ and $u \geq 0$, $\dfrac{1}{(u\alpha+p)s+u} W_{s,\alpha}(A^{-u}|B^p)$ is a decreasing function of s such that $s \geq 1$.*

Theorem 8.2. *Let A and B be positive invertible operators. Then the following assertions are mutually equivalent.*
(I) $A \gg B$ *(i.e., $\log A \geq \log B$).*
(II$_1$) *For each $\alpha \in [0,1]$, and all $p \geq 0$ and all $u \geq 0$,*
$$A^{u(\alpha+1)} \geq \{A^{\frac{u}{2}}(A^{\frac{u\alpha}{2}} B^p A^{\frac{u\alpha}{2}})^s A^{\frac{u}{2}}\}^{\frac{u(\alpha+1)}{(u\alpha+p)s+u}}$$
holds for any $s \geq 1$.
(II$_2$) *For each $\alpha \in [0,1]$, a fixed positive number p_0 and a fixed number s_0 such that $s_0 \geq 1$,*
$$A^{u(\alpha+1)} \geq \{A^{\frac{u}{2}}(A^{\frac{u\alpha}{2}} B^{p_0} A^{\frac{u\alpha}{2}})^{s_0} A^{\frac{u}{2}}\}^{\frac{u(\alpha+1)}{(u\alpha+p_0)s_0+u}}$$
holds for all $u \in [0, u_0]$, where u_0 is a fixed positive number.
(III$_1$) *For each $\alpha \in [0,1]$, all $p \geq 0$ and all $u \geq 0$,*
$$\log A^{(u\alpha+p)s+u} \geq \log\{A^{\frac{u}{2}}(A^{\frac{u\alpha}{2}} B^p A^{\frac{u\alpha}{2}})^s A^{\frac{u}{2}}\}$$
holds for any $s \geq 1$.
(III$_2$) *For each $\alpha \in [0,1]$, a fixed positive number p_0 and a fixed number s_0 such that $s_0 \geq 1$,*
$$\log A^{(u\alpha+p_0)s_0+u} \geq \log\{A^{\frac{u}{2}}(A^{\frac{u\alpha}{2}} B^{p_0} A^{\frac{u\alpha}{2}})^{s_0} A^{\frac{u}{2}}\}$$
holds for all $u \in [0, u_0]$, where u_0 is a fixed positive number.

Theorem 8.3. *Let A, B and C be positive invertible operators. Then the following assertions are mutually equivalent.*

(I) $C \gg A \gg B$ (i.e., $\log C \geq \log A \geq \log B$).

(II$_1$) *For each $\alpha \in [0,1]$, and all $p \geq 0$ and all $u \geq 0$,*
$$\{A^{\frac{u}{2}}(A^{\frac{u\alpha}{2}}C^pA^{\frac{u\alpha}{2}})^sA^{\frac{u}{2}}\}^{\frac{u(\alpha+1)}{(u\alpha+p)s+u}} \geq A^{u(\alpha+1)} \geq \{A^{\frac{u}{2}}(A^{\frac{u\alpha}{2}}B^pA^{\frac{u\alpha}{2}})^sA^{\frac{u}{2}}\}^{\frac{u(\alpha+1)}{(u\alpha+p)s+u}}$$
holds for any $s \geq 1$.

(II$_2$) *For each $\alpha \in [0,1]$, a fixed positive number p_0 and a fixed number s_0 such that $s_0 \geq 1$,*
$$\{A^{\frac{u}{2}}(A^{\frac{u\alpha}{2}}C^{p_0}A^{\frac{u\alpha}{2}})^{s_0}A^{\frac{u}{2}}\}^{\frac{u(\alpha+1)}{(u\alpha+p_0)s_0+u}} \geq A^{u(\alpha+1)} \geq \{A^{\frac{u}{2}}(A^{\frac{u\alpha}{2}}B^{p_0}A^{\frac{u\alpha}{2}})^{s_0}A^{\frac{u}{2}}\}^{\frac{u(\alpha+1)}{(u\alpha+p_0)s_0+u}}$$
holds for all $u \in [0,u_0]$, where u_0 is a fixed positive number.

(III$_1$) *For each $\alpha \in [0,1]$, all $p \geq 0$ and all $u \geq 0$,*
$$\log\{A^{\frac{u}{2}}(A^{\frac{u\alpha}{2}}C^pA^{\frac{u\alpha}{2}})^sA^{\frac{u}{2}}\} \geq \log A^{(u\alpha+p)s+u} \geq \log\{A^{\frac{u}{2}}(A^{\frac{u\alpha}{2}}B^pA^{\frac{u\alpha}{2}})^sA^{\frac{u}{2}}\}$$
holds for any $s \geq 1$.

(III$_2$) *For each $\alpha \in [0,1]$, a fixed positive number p_0 and a fixed number s_0 such that $s_0 \geq 1$,*
$$\log\{A^{\frac{u}{2}}(A^{\frac{u\alpha}{2}}C^{p_0}A^{\frac{u\alpha}{2}})^{s_0}A^{\frac{u}{2}}\} \geq \log A^{(u\alpha+p_0)s_0+u} \geq \log\{A^{\frac{u}{2}}(A^{\frac{u\alpha}{2}}B^{p_0}A^{\frac{u\alpha}{2}})^{s_0}A^{\frac{u}{2}}\}$$
holds for all $u \in [0,u_0]$, where u_0 is a fixed positive number.

(IV$_1$) *For each $\alpha \in [0,1]$, and all $p \geq 0$, and $u \geq 0$,*
$$W_{s,\alpha}(A^{-u}|C^p) \geq W_{s,\alpha}(A^{-u}|A^p) \geq W_{s,\alpha}(A^{-u}|B^p)$$
holds for any $s \geq 1$.

(IV$_2$) *For each $\alpha \in [0,1]$, and a fixed positive nymber p_0, and a fixed number s_0 such that $s_0 \geq 1$,*
$$W_{s_0,\alpha}(A^{-u}|C^{p_0}) \geq W_{s_0,\alpha}(A^{-u}|A^{p_0}) \geq W_{s_0,\alpha}(A^{-u}|B^{p_0})$$
holds for all $u \in [0,u_0]$,where u_0 is a fixed positive number.

Theorem 8.4. *Let A and B be positive invertible operators. For each $\alpha \in [0,1]$ and any positive number x_0 and any $s \geq 1$, the following inequality holds;*
$$(\log x_0 - 1)A + \frac{1}{x_0}(A^{-\alpha/2}BA^{-\alpha/2})^s$$
$$\geq W_{s,\alpha}(A|B)$$
$$\geq (1 - \log x_0)A - \frac{1}{x_0}A(A^{\alpha/2}B^{-1}A^{\alpha/2})^sA,$$
especially , $W_{s,\alpha}(A|B) = 0$ holds if and only if $B^s = A^{1+\alpha s}$ holds.

Theorem 8.5. *Let A and B be positive invertible operators.   Then the following (I), (II), (III) and (IV) hold:*

(I) *For each $\alpha \in [0,1]$, any $s \geq 1$, and any $r \geq 0$,*
$$W_{s,\alpha}(A|A^r) = ((r-\alpha)s - 1)A\log A.$$

(II) *For each* $\alpha \in [0,1]$ *and any* $s \geq 1$,
$$W_{s,\alpha}(A|A^{\frac{\alpha}{2}}BA^{\frac{\alpha}{2}}) = S(A|B^s).$$

(III) *For each* $\alpha \in [0,1]$ *and any* $s \geq 1$,
$$W_{s,\alpha}(A|I) = -(\alpha s + 1)A\log A.$$

(IV) *For each* $\alpha \in [0,1]$, *any* $s \geq 1$ *and any* $t \in [0,1]$, $g_{s,\alpha}(t)$ *is defined by*
$$g_{s,\alpha}(t) = A^{\frac{1}{2}}\{A^{\frac{-1}{2}}(A^{\frac{-\alpha}{2}}BA^{\frac{-\alpha}{2}})^s A^{\frac{-1}{2}}\}^t A^{\frac{1}{2}}.$$
Then the following relation holds:
$$W_{s,\alpha}(A|B) = \text{s-}\lim_{t\to+0}\frac{g_{s,\alpha}(t)-A}{t} = \frac{d}{dt}g_{s,\alpha}(t)\Big|_{t=0}.$$

Remark 8.1. We remark that differential geometric view between the geodestic from A to B and the relative operator entropy $S(A|B)$ is discussed in [CPR] , that is, the geodestic $g(t)$ from A to B is defined by

$$g(t) = A^{\frac{1}{2}}(A^{\frac{-1}{2}}BA^{\frac{-1}{2}})^t A^{\frac{1}{2}}$$

and

$$S(A|B) = A^{\frac{1}{2}}(\log A^{\frac{-1}{2}}BA^{\frac{-1}{2}})A^{\frac{1}{2}} = \text{s-}\lim_{t\to+0}\frac{d}{dt}g(t)\Big|_{t=0}$$

and these relations just correspond to (IV) of Theorem 5.5 in the case $s = 1$ and $\alpha = 0$ since $g(t) = g_{1,0}(A|B)$ and $S(A|B) = W_{1,0}(A|B)$ by (II) of Theorem 5.5.

Theorem 8.2 implies the following Theorem F-1 [F-6] and also Theorem 8.3 yields Theorem F-2 [F-6] and Corollary 8.6 which is an extension of Corollary 2 [F-6].

Theorem F-1 [F-6]. *Let A and B be positive invertible operators. Then the following assertions are mutually equivalent.*

(I) $A \gg B$ *(i.e., $\log A \geq \log B$).*

(II$_1$) $A^u \geq (A^{\frac{u}{2}}B^pA^{\frac{u}{2}})^{\frac{u}{p+u}}$ *for all $p \geq 0$ and all $u \geq 0$,*

(II$_2$) $A^u \geq (A^{\frac{u}{2}}B^{p_0}A^{\frac{u}{2}})^{\frac{u}{p_0+u}}$ *for a fixed positive number p_0 and for all u such that $u \in [0,u_0]$, where u_0 is a fixed positive number.*

(III$_1$) $\log A^{p+u} \geq \log(A^{\frac{u}{2}}B^pA^{\frac{u}{2}})$ *for all $p \geq 0$ and all $u \geq 0$.*

(III$_2$) $\log A^{u+p_0} \geq \log(A^{\frac{u}{2}}B^{p_0}A^{\frac{u}{2}})$ *For a fixed positive number p_0 and for all u such that $u \in [0,u_0]$, where u_0 is a fixed positive number.*

Theorem F-2 [F-6]. *Let A, B and C be positive invertible operators. Then the following assertions are mutually equivalent.*

(I) $C \gg A \gg B$ (i.e., $\log C \geq \log A \geq \log B$).

(II$_1$) $(A^{\frac{u}{2}} C^p A^{\frac{u}{2}})^{\frac{u}{p+u}} \geq A^u \geq (A^{\frac{u}{2}} B^p A^{\frac{u}{2}})^{\frac{u}{p+u}}$ *for all $p \geq 0$ and all $u \geq 0$.*

(II$_2$) $(A^{\frac{u}{2}} C^{p_0} A^{\frac{u}{2}})^{\frac{u}{p_0+u}} \geq A^u \geq (A^{\frac{u}{2}} B^{p_0} A^{\frac{u}{2}})^{\frac{u}{p_0+u}}$ *for a fixed positive number p_0 and for all u such that $u \in [0, u_0]$, where u_0 is a fixed positive number.*

(III$_1$) $\log(A^{\frac{u}{2}} C^p A^{\frac{u}{2}}) \geq \log A^{p+u} \geq \log(A^{\frac{u}{2}} B^p A^{\frac{u}{2}})$ *for all $p \geq 0$ and all $u \geq 0$,*

(III$_2$) $\log(A^{\frac{u}{2}} C^{p_0} A^{\frac{u}{2}}) \geq \log A^{p_0+u} \geq \log(A^{\frac{u}{2}} B^{p_0} A^{\frac{u}{2}})$ *for a fixed positive number p_0 and for all u such that $u \in [0, u_0]$, where u_0 is a fixed positive number.*

(IV$_1$) $S(A^{-u}|C^p) \geq S(A^{-u}|A^p) \geq S(A^{-u}|B^p)$ *for all $p \geq 0$, and $u \geq 0$.*

(IV$_2$) $S(A^{-u}|C^{p_0}) \geq S(A^{-u}|A^{p_0}) \geq S(A^{-u}|B^{p_0})$ *for a fixed positive number p_0, and for all u such that $u \in [0, u_0]$,where u_0 is a fixed positive number.*

Corollary 8.6. *Let A ,B and C be positive invertible operators. If $C \gg A^{-1} \gg B$, then for for each $\alpha \in [0, 1]$*

$$W_{s,\alpha}(A|C) \geq -\{1 + (1 + \alpha)s\} A \log A \geq W_{s,\alpha}(A|B)$$

for any $s \geq 1$.

Corollary 8.7 [F-6]. *Let A and B be positive invertible operators. If $C \gg A^{-1} \gg B$, then $S(A|C) \geq -2A \log A \geq S(A|B)$.*

Corollary 8.8 [F-10]. *Let A and B be positive invertible operators. For any positive number x_0 , the following inequality holds;*

$$(\log x_0 - 1)A + \frac{1}{x_0}B \geq S(A|B) \geq (1 - \log x_0)A - \frac{1}{x_0} AB^{-1}A.$$

Remark 8.2. Several characterizations of operators $\log A \geq \log B$ (A-1) and Generalizations of Ando's theorem (A-2) are given in [F-5],[F-6],[MF-K-1],[FFK-1] [FFK-2] and [FFW]. Results on Generalized Aluthge transformation (A-5) are in [F-17] and [FY] . Order preserving operator inequalities are shown in [BF-1].

We would like to omit to describe the following results; several generalizations of Heinz-Kato (B-1) in [F-7],[F-11] and [F-12], and generalizations of some theorem on norms (B-2) in [F-8] and [F-9] and also generalizations of Pedersen-Takesaki theorem on some operator equation and related results (C-1) in [F-2] and [BF-2].

References

[Al] A.Aluthge, *On p-hyponormal operators for* $0 < p < 1$, Integral Equations and Operator Theory ,**13** (1990),307-315.

[A] T.Ando *On some operator inequality*, Math. Ann.,**279** (1987),157-159.

[A-H] T.Ando and F.Hiai,*Log majorization and complementary Golden-Thompson type inequalities*, Linear Alg. and Its Appl.,**197, 198** (1994),113-131.

[BF-1] E.Bach and T.Furuta, *Order preserving operator inequalities*, J. Operator Theory, **19** (1988),341-346.

[BF-2] E.Bach and T.Furuta, *Counterexample to a question on operator equation* $T(H^{1/n}T)^n = K$, Linear Alg. and Its Appl., **177** (1992),157-162.

[C-K] N.N.Chan and Man Kam Kwong, *Hermitian matrix inequalities and a conjecture*, Amer. Math. Monthly,**92** (1985),533-541.

[C-I] M.Cho and M.Ito, *Putnam's inequality for p-hyponormal operators*, Proc. Amer. Math. Soc.,**123** (1995),2435-2440.

[CPR] G.Corach, H.Porta and L.Recht, *Geodesics and operastor means in the space of positive operators*, International J. Math.,4(1993),193-202.

[D] B.P.Duggal, *On p-hyponormal contraction*, Proc. Amer. Math. Soc., **123**(1995),81-86.

[JI.F-K-1] J.I.Fujii and E.Kamei, *Relative operator entropy in noncommutative information theory*, Math. Japon., **34** (1987), 341-346.

[JI.F-K-2] J.I.Fujii and E.Kamei,*Uhlmann's interpolational method for operator means*, Math. Japon.,**34** (1987), 541-548.

[MF] M.Fujii, *Furuta's inequality and its mean theoretic approach*, J. Operator Theory, **23** (1990), 67-72.

[MF-K-1] M.Fujii and E.Kamei, *Furuta's inequality and its application to Ando's theorem*, Proc. Amer. Math. Soc., **115** (1992), 409-413.

[MF-K-2] M.Fujii and E.Kamei, *Mean theoretic approach to the grand Furuta inequality*, Proc. Amer. Math. Soc., **124** (1996), 2751-2756.

[FFK-1] M.Fujii, T.Furuta and E.Kamei, *Operator functions associated with Furuta's inequality*, Linear Alg. and Its Appl., **149**(1991),91-96.

[FFK-2] M.Fujii, T.Furuta and E.Kamei, *Furuta's inequality and its application to Ando's theorem*, Linear Alg. and Its Appl., **179**(1993),161-169.

[FFW] M.Fujii, T.Furuta and D.Wang, *An application of the Furuta inequality to operator inequalities on chaotic order*, Math. Japon., 40(1994),317-321.

[F-1] T.Furuta, $A \geq B \geq 0$ *assures* $(B^r A^p B^r)^{1/q} \geq B^{(p+2r)/q}$ *for* $r \geq 0, p \geq 0, q \geq 1$ *with* $(1+2r)q \geq p+2r$, Proc. Amer. Math. Soc., **101** (1987),85-88.

[F-2] T.Furuta, *The operator equation* $T(H^{1/n}T)^n = K$, Linear Alg. and Its Appl., **101** (1988),149-152.

[F-3] T.Furuta, *A proof via operator means of an order preserving inequality,* Linear Alg. and Its Appl., **113**(1989),129-130.

[F-4] T.Furuta,*Elementary proof of an order preserving inequality,* Proc. Japan Acad., **65** (1989),126.

[F-5] T.Furuta, *Two operator functions with monotone property,* Proc. Amer. Math. Soc. **111**(1991),511-516.

[F-6] T.Furuta, *Applications of order preserving operator inequalities,* Operator Theory: Advances and Applications, **59**(1992),180-190.

[F-7] T.Furuta, *Generalizations of the Heinz-Kato theorem via the Furuta inequality,* Operator Theory; Advances and Applications, **62**(1993),77-83.

[F-8] T.Furuta, *Some norm inequalities and operator inequalities via the Furuta inequality,* Acta Sci. Math.(Szeged), **57**(1993),139-145.

[F-9] T.Furuta, *Applications of the Furuta inequality to operator inequalities and norm inequalities preserving some order,* Operator Theory, Advances and Applications, Birkhüser , **61**(1993),115-122.

[F-10] T.Furuta, *Furuta's inequality and its application to the relative operator entoropy,* J. Operator Theory ,**30**(1993),21-30.

[F-11] T.Furuta, *Determinant type generalizations of the Heinz-Kato theorem via the Furuta inequality,* Proc. Amer. Math. Soc., **120**(1994),223-231

[F-12] T.Furuta, *An extension of the Heinz-Kato theorem,* Proc. Amer. Math. Soc., **120**(1994), 785-787.

[F-13] T.Furuta, *Extension of the Furuta inequality and Ando-Hiai Log majorization,* Linear Alg. and Its Appl., **219**, (1995),139-155.

[F-14] T.Furuta, *Characterization of operators satisfying* $logA \geq logB$ *and its applications,* Proc. of 15th OT Conference (1995).

[F-15] T.Furuta,*Generalizations of Kosaki trace inequalities and related trace inequalities on chaotic order,* Linear Alg. and Its Appl.,**235** (1995),153-161.

[F-16] T.Furuta, *Applications of order preserving operator inequalities to a generalized relative operator entropy,* General Inequalities 7, Birkhäuser **123**(1997),65-76.

[F-17] T.Furuta, *Generalized Aluthge transformation on p-hyponormal operators*, Proc. Amer. Math. Soc.,**124**(1996),3071-3075.

[F-18] T.Furuta, *Simplified proof of an order preserving operator inequality*, preprint.

[FHI] T.Furuta, M.Hashimoto and M.Itô, *Equivalence relation between generalized Furuta inequality and related operator functions*, to appear in Scientiae Math.

[FY] T.Furuta and M.Yanagida, *Further extensions of Aluthge transformation on p-hyponormal operators*, Integral Equations and Operator Theory.,**29**(1997),122-125.

[FYY-1] T.Furuta,T.Yamazaki and M.Yanagida, *Operator functions implying generalized Furuta inequality*, Mathematical Inequalities and Applications,**1**(1998),123-130.

[FYY-2] T.Furuta,T.Yamazaki and M.Yanagida,*Order preserving operator inequalities via Furuta inequality*, to appear in Math. Japon.

[FW] T.Furuta and D.Wang,*A decreasing operator function associated with the Furuta inequality*, to appear in Proc. Amer. Math. Soc.

[H] E. Heinz, *Beiträge zur Störungstheorie der Spektralzerlegung*, Math. Ann., **123**(1951), 415-438.

[Hu] T.Huruya, *A note on p-hyponormal operators*, Proc. Amer. Math. Soc., **125**(1997), 3617-3624.

[JFK] Jian Fei Jiang ,M.Fujii and E.Kamei, *Operator functions associated with the grand Furuta inequality*, Mathematical Inequalities and Applications, **2** (1998), 000-000.

[K] E.Kamei, *A satellite to Furuta's inequality*, Math. Japon, **33**(1988),883-886.

[K-A] F.Kubo and T.Ando, *Means of positive linear operators*, Math. Ann. **246**(1980),205-224.

[KO] H.Kosaki, *On some trace inequality*, Proc. Centre Math. Anal. Austral. Nat. Univ., **29**(1992),129-134.

[L] K.Löwner, *Über monotone Matrixfunktionen*, Math. Z., **38**(1934),177-216.

[N-U] M.Nakamura and H.Umegaki, *A note on the entropy for operator algebras*, Proc, Japan Acad., **37**(1961),149-154.

[T-1] K.Tanahashi, *Best possibility of the Furuta inequality*, Proc. Amer. Math. Soc., **124**(1996),141-146.

[T-2] K.Tanahashi, *The best possibility of the grand Furuta inequality*, to appear in Proc. Amer. Math. Soc.

[U] H.Umegaki, *Conditional expectation in operator algebra IV,(entropy and information)*, Kodai Math. Sem. Rep., **14**(1962),59-85.

[Xi] D.Xia, *Spectral theory of hyponormal operators*, Birkhäuser Verlag, Boston, 1983.

DEGREE OF CONVERGENCE FOR A CLASS OF LINEAR OPERATORS

N. K. GOVIL
Department of Mathematics, Auburn University
Auburn, Al 36849

R. N. MOHAPATRA
Department of Mathematics, University of Central Florida
Orlando, FL 32817

Abstract. Saturation order and saturation class for operators based on the partial sums of Fourier series have been studied by many authors. Kuttner and Sahney [17] obtained results on the nonuniqueness of the degree of saturation. Subsequently Kuttner, Mohapatra and Sahney [18] obtained general results on saturation order and saturation class of a sequence of operators based on the infinite matrix transformation of any bounded, measurable functions of period 2π. In that paper it was shown that the operator based on harmonic method has a saturation order $O(\frac{1}{n \log n})$ and a saturation class which is contained in the class of all functions f for which the conjugate function $\tilde{f} \in \text{Lip } 1$. In this paper we consider operators based on generalized harmonic method, N_δ, and obtain the saturation order and information regarding its saturation class.

1. Introduction and Statement of Results.

Let $\mathcal{B}$ denote the space of bounded measurable functions with period 2π. Throughout this paper we shall suppose $f \in \mathcal{B}$ unless mentioned otherwise. $\|f\|_p$ will denote the usual L^p norm of f for $1 < p < \infty$ and when $p = \infty$ we shall interprete $\|f\|_\infty = \sup_{x \in \mathbb{R}} |f(x)|$, $\mathbb{R}$ being the set of all real numbers. Let the Fourier series of f at point $x \in \mathbb{R}$ be given by

$$(1) \qquad \frac{1}{2}a_0 + \sum_{k=1}^{\infty}(a_k \cos kx + b_k \sin kx) := \sum_{k=0}^{\infty} A_k(x).$$

Let $\mathcal{D} = (d_{nk})$ $(n, k = 0, 1, \cdots)$ be an infinite matrix.
Let $L_n(f; x)$ be the $\mathcal{D}$ transform of the Fourier Series of f at x i.e.

$$(2) \qquad L_n(f; x) := \sum_{k=0}^{\infty} d_{nk} S_k(x)$$

1991 *Mathematics Subject Classification.* Primary 41A25; Secondary 42A10
Key words and phrases. Saturation of operators and summability methods, saturation order and saturation class, conjugate function, Hermite conjugate function, Lipschitz class.

135

T.M. Rassias and H.M. Srivastava (eds.), Analytic and Geometric Inequalities and Applications, 135–150.
© 1999 *Kluwer Academic Publishers.*

where $S_k(x) = S_k(f; x) = \sum_{j=0}^{k} A_j(x)$. Let us write

$$D_{nk} = \sum_{j=k}^{\infty} d_{nj} \quad (n, k = 0, 1, \cdots).$$

If there is a positive nonincreasing function $\varphi(n)$ and a subspace $\mathcal{K}$ of $\mathcal{B}$ such that

(3) $\|f - L_n(f)\| = o(\varphi(n)) \implies f$ is almost everywhere a constant function;

(4) $$\|f - L_n(f)\| = O(\varphi(n)) \implies f \in \mathcal{K}$$

and

(5) $$f \in \mathcal{K} \implies \|f - L_n(f)\| = O(\varphi(n));$$

then we say that the method of summation $\mathcal{D} = (d_{nk})$ is saturated with the order $\varphi(n)$ and has saturation class $\mathcal{K}$.

After this concept of saturation of an operator was introduced by Favard [6], a number of authors obtained results on the order and class of saturation for linear operators (see Alexits [1], Butzer [2], [3], Favard [6], DeVore [5], Khan [16], Suzuki [30]). Zamansky [31] obtained the order and saturation class of Cesàro-Fejér operator. Local saturation for a class of linear operators has been studied by Mamedov [19], Suzuki [30], Sunouchi [27], [28] and Suuouchi and Watari [29]. Goel, Holland, Nasim and Sahney [8] have obtained order and class of saturation of some Nörlund means. Sunouchi and Watari [29] obtained general conditions under which for suitable class $\mathcal{K}$ of saturation, (3) and (4) hold. However their result did not deal with general conditions under which (5) holds. Mohapatra and Sahney [21] studied a class of operators based on an infinite matrix transformation of the Fourier series of f and generalized the results of Goel, Holland et al [8]. However, their result did not apply to the matirx (d_{nk}) where $d_{nk} = \frac{1}{n+1}$ for $0 \leq k \leq n$ and $d_{nk} = 0$, if $k \geq n + 1$. Later, Kuttner, Mohapatra and Sahney [18] obtained a general class of summability methods which for a suitable class $\mathcal{K}$ of functions satisfy (3), (4) and (5). After obtaining some general results they also applied them to special cases to deduce previously known results for Cesàro, Nörlund and Abel methods and also obtained new results on Riesz and Harmonic methods of summation.

Let $\tilde{f}$ be the conjugate function of f (see Zygmund [32]) which is written as

(6) $$\tilde{f}(x) = -\frac{1}{2\pi} \int_0^\pi [(f(x+t) - f(x-t)] \cot \frac{t}{2} dt$$

where the integral is taken as a Cauchy integral at the origin. It is known that (6) exists almost everywhere.

Let

$$\mathcal{K}_0 = \{f \in \mathcal{B} : \tilde{f} \in \text{Lip } 1\}.$$

The Bernstein theorem (see Joo [11], p. 169) states that if f is a continuous 2π periodic function then $f \in$ Lip α, $0 < \alpha < 1$ is equivalent to the estimate

$$\|\sigma_n f - f\| = O(\frac{1}{n^\alpha})$$

where $\sigma_n f$ denotes the nth Fejér mean of the Fourier series of f. If $f \in$ Lip 1, then $\|\sigma_n f - f\| = O(\log n/n)$ is valid. Alexits [1] and Zamansky [31] showed that the function class for which $\|\sigma_n f - f\| = O(1/n)$ is the class of all functions f for which $\tilde{f} \in$ Lip 1. The notion of Hermite conjugate function was obtained by Muckenhoupt [23] who studied Poisson integrals for Laguerre and Hermite expansions in [22], conjugate functions for Laguerre expansions in [24] and classical expansions of a function and their relation to conjugate harmonic functions in Muckenhoupt and Stein [25]. Among many other things, Muckenhoupt [23] proved the boundedness of the conjugation operator in the weighted L^p space

$$L^p_{e^{-y^2}} := \{f : \|f\|_p = (\int_{-\infty}^{\infty} |f(y)|^p e^{-y^2} dy)^{1/p} < \infty\}.$$

Many authors prefer to use Hermite-weighted L^p space which is given by (see Joo [15], p. 169)

$$L^p(e^{-y^2/2}) := \{f : e^{-y^2/2} f \in L^p(\mathbb{R})\}.$$

In [10], Joo applied this conjugate concept to obtain Alexits type theorems and obtained a saturation theorem for Abel-Poisson means in the weighted L^p space. Harváth [9] obtained both Alexits and Abel-Poisson saturation theorems for the Jacobi and Laguerre expansion. Rate of convergence for Riesz means of Hermite Fourier series of functions belonging to Lipschitz class was studied by Joo [11]. Alexits type theorems for multidimensional trigonometric series are considered in Joo [12]. Concerning the relation between the rate of convergence of an operator in the weighted L^p space and degree of smoothness of f, Joo [13] has shown that n^{th} Riesz means of parameter $1/2$ converge to the function f in the order $O(1/\sqrt{n})$ in the weighted L^p space if and only if its Hermite conjugate has a smoothness property in the sense that the Hermite derivative of the conjugate function belongs to L^p i.e., $\tilde{f} \in$ weighted Lip$(1,p)$ class (see Joo [14]). Freud [7] introduced modulus of continuity for weighted L^p spaces and studied approximation properties with its help. Approximation properties of periodic functions by de la Vallée Poussin sums were studied by Steckin [26]. Milne [20] and later Joo [15] proved the following theorem for weighted L^p space.

Theorem A (see [20], [15, p. 171, Lemma 3]). *If $\mathcal{P}_n$ is the collection of all polynomials of degree less or equal to n and $1 \le p \le \infty$ then for $p_n \in \mathcal{P}_n$,*

$$\|(p_n e^{-x^2/2})'\|_p \le C\sqrt{n}\|p_n e^{-x^2/2}\|_p.$$

Here C is a constant independent of n and $(p_n e^{-x^2/2})'$ is the derivative of $p_n e^{-x^2/2}$.

In [11], Joo mentions that for trigonometric Fourier series conjugation is an involution but in case of Jacobi, Laguerre, Hermite and Walsh Fourier series conjugation does not remain involutorious hence the variants of saturation theorems using f and $\tilde{f}$ need not necessarily be the same (for details see [11]). In all cases conjugate of a function is associated with an operation on the coefficients of the expansion. The simplest case is the Walsh case where the conjugation is a multiplier operator whose continuity follows from a well-known result of Steckin [26]. In the classical orthogonal expansions conjugation makes a shift of the coefficients. For the ultraspherical, Hermite and Laguerre expansions, the conjugation process was introduced in [22], [23], [24] and [25].

When we restrict ourself to trigonometric Fourier series the simplest saturation theorem is given by

Theorem B (see [31]). *The Cesáro-Fejér operator is saturated with order $1/n$ and saturation class $\mathcal{K}_0$.*

It is known that Cesáro method of order $\alpha > 0$, the Abel method and the Nörlund method of certain types have saturation class $\mathcal{K}_0$.

Kuttner and Sahney [17] obtained necessary and sufficient conditions for a summability method to have a certain order of saturation. Precisely, they proved the following.

Theorem C (see [17]). *Let $\mathcal{D} = (d_{nk})$ satisfy*

$$(7) \qquad d_{nk} \geq 0 \ (n, k = 0, 1, 2, \cdots), \ \sum_{k=0}^{\infty} d_{nk} = 1$$

and for all fixed n

$$(8) \qquad \sum_{k=1}^{\infty} d_{nk} \log k < \infty.$$

Let $\varphi(n)$ be a positive function of n. In order that $\mathcal{D}$ is saturated with order $\varphi(n)$ and have same saturation class, it is necessary and sufficient that

$$(9) \qquad 0 < \liminf_{n \to \infty} \frac{\varphi(n)}{d_{no}} < \infty.$$

Remark 1. *In the above, we have used saturation of the summability process $\mathcal{D}$ synonymously with the saturation of operator $\{L_n\}$ which is given by (2).*

During the course of determination of global saturation class corresponding to a general method of summation, Sunouchi and Watari [29] proved

Theorem D (see [29]). *If $D_{no} = 1$ and there exists some positive constant and positive functions $\psi(k)$ and a positive constant C such that*

$$(10) \qquad \lim_{n \to \infty} \frac{1 - D_{nk}}{\varphi(n)} = C \psi(k), \ (k = 1, 2, \ldots,),$$

where $\varphi(n)$ is a positive non-increasing function, then the following hold:

$$\|f - L_n(f)\| = o(\varphi(n)) \implies f \text{ is a constant function almost everywhere,}$$
$$\|f - L_n(f)\| = O(\varphi(n)) \implies \|\textstyle\sum_{k=1}^{N} \psi(k)(1 - \tfrac{k}{N+1})A_k(x)\| = O(1).$$

Kuttner, Mohapatra and Sahney [18] generalized above results and proved the following:

Theorem E ([18], Theorem 1). *Let $\mathcal{D} = (d_{nk})$ satisfy (7) and (8). Then the following hold:*

(a) $\|f - L_n(f)\| = o(d_{no}) \implies f$ is almost everywhere a constant;

(b) If, further,

$$(11) \qquad \frac{d_{nk}}{d_{no}} \to \lambda_k \text{ as } n \to \infty \text{ for each fixed } k \geq 0,$$

then

$$(12) \qquad \|f - L_n(f)\| = O(d_{no}) \implies \|\sum_{k=1}^{N} \mu_k(1 - \frac{k}{N+1})A_k(x)\| = O(1)$$

and hence that

$$(13) \qquad \sum_{k=1}^{\infty} \mu_k A_k(x)$$

is the Fourier series of a bounded function, where

$$(14) \qquad \mu_k = \sum_{r=0}^{k-1} \lambda_r.$$

Concerning saturation order, they proved

Theorem F (see [18], Theorem 3). *Let $D = (d_{nk})$ satisfy (7), (8) and (11). Let, further, $\lambda_k \nrightarrow 0$ as $k \to \infty$ and*

$$(15) \qquad \sum_{k=0}^{\infty} |d_{nk} - d_{n,k+1}| = O(d_{no}).$$

Then $\|f - L_n(f)\| = O(d_{no})$ if and only if $f \in \mathcal{K}_0$.

Remark 2. *From Theorem E and Theorem F we can see that if $D = (d_{nk})$ satisfy requirements of Theorem F, then the summability method D is saturated with order d_{no} and has saturation class $\mathcal{K}_0$.*

Next we apply Theorems E and F to special methods like Nörlund and Harmonic methods and mention related results from [18].

Let $\{p_n\}$ be a sequence of real numbers with $P_n = p_0 + p_1 + \cdots + p_n$. Let $N_n(f)$ be the operator $L_n(f)$ where $d_{nk} = \frac{p_{n-k}}{P_n}$, $k = 0, 1, \cdots, n$ and $d_{nk} = 0$, otherwise. The corresponding method of summability is called (N, p_n) or Nörlund method. When $p_n = \frac{1}{n+1}$, $n = 0, 1, \ldots$, the (N, p_n) method reduces to Harmonic method of summation and in this case the operator $N_n(f)$ is written as $H_n(f)$.

From Theorem E and Theorem F and our remark following those we have the following:

Theorem G. *Let $\{p_n\}$ be a sequence of positive constants satisfying*

$$(16) \qquad \frac{p_{n-1}}{p_n} \to \lambda \ (\lambda \geq 1), \ as \ n \to \infty$$

$$(17) \qquad \sum_{k=0}^{n} |p_{n-k} - p_{n-k-1}| = O(p_n) \ where \ p_{-1} = 0$$

Then the method (N, p_n) is saturated with order (p_n/P_n) and class $\mathcal{K}_0$.

If $\lambda = 1$ one gets the result of Goel, Holland, Nassim and Sahney [8].

If $\{p_n\}$ is such that $p_n = A_n^{\alpha-1}$ where $A_n^{\alpha} = \binom{n+\alpha}{\alpha}$, $\alpha > 0$, then (N, p_n) is the (C, α) method of summation. If $0 < \alpha < 1$ then (17) is not satisfied and hence Theorem G cannot be used. However, Sunouchi and Watari's result in [29] covers the saturation result for method (C, α).

Kuttner, Mohapatra and Sahney [18] proved the next result which complements Theorem G. Precisely they proved

Theorem H ([18]). *(a) Suppose that $p_n > 0$ and that $p_{n-1}/p_n \to 1$ as $n \to \infty$. Then (N, p_n) is saturated with order (p_n/P_n) and some class F. This class F is equal to $\mathcal{K}_0$ if and only if,*

$$(18) \qquad \int_0^{\pi} |G_n(t)| dt = O(\frac{p_n}{P_n})$$

where

$$(19) \qquad G_n(t) = \int_t^{\pi} M_n(u) du$$

with

$$(20) \qquad M_n(t) = \sum_{k=0}^{n} \frac{p_{n-k}}{P_n} \frac{\cos(k + 1/2)t}{\sin \frac{1}{2}t}.$$

If (18) does not hold, then the class of saturation is a proper subclass of $\mathcal{K}_0$.
(b) The condition (18) does not hold, in particular, if

$$(21) \qquad \liminf_{n \to \infty} \left(\frac{np_n}{P_n} \right) = 0.$$

Remark 3. *By Theorem H when* $p_n = \frac{1}{n+1}$, $n = 0, 1, 2, \ldots$, *the corresponding Nörlund method, which is the Harmonic method, is saturated with order* $1/(n \log n)$ *and has a saturation class which is a subclass of* $\mathcal{K}_0$. *In* [18] *a subclass has been investigated.*

Let $\Lambda(t)$ be any positive non-decreasing function in $(0, \pi)$ such that

$$(22) \qquad \int_0^\pi \{\Lambda(t) \frac{\log(k/t)}{t}\} dt < \infty$$

(k being a constant greater than π). Let $\mathcal{F}$ be the collection of all Lebesgue integrable functions with period 2π with conjugate function $\tilde{f}$. The conjugate function of f (as given in [32]) is an indefinite integral of a function $\hat{f}'(t)$ satisfying

$$(23) \qquad \hat{f}'(x + t) - \hat{f}'(x) = O(\Lambda(t)),$$

uniformly for $x \in (0, \pi)$ and all $t \in (0, \pi)$.

It is not difficult to see that $\mathcal{F} \subset \mathcal{K}_0$ and the inclusion is strict.

Using $\mathcal{F}$ we can now deduce a result on the method $(N, \frac{1}{n+1})$ or the operator $H_n(f)$ which shows that the saturation class for the Harmonic method or the operator H_n includes $\mathcal{F}$ and is a proper subclass of $\mathcal{K}_0$. Precisely, the following is true:

Theorem I. *If* $f \in \mathcal{F}$ *then*

$$(24) \qquad \|f - H_n(f)\| = O\left(\frac{1}{n \log(n+1)}\right).$$

The saturation class of the Harmonic method $(N, \frac{1}{n+1})$ *corresponding to the order* $1/n \log(n+1)$ *is a proper subclass of* $\mathcal{K}_0$ *but it includes the class* $\mathcal{F}$.

Now we state our objective of proving a result concerning generalized Harmonic method introduced first by Iyenger and later considered by Das and Mohapatra [4] where several features of this method have been discussed. (See [4] for all pertinent references for this summability method).

Let us write for any real δ,

$$(25) \qquad \left(\frac{1}{z} \log \frac{1}{1-z}\right)^\delta = \left(\sum_{n=0}^\infty \frac{z^n}{n+1}\right)^\delta = \sum_{n=0}^\infty b_n^\delta z^n,$$

which exists for $|z| \leq 1$, $z \neq 1$. The (N_δ) mean is the Nörlund mean (N, p_n) with $p_n = b_n^\delta$. It is obvious that $(N_1) = (N, \frac{1}{n+1})$.

From [4, p. 219, (1.12)], we have, for any real δ,

$$(26) \qquad p_n = b_n^\delta \sim \frac{\delta(\log n)^{\delta-1}}{n}, \quad P_n = \sum_{k=0}^{n} b_k^\delta \sim (\log n)^\delta.$$

From (26) we note that for the corresponding Nörlund method (N, p_n), $\{p_n\}$ is a positive, monotonically nonincreasing sequence with $p_n/P_n \sim \frac{1}{n \log n}$.

For $\delta = 1, 2, 3, \ldots$, this method was first considered by Iyenger who used it for obtaining a tauberian theorem for weighted or Harmonic summability. The (N_δ) method for $\delta > 0$ is due to Das and Mohapatra [4] where they also applied this to the Harmonic summability of Fourier series. For the remainder of the paper we shall only consider (N, p_n) method with $p_n = b_n^\delta$ given in (25) and (26). Since it is a Nörlund method by Theorem H the saturation class is a subclass of $\mathcal{K}_0$ since

$$\liminf_{n \to \infty} \left(\frac{n p_n}{P_n} \right) = 0,$$

provided $p_n = b_n^\delta$ satisfies all the conditions of Theorem H. A careful look at [18] shows that the derivation of the subclass of $\mathcal{K}_0$ which includes $\mathcal{F}$ was achieved by taking $p_n = \frac{1}{n+1}$ and using some special lemmas which played a crucial role. We shall modify the lemmas and obtain some estimates which will allow us to prove a result concerning saturation order and saturation class of the generalized Harmonic method.

We denote by H_n^δ,

$$(27) \qquad H_n^\delta(f; x) = \frac{1}{P_n} \sum_{k=0}^{n} p_{n-k} S_k(f; x)$$

where $p_n = b_n^\delta$, $\delta > 0$, and $\mathcal{F}^\delta$ is the class of functions f such that

$$\int_0^\pi \{\Lambda(t) \frac{(\log(k/t))^\delta}{t}\} \, dt < \infty,$$

with Λ as in (23).

Our result is the following

Theorem. *If $f \in \mathcal{F}$ then*

$$(28) \qquad \|f - H_n^\delta(f)\| = O\left(\frac{1}{n \log(n+1)} \right), \quad \delta = 1, 2, 3, \ldots.$$

The saturation class corresponding to the order $\frac{1}{n \log n}$ is a proper subclass of $\mathcal{K}_0$ but it includes the class $\mathcal{F}$.

Remark 4. (i) *The case $\delta = 1$ is Theorem I.*

(ii) *The case for non-integer δ is still open.*

(iii) *Precise class of saturation concerning the saturation order $\frac{1}{n \log(n+1)}$ is still not known*

2. Lemmas.

We shall need the following lemmas for the proof of our theorem.

Lemma 1. *Let $H_n^\delta(f; x)$ be defined by (27) and $\mathcal{G}_n(t)$ be as in (19) with $p_n = b_n^\delta$ which is defined in (25). Then, we have, uniformly in $0 < t \leq \pi$,*

$$(29) \qquad \mathcal{G}_n(t) = O(\frac{(\log(k/t))^\delta}{n(\log n)t}), \quad \delta = 1, 2, \ldots,$$

where k is a constant greater than π.

Proof: Consider $\mathcal{G}_n(t)$ and replace k by $(n-k)$ in the sum. We see that

$$\sum_{k=0}^{n} p_k \cos(n - k + \frac{1}{2})u = \operatorname{Re}\{e^{(n+\frac{1}{2})iu} \sum_{k=0}^{n} p_k e^{-iku}\}$$

$$(30) \qquad = \operatorname{Re}\left[e^{(n+\frac{1}{2})iu}\{\sum_{k=0}^{\infty} p_k e^{-iku} - \sum_{k=n+1}^{\infty} p_k e^{-iku}\} \right].$$

Replacing the expression in (30) in the inner sum of $\mathcal{G}_n(t)$, we can write

$$(31) \qquad \mathcal{G}_n(t) = \mathcal{G}_{n1}(t) - \mathcal{G}_{n2}(t),$$

where

$$(32) \qquad \mathcal{G}_{n1}(t) = \int_t^\pi (P_n \sin\frac{u}{2})^{-1} \operatorname{Re}\{e^{i(n+\frac{1}{2})u} p(e^{-iu})\} du$$

and

$$(33) \qquad \mathcal{G}_{n2}(t) = \int_t^\pi (P_n \sin\frac{u}{2})^{-1} \operatorname{Re}\{e^{i(n+\frac{1}{2})u} \sum_{k=n+1}^{\infty} p_k e^{-iku}\} du$$

with $p(x) = \sum_{k=0}^{\infty} p_k x^k$, $|x| < 1$.

From Das and Mohapatra [4], we know that $\{p_n\}$ satisfies (26) and is monotonically decreasing and positive. Since $\{p_k\}$ is positive, decreasing and $p_n \to 0$ as $n \to \infty$, we observe that

$$|\sum_{k=n+1}^{\infty} p_k e^{-iku}| \leq p_{n+1} \sup_{j \geq n+1} |\sum_{k=n+1}^{j} e^{-iku}| = O(\frac{p_{n+1}}{u}).$$

Hence,

144

$$(34) \qquad \mathcal{G}_{n2}(t) = \int_t^\pi (P_n \sin \frac{1}{2}u)^{-1} O(\frac{p_{n+1}}{u})\, du$$

$$= O\{\int_t^\pi (\frac{p_n}{P_n u^2})du\} = O\{\frac{1}{n \log n} \int_t^\pi \frac{du}{u^2}\}$$

$$= O\{\frac{1}{n(\log n)t}\}.$$

Before we evaluate $\mathcal{G}_{n1}(t)$, let us note that for $0 < t \leq \pi$, (25) suggests that

$$(35) \qquad p(e^{-it}) = \{-e^{it} \log(1 - e^{-it})\}^\delta = (-1)^\delta e^{i\delta t} \{\log(1 - e^{-it})\}^\delta,$$

$\delta = 1, 2, 3, \ldots$.

Denoting $\{\log(1 - e^{-it})\}^\delta$ by $re^{i\theta}$, we see that

$$(36) \qquad \log(2\sin(\frac{t}{2})) + i(\frac{\pi}{2} - \frac{t}{2}) = r^{1/\delta} e^{i\theta/\delta}.$$

Clearly, from (36),

$$(37) \qquad r^{1/\delta} = \left\{ (\log(2\sin t/2))^2 + \left(\frac{\pi}{2} - \frac{t}{2}\right)^2 \right\}^{\frac{1}{2}},$$

and

$$(38) \qquad \tan(\theta/\delta) = \frac{\pi/2 - t/2}{\log(2\sin \frac{t}{2})}.$$

Now, from (35),

$$(39) \qquad \mathbb{Re}\{e^{i(n+\frac{1}{2})t} p(e^{-it})\}$$

$$= \mathbb{Re}\{e^{i(n+\frac{1}{2})t} (-1)^\delta e^{i\delta t} r e^{i\theta}\}$$

$$= (-1)^\delta r \mathbb{Re}\{e^{i\{(n+\frac{1}{2})t + \delta t + \theta\}}\}$$

$$= (-1)^\delta r \cos\{(n + \frac{1}{2})t + \delta t + \theta\}$$

$$= (-1)^\delta r \{\cos(n + \frac{1}{2} + \delta)t \cos\theta - \sin(n + \frac{1}{2} + \delta)t \sin\theta\}$$

where r and θ are as in (37) and (38).

Let us write

$$(40) \qquad \eta = \arctan \left(\frac{\frac{\pi}{2} - t/2}{\log 2\sin t/2} \right).$$

Then

(41)
$$\theta = \delta\eta$$

Using $(\cos\eta + i\sin\eta)^\delta = \cos\delta\eta + i\sin\delta\eta = \cos\theta + i\sin\theta$ and binomial theorem to the expression $(\cos\eta + i\sin\eta)^\delta$, we get, when $\delta = 2\beta$, $\beta = 1, 2, \ldots$,

(42)
$$\cos 2\beta\eta = \sum_{r=0}^{\beta}(-1)^r \binom{2\beta}{2r} \cos^{2\beta-2r}\eta \sin^{2r}\eta$$

and

(43)
$$\sin 2\beta\eta = \sum_{r=1}^{\beta}(-1)^{r-1} \binom{2\beta}{2r-1} \cos^{2\beta-2r+1}\eta \sin^{2r-1}\eta.$$

When δ is odd, viz. $\delta = 2\rho + 1$, $\rho = 0, 1, 2, \ldots$, then

(44)
$$\cos(2\rho+1)\eta = \sum_{r=0}^{\rho} \binom{2\rho+1}{2r}(-1)^r \cos^{2\rho+1-2r}\eta \sin^{2r}\eta$$

and

(45)
$$\sin(2\rho+1)\eta = \sum_{r=0}^{\rho}(-1)^r \binom{2\rho+1}{2r+1} \cos^{2\rho-2r}\eta \sin^{2r+1}\eta.$$

Now we shall evaluate $\mathcal{G}_{n1}(t)$, when $\delta = 2\beta$, $\beta = 1, 2, \ldots$ and the case when $\delta = 2\rho+1$, $\rho = 0, 1, \ldots$, will follow along the same lines with the help of (44) and (45) instead of (42) and (43) respectively.

Note that $\cos\theta = \cos\delta\eta = \cos 2\beta\eta$ when $\delta = 2\beta$. From (40)

(46)
$$\cos^{2\beta-2r}\eta = \left\{\frac{\log(2\sin(t/2))}{\Delta}\right\}^{2\beta-2r}$$

and

(47)
$$\sin^{2r}\eta = \left\{\frac{(\frac{\pi}{2}-\frac{t}{2})}{\Delta}\right\}^{2r},$$

where

(48)
$$\Delta^2 = (\frac{\pi}{2}-\frac{t}{2})^2 + (\log(2\sin\frac{t}{2}))^2.$$

Thus

(49)
$$\cos^{2\beta-2r}\eta \sin^{2r}\eta = \frac{[\log(2\sin\frac{t}{2})]^{2\beta-2r}(\frac{\pi}{2}-\frac{t}{2})^{2r}}{\Delta^{2\beta}}$$

with Δ given by (48).

Contribution of a typical term to $\mathcal{G}_{n1}(t)$ is

(50)
$$\int_t^\pi \frac{(\log 2 \sin \frac{u}{2})^{2\beta-2r}(\frac{\pi}{2} - \frac{u}{2})^{2r} \cos(n + \frac{1}{2} + 2\beta)u}{P_n \sin(\frac{u}{2})} du.$$

Since

(51)
$$(\log(2 \sin \tfrac{u}{2}))^{2\beta-2r} = \left(\log 2 + \log \sin(\tfrac{u}{2})\right)^{2\beta-2r}$$
$$= \textstyle\sum_{j=0}^{2\beta-2r} \binom{2\beta-2r}{j} (\log 2)^{2\beta-2r-j} (\log \sin \tfrac{u}{2})^j,$$

the expression in (50) can be written as

(52)
$$\tfrac{1}{P_n} \textstyle\sum_{j=0}^{2\beta-2r} \binom{2\beta-2r}{j} (\log 2)^{2\beta-2r-j} \int_t^\pi \frac{(\log \sin \frac{u}{2})^j}{\sin \frac{u}{2}}$$
$$\times(\tfrac{\pi}{2} - \tfrac{u}{2})^{2r} \cos(n + \tfrac{1}{2} + 2\beta)u\, du.$$

If j is even, then $(\log \sin \frac{u}{2})^j / \sin(\frac{u}{2})$ is nonnegative and decreasing in $t \le u \le \pi$. If j is odd, then $|(\log(\sin \frac{u}{2}))^j| /(\sin u/2)$ is nonnegative and decreasing in $t \le u \le \pi$. Also $(\frac{\pi}{2} - \frac{u}{2})^{2r}/(\sin u/2)$, $r = 0, 1, 2, \ldots, 2\beta$ is nonnegative and decreasing in the same range of u. Hence by Second Mean Value theorem for integrals and an analysis similar to the one used in [18, Lemma 3] we can show that

(53)
$$\mathcal{G}_{n1}(t) = O\left(\frac{(\log(k/t))^{2\beta}}{n(\log n)t}\right).$$

Substituting the estimates from (34) and (53) in (31), we see that

(54)
$$\mathcal{G}_n(t) = O\left(\frac{(\log(k/t))^{2\beta}}{n(\log n)t}\right), \quad \text{when } \delta = 2\beta.$$

Using similar arguments for the case when $\delta = 2\rho + 1$, $\rho = 0, 1, \ldots$, we conclude that

(55)
$$\mathcal{G}_n(t) = O\left(\frac{(\log(k/t))^{2\rho+1}}{n(\log n)t}\right),$$

and this completes the proof of Lemma 1.

Lemma 2. *Let $\mathcal{E}_n$ be the error in taking the nth generalized harmonic mean of the series*

$$\sum_{k=1}^\infty \frac{(-1)^{k-1}}{k},$$

which is same as saying

$$(56) \qquad \mathcal{E}_n = \frac{1}{P_n} \sum_{k=1}^{n} P_{n-k} \frac{(-1)^{k-1}}{k} - \log 2,$$

where

$$P_n = \sum_{k=0}^{n} b_k^\delta,$$

with b_n^δ given by (25). Then, as $n \to \infty$,

$$(57) \qquad \mathcal{E}_n = 0(\frac{1}{n \log n}).$$

This lemma can be proved by modifying the proof of Lemma 4 in [18], we omit the details.

3. Proof of the Theorem.

Let $\tilde{S}_n(\tilde{f}; x)$ be the nth partial sum of the conjugate series associated with $\tilde{f}(x)$ (see (6)). Hence, from the definition of H_n^δ, $\delta = 1, 2, \ldots$, we have,

$$
\begin{aligned}
H_n^\delta(\tilde{S}_n(\tilde{f}; x)) &= (P_n)^{-1} \sum_{k=0}^{n} p_{n-k} \tilde{S}_k(\tilde{f}; x) \\
&= \frac{1}{2\pi P_n} \sum_{k=0}^{n} p_{n-k} \int_0^\pi \{\tilde{f}(x-t) - \tilde{f}(x+t)\} \cot \frac{t}{2} dt \\
&\quad - \frac{1}{2\pi P_n} \sum_{k=0}^{n} p_{n-k} \int_0^\pi \{\tilde{f}(x-t) - \tilde{f}(x+t)\} \frac{\cos(k+\frac{1}{2})t}{\sin t/2} dt.
\end{aligned}
$$

Since $f \in \mathcal{B}$ implies $(-f + \frac{1}{2}a_0)$ is equivalent to $\tilde{\tilde{f}}$, we have, for almost all x,

$$(58) \qquad f(x) - H_n^\delta(x) = \frac{1}{2\pi} \int_0^{2\pi} [\tilde{f}(x+t) - \tilde{f}(x-t)] M_n(t) dt,$$

with $M_n(t)$ given by (20).

Since $f \in \mathcal{F}^\delta$ implies $f \in \mathcal{K}_0$, $\tilde{f}$ is equivalent to an indefinite integral of a bounded function, say $\tilde{f}'(x)$. Hence as in [18], we get on integration by parts,

$$(59) \qquad f(x) - H_n^\delta(x) = \frac{1}{2\pi} \int_0^\pi [\tilde{f}'(x+t) + \tilde{f}'(x-t)] \mathcal{G}_n(t) dt$$

where $\mathcal{G}_n(t)$ is given by (19).

Adding and subtracting $-2\tilde{f}'(x)$ in the square bracket in the right side of (59), we can express

$$(60) \qquad f(x) - H_n^\delta(x) = R_n(x) + S_n(x),$$

where

$$(61) \qquad R_n(x) = \frac{1}{2\pi} \int_0^\pi [\tilde{f}'(x+t) - 2\tilde{f}'(x) + \tilde{f}'(x-t)]\mathcal{G}_n(t)dt$$

and

$$(62) \qquad S_n(x) = \frac{1}{\pi}\tilde{f}'(x) \int_0^\pi \mathcal{G}_n(t)dt.$$

Since under the hypothesis of our theorem

$$[\tilde{f}'(x+t) - 2\tilde{f}'(x) + \tilde{f}'(x-t)] = O(\Lambda(t)),$$

uniformly in x, using Lemma 1, we conclude that

$$\int_0^\pi \Lambda(t)|\mathcal{G}_n(t)|dt = O(\frac{1}{n\log n}),$$

and consequently

$$(63) \qquad \|R_n(x)\| = O(\frac{1}{n\log n}).$$

By (59),

$$\frac{1}{\pi} \int_0^\pi \mathcal{G}_n(t)dt$$

is equal to the value of $f(0) - H_n^\delta(0)$ in the special case when $\tilde{f}(t) = t$, $-\pi < t < \pi$. In this case the Fourier series of $\tilde{f}(t)$ is

$$2\sum_{n=1}^\infty (-1)^{n-1}\left(\frac{\sin nt}{n}\right),$$

and thus the Fourier series of $f(t)$ is

$$(64) \qquad 2\sum_{n=1}^\infty (-1)^{n-1}\left(\frac{\cos nt}{n}\right)$$

apart from a constant term.

Thus $f(0) - H_n^\delta(0) = -2\mathcal{E}_n$, and

$$(65) \qquad \frac{1}{\pi} \int_0^\pi \mathcal{G}_n(t)dt = O(\frac{1}{n\log n}),$$

follows from Lemma 2. Since $\tilde{f}'(x)$ is bounded

$$(66) \qquad \|S_n(x)\| = O(\frac{1}{n\log n}).$$

Now substituting the estimates (63) and (66) in (65), the proof of the theorem becomes complete.

References

[1] G. Alexits, *On the order of approximation of Fejér means*, Acta Math. Hung. 3 (1944), 20-25.

[2] P. L. Butzer, *On the singular integral of de la Vallée Poussin*, Archiv. Math. 7 (1956), 295-303.

[3] P. L. Butzer, *Über den Grad der Approximation des Identitäsoperators durch Halbgruppen van linearen operatoren und anwendungen auf die theorie der singularen integrale*, Math. Ann. 133 (1957), 97-110.

[4] G. Das and P. C. Mohopatra, *Necessary and sufficient conditions for absolute Nörlund summability of Fourier series*, Proc. London Math. Soc., 41 (1980), 217-253.

[5] R. DeVore, *The approximation of continuous functions by positive linear operators*, Lecture Notes in Mathematics 293, Springer Verlag, Berlin, 1972.

[6] J. Favard, *Sur la saturation des precédés de sommation*, J. Math. 36 (1957), 359-372.

[7] G. Freud, *On polynomial approximation with the weight* $\exp\{-\frac{1}{2}x^{2k}\}$, Acta Math. Acad. Sci. Hung. 24 (1973), 363-371.

[8] D. S. Goel, A. S. B. Holland, C. Nasim, and B. N. Sahney, *Best approximation by a saturation class of polynomial operators*, Pacific J. Math., 55 (1974), 149-155.

[9] M. Horváth, *Some saturation theorems for classical orthogonal expansions*, preprint of Math. Inst. 76, Budapest 1987.

[10] I. Joo, *On the order of approximation by Fejér means of Hermite-Fourier and Laguerre-Fourier series*, Acta Math. Hung. 51 (1988), 365-370.

[11] I. Joo, *Saturation theorems for Hermite-Fourier series*, Acta Math. Hung. 57 (1991), 169-179.

[12] I. Joo, *Alexits-type theorem for multidimensional trigonometric series*, Annales Univ. Sci. Budapest, 34 (1991), 173-180.

[13] I. Joo, *On Riesz means of Hermite-Fourier series of functions from Lipschitz class*, Annales Univ. Sci. Budapest, 35 (1992), 69-76.

[14] I. Joo, *On Hermite-Fourier series*, Periodica Math. Hung., 24 (1992), 87-118.

[15] I. Joo, *Saturation Theorems for Hermite-Fourier series*, Acta Math. Hung. 57 (1991), 169-179.

[16] H. H. Khan, and S. M. Rizvi, *On the saturation of functions by* (N, p_n, q_n) *method*, Indian J. Pure and Applied Math. 6 (1974), 1262-1269.

150

[17] B. Kuttner, and B. N. Sahney, *On nonuniqueness of the degree of saturation*, Math. Proc. Cambridge Phil. Soc. 84 (1978), 113-116.

[18] B. Kuttner, R. N. Mohaptra, B. N. Sahney, *Saturation results for a class of linear operators*, Math. Proc. Cambridge Phil. Soc., 94 (1983), 133-148.

[19] R. G. Mamedov, *Local saturation of a family of linear operators*, Doklady Akad. Nauk. 155 (1964), 499-502.

[20] W. E. Milne, *On the maximum absolute value of the derivative of* $exp(-x^2)P_n(x)$, Trans. Amer. Math. Soc. 139 (1969), 231-242.

[21] R. N. Mohapatra, and B. N. Sahney, *Approximation by a class of linear operators involving a lower triangular matrix*, Studia Sci. Math. Hung. 14 (1979), 87-94.

[22] B. Muckenhoupt, *Poisson integrals for Hermite and Laguerre expansions*, Trans. Amer. Math. Soc. 139 (1969), 231-242.

[23] B. Muckenhoupt, *Hermite conjugate expansions*, Trans. Amer. Math. Soc, 139 (1969), 243-260.

[24] B. Muckenhoupt, *Conjugate functions for Laguerre expansions*, Trans. Amer. Math. Soc, 147 (1970), 403-418.

[25] B. Muckenhoupt, and E. M. Stein, *Classical expansions and their relation to conjugate harmonic functions*, Trans. Amer. Math. Soc., 118 (1956), 17-92.

[26] S. B. Steckin, *On the approximation of periodic functions by de la Vallée-Poussin sums*, Anal. Math. 4 (1978), 61-74.

[27] G. Sunouchi, *On the class of saturation in the theory of approximation II, III*, Tôhoku Math. J. 13 (1961), 112-118; 320-328.

[28] G. Sunouchi. *Saturation in the local approximation*, Tôhoku Math. J. 17 (1965), 16-28.

[29] G. Sunouchi, and C. Watari, *On determination of the class of saturation in the theory of approximation of functions II*, Tôhoku Math. J. 11 (1959), 480-488.

[30] Y. Suzuki, *Saturation of local approximation by linear positive operators*, Tôhoku Math. J. 17 (1965), 210-221.

[31] M. Zamansky, *Classes de saturation de certaines procédés d'approximation des séries de Fourier des fonctions continues*, Ann. Sci Ecole Normale sup. 66 (1949), 19-93.

[32] A. Zygmund, *Trigonometric series* I & II, Cambridge University Press, 1959.

SOME COMMENTS AND A BIBLIOGRAPHY ON THE LAGUERRE-SAMUELSON INEQUALITY WITH EXTENSIONS AND APPLICATIONS IN STATISTICS AND MATRIX THEORY

SHANE T. JENSEN and GEORGE P. H. STYAN

Department of Mathematics and Statistics, McGill University,
805 rue Sherbrooke ouest, Montréal, Québec, Canada H3A 2K6

Abstract. We examine an 1880 theorem of Laguerre [50] concerning polynomials with all real roots and a 1968 inequality of Samuelson [117] for the maximum and minimum deviation from the mean, and establish their equivalence. The bounds provided by Laguerre's Theorem involve the first three coefficients of an n-th degree polynomial while Samuelson's Inequality is in terms of the standard deviation (and the mean) of a set of n real numbers (observations). We present eight proofs of this Laguerre-Samuelson inequality and survey the literature; we also give various extensions and applications in statistics and matrix theory. We include some historical and biographical information and present an extensive bibliography with over 100 entries.

1. Introduction and mise-en-scène

1.1. The Laguerre-Samuelson Inequality. Throughout this paper x_1, $x_2, \ldots, x_n$ will denote n real numbers with (arithmetic) mean

$$(1.1) \qquad \bar{x} = \frac{1}{n} \sum_{i=1}^{n} x_i$$

and standard deviation (with divisor n):

$$(1.2) \qquad s = \sqrt{\frac{1}{n} \sum_{i=1}^{n} (x_i - \bar{x})^2} = \sqrt{\frac{1}{n} \left(\sum_{i=1}^{n} x_i^2 - n\bar{x}^2 \right)}.$$

1991 *Mathematics Subject Classifications:* Primary 15A45, Secondary 01A60, 15-03, 15A24, 26C05, 26D99, 62G30.

Key words and phrases: Boyd-Hawkins inequalities, Brunk inequalities, Cauchy-Schwarz inequality, deviant observations, extreme deviations inequality, Laguerre's theorem, matrix inequalities, order statistics, outliers, Samuelson's inequality, Scott inequalities, Studentized deviations.

T.M. Rassias and H.M. Srivastava (eds.), Analytic and Geometric Inequalities and Applications, 151–181.

152

Then

(1.3) $\quad \bar{x} - s\sqrt{n-1} \leq x_j \leq \bar{x} + s\sqrt{n-1} \qquad$ for all $j = 1, 2, \ldots, n$

or equivalently

(1.4) $\qquad (x_j - \bar{x})^2 \leq (n-1)s^2 \qquad$ for all $j = 1, 2, \ldots, n.$

Equality holds in (1.4) if and only if all the x_i other than x_j are equal and so then x_j is either the largest or the smallest of the x_i; equality holds on the left (right) of (1.3) if and only if the $n-1$ largest (smallest) x_i are all equal.

We see, therefore, that given the mean and standard deviation of a set of real numbers, their minimum is bounded below and their maximum bounded above. These bounds are often referred to as "Samuelson's Inequality" in the statistical literature[1] in view of the inequalities established in 1968 by the American economist and Nobel laureate Paul Anthony Samuelson[2] (b. 1915) in the *Journal of the American Statistical Association* [117].

The inequalities (1.3) were (almost certainly first) established in 1880 by the well-known French mathematician Edmond Nicolas Laguerre[3] (1834–1886) in the *Nouvelles Annales de Mathématiques (Paris)* [50]. Laguerre's results were obtained in a completely different notation and context[4].

Laguerre's interest focused on n-th degree polynomials with all roots real. Let $x_1, x_2, \ldots, x_n$ denote the roots, all of which we will assume to be real, of the n-th degree polynomial equation with $n \geq 2$:

(1.5) $\qquad f(x) = a_0 x^n + a_1 x^{n-1} + a_2 x^{n-2} + \cdots + a_{n-1} x + a_n = 0.$

Since we will assume that this polynomial has degree n we will now suppose, without loss of generality, that

(1.6) $\qquad\qquad\qquad\qquad a_0 = 1.$

---

[1] Cf. e.g., Arnold [3], Borwein, Styan and Wolkowicz [16], Chaganty and Vaish [26], Farnum [31], Kabe [46], Mărgăritescu [58], Mathew and Nordström [60], Murty [83], Patel, Kapadia and Owen [99] (p. 263), Puntanen [106] (Example 6.16, pp. 275–276), and Wolkowicz and Styan [135].

[2] For an "autobiographical account of his career" see Samuelson [119].

[3] For a biographical account see Brezinski [20].

[4] While several authors in the mathematical literature refer to Laguerre (cf. e.g., Lupaş [53], Madhava Rao and Sastry [55], Mitrinović [78], pp. 210–211, Popoviciu [104], Sz.-Nagy [126], [127], [128], and Weber [134], pp. 364–371, the only author who we could find in the statistical literature to do so was Rodica-Cristina Vodă [132] in 1983 (in Romanian), who also references Mihăileanu [71].

Let

(1.7)
$$t_1 = \sum_{i=1}^{n} x_i \qquad \text{and} \qquad t_2 = \sum_{i=1}^{n} x_i^2.$$

Then

(1.8)
$$a_1 = -\sum_{i=1}^{n} x_i = -t_1 \qquad \text{and} \qquad a_2 = \sum_{i<j} x_i x_j = \tfrac{1}{2}(t_1^2 - t_2).$$

Laguerre [50] proved that

(1.9)
$$-\frac{a_1}{n} - b\sqrt{n-1} \le x_j \le -\frac{a_1}{n} + b\sqrt{n-1} \qquad \text{for all } j = 1, 2, \ldots, n,$$

where

(1.10)
$$b = \sqrt{\frac{(n-1)a_1^2}{n^2} - \frac{2a_2}{n}} = \frac{\sqrt{nt_2 - t_1^2}}{n}$$

using (1.8). It follows at once that

(1.11)
$$-\frac{a_1}{n} = \bar{x} \qquad \text{and} \qquad b = s,$$

respectively the mean and the standard deviation defined in (1.1) and (1.2) above, and so the inequalities (1.9) coincide with (1.3).

Laguerre [50], however, did not observe that $-a_1/n$ and b were in fact the mean and standard deviation[5] of the roots x_i; his interest was in obtaining bounds for the roots, whenever they are all real, of an n-th degree polynomial given the first three coefficients—in our formulation the first of these: $a_0 = 1$, cf. (1.6)[6].

In this paper we will, therefore, refer to the inequalities (1.3) or (1.4) as the "Laguerre-Samuelson Inequality".

While "Samuelson's Inequality" is certainly the most popular name for (1.3), the name "Extreme Deviations Inequality" is also used in the (relatively recent) statistical literature[7]; in 1974 Arnold used "extreme deviance" in

---

[5] The term "standard deviation" was introduced in 1893 (by Karl Pearson (1857–1936) "in a lecture to the Royal Society", cf. Hart [37], p. 626; Stigler [124], p. 328, "although the idea was by then nearly a century old", cf. Abbott [1], p. 105.

[6] Laguerre [50] did not assume that $a_0 = 1$ and so his results involve a_1/a_0 and a_2/a_0 instead of our a_1 and a_2.

[7] Cf. Dwass (1975) [30], O'Reilly (1976) [98], and Quesenberry (1974) [107].

the title of his paper [3], while "How deviant can you be?" is the title of the seminal paper by Samuelson (1968) [117]; the 1992 survey paper by Olkin [96] is entitled "A matrix formulation on how deviant an observation can be". Much earlier, however, the term "extreme deviate" appears in the title of the 1948 paper by Nair [86] and "extreme observation" in the titles of the papers by Hartley and David (1954) [38] and McKay (1935) [61]. In the hydrology journal *Water Resources Research*, Kirby (1974) [48] uses "standardized maximum deviate".

Wolkowicz and Styan (1988) call (1.3) the "Samuelson-Nair Inequality" in their *Encyclopedia of Statistical Sciences* entry [139], while Arnold and Balakrishnan in their 1989 monograph *Relations, Bounds and Approximations for Order Statistics* [4] present many inequalities related to and including the Laguerre-Samuelson Inequality in their Section 3.2 entitled "Variations on the Samuelson-Scott theme"[8].

The Indian statistician Keshavan Raghavan Nair[9] (b. 1910) established the Laguerre-Samuelson Inequality (1.3) in his 1947 Ph.D. thesis [84], publishing his proof a year later in 1948 in the *Journal of the Indian Society of Agricultural Statistics (Delhi)* [85], cf. also Nair [89], [90]. J. M. C. Scott[10] established several inequalities (see §1.4 below) on ordered absolute deviations $|x_j - \bar{x}|$ in the Appendix to the 1936 paper [101] by Egon Sharpe Pearson (1895–1980), assisted by C. Chandra Sekar in *Biometrika (London)*; as noted by Arnold and Balakrishnan [4] (Theorem 3.2, p. 44) the Laguerre-Samuelson Inequality is a special case of one of Scott's inequalities.

1.2. The Brunk Inequalities. Now let us arrange the x_i's in nondecreasing order:

$$(1.12) \qquad x_{\min} = x_{(n)} \leq x_{(n-1)} \leq \cdots \leq x_{(2)} \leq x_{(1)} = x_{\max}$$

so that $x_{(j)}$ is the j-th largest. Then:

$$(1.13) \qquad \bar{x} + \frac{s}{\sqrt{n-1}} \leq x_{\max} = x_{(1)} \leq \bar{x} + s\sqrt{n-1}$$

and

$$(1.14) \qquad \bar{x} - s\sqrt{n-1} \leq x_{\min} = x_{(n)} \leq \bar{x} - \frac{s}{\sqrt{n-1}}.$$

---

[8] Cf. [4], Theorem 3.3, pp. 45–46, for six proofs of the Laguerre-Samuelson Inequality; a further proof using the arithmetic-geometric mean inequality is proposed in [4] as Exercise 7, p. 62.

[9] For an "autobiographical article" see Nair [90].

[10] We believe J. M. C. Scott was at the Cavendish Laboratory, Cambridge, England, in the mid-50s, but have no further biographical information.

The right-hand inequality in (1.13) and the left-hand inequality in (1.14) are the Laguerre-Samuelson inequality (1.3). The left-hand inequality in (1.13) and the right-hand inequality in (1.14) were established (possibly for the first time) in 1959 by Hugh Daniel Brunk (b. 1919), also in the *Journal of the Indian Society of Agricultural Statistics* [22], and so we will refer to them as the "Brunk Inequalities". Unaware of Brunk's results these inequalities were established again by Boyd (1971) [18], Hawkins (1971) [41] and Wolkowicz and Styan (1979) [135].

Equality holds on the left of (1.13) if and only if equality holds on the left of (1.14) if and only if:

$$x_{(1)} = \cdots = x_{(n-1)} = \bar{x} + \frac{s}{\sqrt{n-1}} \qquad \text{and} \qquad x_{(n)} = \bar{x} - s\sqrt{n-1};$$

equality holds on the right of (1.13) if and only if equality holds on the right of (1.14) if and only if:

$$x_{(1)} = \bar{x} + s\sqrt{n-1} \qquad \text{and} \qquad x_{(2)} = \cdots = x_{(n)} = \bar{x} - \frac{s}{\sqrt{n-1}}.$$

1.3. The Boyd-Hawkins Inequalities. For the k-th largest observation or "order statistic" $x_{(k)}$ we have the following inequalities

$$(1.15) \qquad \bar{x} - s\sqrt{\frac{k-1}{n-k+1}} \leq x_{(k)} \leq \bar{x} + s\sqrt{\frac{n-k}{k}} \qquad \text{for } k = 2, \ldots, n-1;$$

equality holds on the left of (1.15) if and only if

$$x_{(1)} = \cdots = x_{(k-1)} = \bar{x} + s\sqrt{\frac{n-k+1}{k-1}} \qquad \text{and} \qquad x_{(k)} = \cdots = x_{(n)} = \bar{x} - s\sqrt{\frac{k-1}{n-k+1}}$$

and on the right of (1.15) if and only if

$$x_{(1)} = \cdots = x_{(k)} = \bar{x} + s\sqrt{\frac{n-k}{k}} \qquad \text{and} \qquad x_{(k+1)} = \cdots = x_{(n)} = \bar{x} - s\sqrt{\frac{k}{n-k}}.$$

If we put $k = 1$ in (1.15) then we obtain the same upper bound for $x_{\max} = x_{(1)}$ as in (1.13) but a weaker lower bound. Similarly, if we put $k = n$ in (1.15) then we obtain the same lower bound for $x_{\min} = x_{(n)}$ as in (1.14) but a weaker upper bound.

The inequalities (1.15) were established (possibly for the first time[11]) in 1971 by A. V. Boyd [18] in the *Publikacije Elektrotehničkog Fakulteta Univerziteta u Beogradu, Serija Matematika i Fizika (Belgrade)*[12] (in English)

---

[11] Rodica-Cristina Vodă [132], p. 547, comments (in Romanian) that (1.15) "este şi el inclus parţial in rezultatul lui Laguerre" (p. 547) or (in English) "can be partially derived from an old inequality due to Laguerre" (p. 548): no further details are given.

[12] The masthead of this journal also carries the French subtitle: *Publications de la Faculté d'Électrotechnique de l'Université à Belgrade, Série Mathématiques et Physique.*

and, also in 1971, by Douglas M. Hawkins [41] in the *Journal of the American Statistical Association*; see also Wolkowicz and Styan [135], [136], [137]. As observed by Arnold and Balakrishnan [4] (p. 49) and Wolkowicz and Styan [135], the inequalities (1.15) are "implicit" in the papers by Mallows and Richter (1969) [56] and Arnold and Groeneveld (1979) [7], while Scott (1936) [122] gives (without proof) the inequality

$$(1.16) \qquad x_{(2)} \leq \bar{x} + s\sqrt{\frac{n-2}{2}},$$

the special case of the upper bound in (1.15) for $k = 2$.

We will call (1.15) the "Boyd-Hawkins Inequalities".

1.4. The Scott Inequalities. The first (explicit) proof of the Laguerre-Samuelson Inequality in the statistical literature was almost certainly that given in 1936 by J. M. C. Scott [122] in the Appendix to the paper by Pearson and Chandra Sekar [101]; the Laguerre-Samuelson Inequality appears there as a special case of (1.19a), the first of three inequalities below, cf. Arnold and Balakrishnan [4], Theorem 3.2, p. 44, where it is observed that "Scott's ingenious constructive proof is apparently the only proof available in the literature."

Let us define the absolute deviations:

$$(1.17) \qquad \delta_i = |x_i - \bar{x}|; \qquad i = 1, \ldots, n,$$

and let $\delta_{(i)}$ denote the i-th largest absolute deviation so that

$$(1.18) \qquad \delta_{(n)} \leq \delta_{(n-1)} \leq \cdots \leq \delta_{(1)}.$$

Of course the i-th largest absolute deviation $\delta_{(i)}$ will not, in general, be equal to $|x_{(i)} - \bar{x}|$.

Then

$$(1.19a) \qquad \delta_{(j)} \leq s\sqrt{\frac{n(n-j)}{j(n-j)+1}} \quad \text{for } j \text{ odd and } j \neq n,$$

$$(1.19b) \qquad \delta_{(n)} \leq s\sqrt{\frac{n-1}{n(n+1)}} \qquad \text{for } n \text{ odd,}$$

$$(1.19c) \qquad \delta_{(j)} \leq s\sqrt{\frac{1}{j}} \qquad \text{for } j \text{ even.}$$

We note that $j = 1$ in (1.19a) corresponds to the Laguerre-Samuelson Inequality (1.4). The inequality (1.19b) is, of course, quite different to the

Brunk Inequality, cf. (1.14):

$$(1.20) \qquad x_{\min} \leq \bar{x} - \frac{s}{\sqrt{n-1}}.$$

Indeed, we obtain equality in (1.19b) when $(n-1)/2$ of the x_i are equal to b and all other x_i are equal to $-1/b$, where

$$(1.21) \qquad b = \sqrt{\frac{n+1}{n-1}}.$$

On the other hand equality holds in (1.20) if and only if the largest $n-1$ of the x_i are equal.

1.5. Purpose and Overview. Our main purpose in this paper is to survey the literature associated with the Laguerre-Samuelson, Brunk, and Boyd-Hawkins Inequalities, and to give several proofs. As observed by Arnold and Balakrishnan [4] (in their introduction to Chapter 3) the publication by Samuelson [117] "... spawned a torrent of generalizations, several of which referred to bounds on order statistics. It also spawned a flurry of rediscoveries of earlier notes on these topics. Ultimate priority seems hard to pin down ..."

In Section 2 we present eight different proofs of the "Laguerre-Samuelson Inequality" (1.3):

- 2.1. Laguerre (1880), Madhava Rao & Sastry (1940), Mitrinović (1970)

- 2.2. Thompson (1935)

- 2.3. Nair (1947, 1948), Kempthorne (1973), Arnold & Balakrishnan (1989)

- 2.4. Arnold (1974), Dwass (1975), Arnold & Balakrishnan (1989)

- 2.5. Arnold (1974), O'Reilly (1975, 1976), Arnold & Balakrishnan (1989), Murty (1990)

- 2.6. Wolkowicz and Styan (1979, 1980)

- 2.7. Smith(1980), Arnold & Balakrishnan (1989)

- 2.8. Olkin (1992).

Arnold and Balakrishnan [4], pp. 45–46, present six proofs, of which five (all but their first) are included in four (§2.3–2.5, 2.7) of our eight. As Arnold and

Balakrishnan [4] point out (p. 45): "It is instructive to ... consider several alternative proofs. The alternative proofs often suggest different possible extensions ... The Schwarz inequality[13] may be perceived to be lurking in the background of many of the proofs."

We also present several related inequalities and some applications in statistics and matrix theory. We include some historical and biographical information and present an extensive bibliography of over 100 entries from both the mathematical and the statistical literature. References to *Jahrbuch für die Fortschritte der Mathematik* are denoted by JFM (for reviews published in 1868–1930), *Mathematical Reviews* by MR (for reviews published since 1940), and to *Zentralblatt für Mathematik* [140] by Zbl (for reviews published since 1931).

2. The Laguerre-Samuelson Inequality: Eight Proofs

2.1. Laguerre (1880), Madhava Rao & Sastry (1940), Mitrinović (1970).

Our first proof is that given in 1880 by Edmond Nicolas Laguerre [50], cf. also Madhava Rao and Sastry [55] and Mitrinović [78], pp. 210–211.

For any real scalar u, we have the sum of squares expansion:

$$(2.1) \quad \sum_{i=1}^{n}(u - x_i)^2 = nu^2 - 2t_1 u + t_2 \geq (u - x_j)^2 = u^2 - 2x_j u + x_j^2$$

for any particular x_j, since a sum of squared terms is always greater than or equal to any one of its summands. Here t_1 and t_2 are as in (1.7).

Rearranging (2.1), we see that for any real u,

$$(2.2) \quad (n - 1)u^2 + 2(x_j - t_1)u + (t_2 - x_j^2) \geq 0.$$

Since this quadratic function in u is nonnegative, its discriminant must be non-positive:

$$(2.3) \quad 4(x_j - t_1)^2 - 4(n - 1)(t_2 - x_j^2) \leq 0.$$

Rearranging and simplifying (2.3) as a quadratic in x_j yields:

$$(2.4) \quad nx_j^2 - 2t_1 x_j + t_1^2 - (n - 1)t_2 \leq 0$$

---

[13] Named after [Karl] Hermann Amandus Schwarz (1843–1921) for the inequality he established in 1888 in [121], pp. 343–345; the inequality was established, however, already in 1821 by [Baron] Augustin-Louis Cauchy (1789–1857) in [23], pp. 373–374, and in 1859 by Viktor Yakovlevich Bouniakowsky [Buniakovski, Bunyakovsky] (1804–1899) in [17], pp. 3–4. In this paper we will call it the Cauchy-Schwarz Inequality, cf. (2.14) below.

and so x_j must lie in the closed interval $[\alpha_1, \alpha_2]$, where α_1, α_2 are the roots of

(2.5)
$$nx_j^2 - 2t_1 x_j + t_1^2 - (n-1)t_2 = 0.$$

These roots α_1, α_2 are:

(2.6)
$$\frac{2t_1 \pm \sqrt{4t_1^2 - 4n(t_1^2 - (n-1)t_2)}}{2n} = \frac{-a_1}{n} \pm b\sqrt{n-1}$$

using (1.10) and so (1.9) is established. □

We may arrive at the inequality (2.4) more easily, however, cf. Madhava Rao and Sastry [55], since

$$-\{nx_j^2 - 2t_1 x_j + t_1^2 - (n-1)t_2\} = (n-1)(t_2 - x_j^2) - (t_1 - x_j)^2$$

$$= (n-1)\sum_{i \neq j} x_i^2 - \left(\sum_{i \neq j} x_i\right)^2$$

$$= (n-1)\sum_{i \neq j}(x_i - \hat{x})^2 \geq 0,$$

cf. (1.2), where

(2.7)
$$\hat{x} = \frac{1}{n-1}\sum_{i \neq j} x_i$$

is the "reduced" mean of the $n-1$ roots $x_1, ..., x_n$ excluding x_j. □

2.2. Thompson (1935).

Almost certainly the first proof in a statistical context is the following proof which is implicit in the 1935 paper of William R. Thompson[130].

Let $\hat{x}$ denote the "reduced" mean of the $n-1$ real numbers $x_1, ..., x_n$ excluding x_j, cf. (2.7), and let $\bar{x}$ and s denote the mean and standard deviation, respectively, of all n observations, cf. (1.1) and (1.2). Then

(2.8)
$$\bar{x} - \hat{x} = \frac{1}{n}(x_j - \hat{x}) = \frac{1}{n-1}(x_j - \bar{x})$$

and so

$$ns^2 = \sum_{i=1}^{n}(x_i - \hat{x} + \hat{x} - \bar{x})^2$$

$$= \sum_{i \neq j}(x_i - \hat{x})^2 + (x_j - \hat{x})^2 - n(\hat{x} - \bar{x})^2$$

$$= \sum_{i \neq j}(x_i - \hat{x})^2 + n(n-1)(\hat{x} - \bar{x})^2$$

$$(2.9) \qquad = \sum_{i \neq j}(x_i - \hat{x})^2 + \frac{n}{n-1}(x_j - \bar{x})^2$$

$$(2.10) \qquad \geq \frac{n}{n-1}(x_j - \bar{x})^2,$$

using (2.8). The inequality (1.4) follows at once.

This proof also shows that equality holds in (1.4) if and only if equality holds in (2.10) and this is so if and only if $x_i = \hat{x}$ for all $i \neq j$. Hence equality holds in (1.4) if and only if all the x_i other than x_j are equal. $\qquad \square$

Thompson [130] obtains (2.9) explicitly—cf. his (6) on p. 215—but apparently does not obtain the inequality (2.10). Thompson's interest focused on the distribution of the "Studentized deviations" $(x_j - \bar{x})/s$ when the "'observations" $x_1, ..., x_n$ are independently and identically distributed as a normal random variable with unknown mean and variance.

2.3. Nair (1947, 1948), Kempthorne (1973), Arnold & Balakrishnan (1989). We consider the $n \times n$ orthogonal matrix $\mathbf{E} =$

$$\begin{pmatrix} \frac{1}{\sqrt{n}} & \frac{1}{\sqrt{n}} & \frac{1}{\sqrt{n}} & \cdots & \frac{1}{\sqrt{n}} & \frac{1}{\sqrt{n}} \\ \frac{1}{\sqrt{2}} & \frac{-1}{\sqrt{2}} & 0 & \cdots & 0 & 0 \\ \vdots & \vdots & \vdots & \cdots & \vdots & \vdots \\ \frac{1}{\sqrt{(n-1)(n-2)}} & \frac{1}{\sqrt{(n-1)(n-2)}} & \frac{1}{\sqrt{(n-1)(n-2)}} & \cdots & \frac{-(n-2)}{\sqrt{(n-1)(n-2)}} & 0 \\ \frac{1}{\sqrt{n(n-1)}} & \frac{1}{\sqrt{n(n-1)}} & \frac{1}{\sqrt{n(n-1)}} & \cdots & \frac{1}{\sqrt{n(n-1)}} & \frac{-(n-1)}{\sqrt{n(n-1)}} \end{pmatrix},$$

the so-called Helmert matrix[14] and let $\mathbf{x} = \{x_i\}$ and $\mathbf{y} = \mathbf{Ex} = \{y_i\}$. Then

(2.11)
$$\sum_{i=1}^{n} x_i^2 = \mathbf{x}'\mathbf{x} = \mathbf{x}'\mathbf{E}'\mathbf{Ex} = \mathbf{y}'\mathbf{y} = \sum_{i=1}^{n} y_i^2 \geq y_1^2 + y_n^2.$$

Since

(2.12)
$$y_1 = \frac{1}{\sqrt{n}} \sum_{i=1}^{n} x_i \quad \text{and} \quad y_n = \sqrt{\frac{n}{n-1}}(\bar{x} - x_n)$$

it follows at once from (2.11) that

(2.13)
$$\sum_{i=1}^{n} x_i^2 \geq \frac{1}{n} \left(\sum_{i=1}^{n} x_i \right)^2 + \frac{n}{n-1}(\bar{x} - x_n)^2.$$

If we rearrange the components of the vector $\mathbf{x}$ so that x_j is in the n-th position then, with x_n replaced by x_j, (2.13) becomes (1.4).

Equality holds in (2.13) if and only if equality holds in (2.11) and this is so if and only if $y_2 = \cdots = y_{n-1} = 0$, i.e., all the x_i are equal except for x_n (which we now choose to be x_j). $\qquad\square$

This is the third proof given by Arnold and Balakrishnan [4], p. 45, and follows that given by K. R. Nair in "a small section of the third part" of his 1947 Ph.D. thesis [84] and published in 1948 [85], and by Oscar Kempthorne in a 1973 "Personal communication" [47] to Barry C. Arnold[15].

2.4. Arnold (1974), Dwass (1975), Arnold & Balakrishnan (1989).
Barry C. Arnold [3] and Meyer Dwass [30] proved (1.4) using the Cauchy-Schwarz inequality:

(2.14)
$$(\mathbf{a}'\mathbf{b})^2 \leq \mathbf{a}'\mathbf{a} \cdot \mathbf{b}'\mathbf{b}$$

for any $n \times 1$ real vectors $\mathbf{a}$ and $\mathbf{b}$. This is the second proof given by Arnold and Balakrishnan [4], p.45.

---

[14] Named after Friedrich Robert Helmert (1843–1919) for the matrix he introduced in 1876 [42], cf. also Harville [40], pp. 85–86, Lancaster [51], Read [116], and Stuart and Ord [125], Example 11.3.

[15] Cf. Arnold and Balakrishnan [4], pp. 45 & 158, and Arnold [3] where, in an acknowledgement, it is observed that: "Upon seeing an earlier draft of this note, Oscar Kempthorne supplied me with three of several alternative proofs that he derived for Samuelson's inequality".

162

Since $\sum_{i=1}^{n}(x_i - \bar{x}) = 0$, it follows that

$$(2.15) \qquad\qquad x_j - \bar{x} = -\sum_{i \neq j}(x_i - \bar{x})$$

and so

$$
\begin{aligned}
(x_j - \bar{x})^2 &= \left(\sum_{i \neq j}(x_i - \bar{x})\right)^2 \\
&\leq (n-1)\sum_{i \neq j}(x_i - \bar{x})^2 \\
&= (n-1)\sum_{i=1}^{n}(x_i - \bar{x})^2 - (n-1)(x_j - \bar{x})^2
\end{aligned}
$$

from (2.14) with the vectors $\mathbf{a} = \{x_i - \bar{x}\}_{i \neq j}$ and $\mathbf{b} = (1, 1, \ldots, 1)'$ both $(n-1) \times 1$. Hence

$$(x_j - \bar{x})^2 \leq \frac{n-1}{n}\sum_{i=1}^{n}(x_i - \bar{x})^2 = (n-1)s^2$$

from (2.14), and so (1.4) follows immediately. Equality holds if and only if the vectors $\mathbf{a}$ and $\mathbf{b}$ are proportional, i.e., all the x_i except for x_j are equal. □

2.5. Arnold (1974), O'Reilly (1975, 1976), Arnold & Balakrishnan (1989), Murty (1990).

Barry C. Arnold (1974) gave a second proof in [3] which used the "hat" matrix from linear regression analysis; see also O'Reilly [97], [98], and Murty [83].

In the usual full-rank Gauss-Markov linear statistical model

$$(2.16) \qquad\qquad \mathbf{Ey} = \mathbf{X\beta},$$

where E denotes (mathematical) expectation and the "model" or "design" matrix $\mathbf{X}$ is $n \times p$ with rank $p < n$. Then it is well known that the $n \times n$ "hat matrix"

$$(2.17) \qquad\qquad \mathbf{H} = \mathbf{X}(\mathbf{X'X})^{-1}\mathbf{X'}$$

is symmetric and idempotent, and hence nonnegative definite, as is the residual matrix $\mathbf{M} = \mathbf{I} - \mathbf{H}$.

We now let $p = 2$ and $\mathbf{X} = (\mathbf{e} : \mathbf{Cx})$ as in (centered) simple linear regression; here the $n \times 1$ sum vector

(2.18)
$$\mathbf{e} = (1, 1, \ldots, 1)',$$

while the $n \times n$ centering matrix

(2.19)
$$\mathbf{C} = \mathbf{I}_n - \frac{1}{n}\mathbf{ee}'$$

is symmetric and idempotent. Hence

(2.20)
$$\mathbf{H} = \frac{1}{n}\mathbf{ee}' + \frac{1}{\mathbf{x}'\mathbf{Cx}}\mathbf{Cxx}'\mathbf{C} = \frac{1}{n}\mathbf{ee}' + \frac{n}{s^2}\mathbf{Cxx}'\mathbf{C}$$

and so the j-th diagonal element of $\mathbf{M} = \mathbf{I} - \mathbf{H}$:

(2.21)
$$m_{jj} = 1 - \frac{1}{n} - \frac{(x_j - \bar{x})^2}{ns^2} \geq 0,$$

since $\mathbf{M}$ is nonnegative definite; the Laguerre-Samuelson Inequality (1.4) follow at once.

Equality holds in (1.4) if and only if equality holds throughout (2.21) and this is so if and only if all the elements in the j-th row (and column) of $\mathbf{M}$ are zero, i.e., all the x_i except for x_j are equal. □

The proof given by O'Reilly [97], [98], is similar but uses the model matrix $\mathbf{X} = (\mathbf{e} : \mathbf{x})$ as in uncentered simple linear regession. This O'Reilly proof is the fifth proof of the Laguerre-Samuelson Inequality given by Arnold and Balakrishnan [4], p. 46, while the Arnold-Murty proof is their fourth.

2.6. Wolkowicz & Styan (1979, 1980). The proof given by Henry Wolkowicz and George P. H. Styan (1979, 1980) [135], [137], cf. also Bancroft [10], Chaganty [24], Chaganty and Vaish [25], [26], Neudecker and Liu [91], Puntanen [106] (Example 6.16, pp. 275–276), and Trenkler [131], essentially uses the following result (Lemma 2.1 in [137], p. 475):

Lemma 2.6.1. *Let* $\mathbf{w}$ *and* $\mathbf{x}$ *be real nonnull* $n \times 1$ *vectors and let* $\bar{x}$ *and* s *be defined as in (1.1) and (1.2) above, so that* $\bar{x} = \mathbf{x}'\mathbf{e}/n$ *and* $s^2 = \mathbf{x}'\mathbf{Cx}/n$, *where the centering matrix* $\mathbf{C} = \mathbf{I} - \mathbf{ee}'/n$ *as in (2.19), with* $\mathbf{e}$ *the* $n \times 1$ *vector of ones. Then*

(2.22)
$$-s\sqrt{n}\,\mathbf{w}'\mathbf{Cw} \leq \mathbf{w}'\mathbf{Cx} \leq s\sqrt{n}\,\mathbf{w}'\mathbf{Cw}.$$

Equality holds on the left (right) of (2.22) *if and only if*

$$(2.23) \qquad\qquad \mathbf{x} = c\mathbf{w} + d\mathbf{e}$$

for some scalars c and d with c < 0 (c > 0).

Proof. The inequality string (2.22) follows at once from the Cauchy-Schwarz Inequality (2.14) with $\mathbf{a} = \mathbf{Cw}$ and $\mathbf{b} = \mathbf{Cx}$. $\qquad\qquad\square$

If in (2.22) we now substitute

$$(2.24) \qquad\qquad \mathbf{w} = \mathbf{e}_j - \mathbf{e}/n = \mathbf{h}_j,$$

say, where

$$(2.25) \qquad\qquad \mathbf{e}_j = (0,\ldots,0,1,0,\ldots,0)'$$

with 1 in the j-th position, then (2.22) becomes (1.3). The equality condition $\mathbf{x} = c\mathbf{w} + d\mathbf{e} = c\mathbf{e}_j + d\mathbf{e}$ shows that equality holds in (1.4) if and only if all the x_i are equal except for x_j. $\qquad\qquad\square$

2.7. Smith (1980), Arnold & Balakrishnan (1989).

Arnold and Balakrishnan [4], p. 46, give the following proof credited to William P. Smith [123], as their sixth (and last) proof of the Laguerre-Samuelson Inequality. This proof is based on the Cantelli Inequality[16], cf. e.g., Patel, Kapadia and Owen [99], p. 51.

Let X denote a random variable with mean 0 and variance 1. Then

$$(2.26) \qquad\qquad \text{Prob}(X \leq u) \ \leq \ \frac{1}{1+u^2} \qquad \text{if } u \leq 0$$

$$(2.27) \qquad\qquad \text{Prob}(X \geq u) \ \leq \ \frac{1}{1+u^2} \qquad \text{if } u \geq 0.$$

We now suppose that X is a discrete uniform random variable with

$$(2.28) \qquad \text{Prob}\left(X = \frac{x_i - \bar{x}}{s}\right) = \frac{1}{n} \qquad \text{for all } i = 1,\ldots,n.$$

Then X has expectation $\mathsf{E}X = 0$ and variance $\text{var}\,X = 1$.

---

[16] Named after Francesco Paolo Cantelli (1875–1966); for a biographical account see Benzi [15].

If we substitute $u = (x_{\min} - \bar{x})/s < 0$ in (2.26) then it becomes

$$\frac{1}{n} \leq 1 \Big/ \left\{ 1 + \left(\frac{x_{\min} - \bar{x}}{s} \right)^2 \right\}$$

and so

(2.29)
$$\left(\frac{x_{\min} - \bar{x}}{s} \right)^2 \leq n - 1.$$

Substituting $u = (x_{\max} - \bar{x})/s > 0$ in (2.27) gives

$$\frac{1}{n} \leq 1 \Big/ \left\{ 1 + \left(\frac{x_{\max} - \bar{x}}{s} \right)^2 \right\}$$

and so

(2.30)
$$\left(\frac{x_{\max} - \bar{x}}{s} \right)^2 \leq n - 1.$$

Combining (2.29) and (2.30) yields the Laguerre-Samuelson Inequality (1.4).
□

2.8. Olkin (1992). Ingram Olkin, in his 1992 survey paper [96], used the following result:

(2.31)
$$c(x_j - \bar{x})^2 \leq \sum_{i=1}^{n} (x_i - \bar{x})^2 \text{ for all } j = 1, \ldots, n \quad \Longleftrightarrow \quad 0 \leq c \leq \frac{n}{n-1}.$$

To prove (2.31) we express both sides of its right-hand side as quadratic forms. Let $\mathbf{x} = (x_1, \ldots, x_n)'$, $\mathbf{e} = (1, \ldots, 1)'$ and where, cf. (2.25), $\mathbf{e}_j = (0, \ldots, 0, 1, 0, \ldots, 0)'$ with 1 in the j-th position—all $n \times 1$. We may write

(2.32)
$$x_j - \bar{x} = \mathbf{x}'\mathbf{h}_j \quad \text{with} \quad \mathbf{h}_j = \mathbf{e}_j - \frac{1}{n}\mathbf{e},$$

cf. (2.24) above, and so the right-hand side of (2.31) becomes

(2.33)
$$c(x_j - \bar{x})^2 = c\mathbf{x}'\mathbf{h}_j\mathbf{h}_j'\mathbf{x} \leq \mathbf{x}'\mathbf{C}\mathbf{x} = \sum_{i=1}^{n} (x_i - \bar{x})^2,$$

where the centering matrix $\mathbf{C}$ is defined as in (2.19). Then (2.33) holds if and only if

$$(2.34) \qquad \mathbf{C} - c\,\mathbf{h}_j\mathbf{h}_j' = \mathbf{I}_n - \frac{1}{n}\mathbf{ee}' - c\,\mathbf{h}_j\mathbf{h}_j' = \mathbf{I}_n - \mathbf{AA}'$$

is nonnegative definite; here $\mathbf{A} = (\mathbf{e}/\sqrt{n} : \sqrt{c}\,\mathbf{h}_j)$. Since the nonzero eigenvalues of the matrices $\mathbf{AA}'$ and $\mathbf{A}'\mathbf{A}$ coincide, it follows at once that $\mathbf{C} - c\,\mathbf{h}_j\mathbf{h}_j'$ is nonnegative definite whenever

$$\mathbf{I}_2 - \mathbf{A}'\mathbf{A} = \mathbf{I}_2 - \begin{pmatrix} \mathbf{e}'/\sqrt{n} \\ \sqrt{c}\,\mathbf{h}_j' \end{pmatrix} (\mathbf{e}/\sqrt{n} : \sqrt{c}\,\mathbf{h}_j) = \begin{pmatrix} 0 & 0 \\ 0 & 1 - c(n-1)/n \end{pmatrix}$$

is nonnegative definite. The result (2.31) follows at once.

Substituting $c = n/(n-1)$ in the right-hand side of (2.31) gives the Laguerre-Samuelson Inequality (1.4). $\qquad\square$

Some discussion of this proof is given in [10], [24], [25], [91], and [131]—see §4.2 below for additional commentary.

3. Proofs of Some Inequalities Closely Related to the Laguerre-Samuelson Inequality

3.1. The Brunk Inequalities. Let us arrange the n real numbers $x_1, \ldots, x_n$ in nondecreasing order as in (1.12):

$$(3.1) \qquad x_{\min} = x_{(n)} \leq x_{(n-1)} \leq \cdots \leq x_{(2)} \leq x_{(1)} = x_{\max}$$

so that $x_{(j)}$ is the j-th largest. Then:

$$(3.2) \qquad \bar{x} + \frac{s}{\sqrt{n-1}} \leq x_{\max} = x_{(1)} \leq \bar{x} + s\sqrt{n-1}$$

and

$$(3.3) \qquad \bar{x} - s\sqrt{n-1} \leq x_{\min} = x_{(n)} \leq \bar{x} - \frac{s}{\sqrt{n-1}}.$$

The right-hand inequality in (3.2) and the left-hand inequality in (3.3) are the Laguerre-Samuelson inequality (1.3). As announced in §1.3 above, we will refer to the left-hand inequality in (3.2) and the right-hand inequality in (3.3) as the "Brunk inequalities" since we believe that they were established for the first time in 1959 by H. D. Brunk [22].

Equality holds on the left of (3.2) if and only if equality holds on the left of (3.3) if and only if:

$$(3.4) \qquad x_{(1)} = \cdots = x_{(n-1)} = \bar{x} + \frac{s}{\sqrt{n-1}} \quad \text{and} \quad x_{(n)} = \bar{x} - s\sqrt{n-1};$$

equality holds on the right of (3.2) if and only if equality holds on the right of (3.3) if and only if:

$$(3.5) \qquad x_{(1)} = \bar{x} + s\sqrt{n-1} \quad \text{and} \quad x_{(2)} = \cdots = x_{(n)} = \bar{x} - \frac{s}{\sqrt{n-1}}.$$

3.1.1. Brunk (1959). To prove the "Brunk inequalities" Brunk used the following result ([22], Corollary 1), which we find to be interesting in its own right:

Lemma 3.1.1. *Let the random variable Z be distributed over the closed interval $[0,1]$ and let p be a nonnegative constant so that $p \le \text{Prob}(Z = 1)$. Then*

$$(3.6) \qquad\qquad p \, \mathsf{E}Z^2 \le (\mathsf{E}Z)^2,$$

with equality if and only if

$$(3.7) \qquad\qquad \text{Prob}(Z = 0) = 1 - p \quad \text{and} \quad \text{Prob}(Z = 1) = p.$$

Proof. Since $0 \le Z \le 1$ we have $Z^2 \le Z$ with probability one and so $\mathsf{E}Z^2 \le \mathsf{E}Z$ and $\text{Prob}(Z = 1) \le \mathsf{E}Z$. Combining these two inequalities yields

$$(3.8) \qquad\qquad p \, \mathsf{E}Z^2 \le \text{Prob}(Z = 1) \cdot \mathsf{E}Z \le (\mathsf{E}Z)^2,$$

and (3.6) is established. Equality holds in (3.6) if and only if equality holds throughout (3.8) if and only if $p = \text{Prob}(Z = 1)$ and $Z = Z^2$ with probability one, and so the equality condition (3.7) follows at once. □

To prove the "Brunk inequalities" we now let the random variable X assume each of the n values in (3.1) with probability $1/n$. Then the random variable $Z = (x_{\max} - X)/r$, where the range $r = x_{\max} - x_{\min}$, is distributed over $[0,1]$. The expectation $\mathsf{E}X = \bar{x}$ and the variance $\text{var}X = s^2$. Hence

$$(3.9) \qquad\qquad \mathsf{E}Z^2 = \text{var}Z + (\mathsf{E}Z)^2 = \frac{s^2 + (x_{\max} - \bar{x})^2}{r^2}$$

168

and so from Lemma 3.1.1:

(3.10) $$\frac{1}{n}EZ^2 = \frac{s^2 + (x_{\max} - \bar{x})^2}{nr^2} \le \frac{(x_{\max} - \bar{x})^2}{r^2},$$

which simplifies to

(3.11) $$s^2 \le (n-1)(x_{\max} - \bar{x})^2,$$

from which the left-hand inequality in (3.2) follows at once. Equality holds in (3.10) if and only if (3.7) holds and here this becomes (3.4).

To establish the right-hand inequality in (3.3) we repeat the above argument with $Z = (X - x_{\min})/r$. □

3.1.2. Wolkowicz and Styan (1979).

Wolkowicz and Styan [135] provided a completely algebraic (non-statistical) proof of the Brunk inequalities. Since $n(x_{\max} - \bar{x}) = \sum_{i=1}^{n}(x_{\max} - x_i)$ it follows that

$$
\begin{aligned}
n^2(x_{\max} - \bar{x})^2 &= \left\{\sum_{i=1}^{n}(x_{\max} - x_i)\right\}^2 \\
&= \sum_{i=1}^{n}(x_{\max} - x_i)^2 + \sum_{i \neq i'}(x_{\max} - x_i)(x_{\max} - x_{i'}) \\
&\ge \sum_{i=1}^{n}(x_{\max} - x_i)^2 \\
&= \sum_{i=1}^{n}(x_{\max} - \bar{x} + \bar{x} - x_i)^2 = n\{(x_{\max} - \bar{x})^2 + s^2\},
\end{aligned}
$$

from which the left-hand inequality in (3.2) follows at once, with equality if and only if $x_{\max} = x_{(1)} = \ldots = x_{(n-1)}$ or (3.4) holds.

If $n^2(\bar{x} - x_{\min})^2$ is expanded similarily, then the right-hand inequality in (3.3) follows at once, with equality if and only if $x_{(2)} = \cdots = x_{(n)} = x_{\min}$ or (3.5) holds. □

3.2. The Boyd-Hawkins Inequalities.

As observed above in §1.3 the k-th largest observation or "order statistic" $x_{(k)}$ satisfies the following "'Boyd-Hawkins inequalities":

(3.12) $$\bar{x} - s\sqrt{\frac{k-1}{n-k+1}} \le x_{(k)} \le \bar{x} + s\sqrt{\frac{n-k}{k}} \qquad \text{for } k = 2, \ldots, n-1;$$

equality holds on the left of (3.12) if and only if

$$x_{(1)} = \cdots = x_{(k-1)} = \bar{x} + s\sqrt{\tfrac{n-k+1}{k-1}} \quad \text{and} \quad x_{(k)} = \cdots = x_{(n)} = \bar{x} - s\sqrt{\tfrac{k-1}{n-k+1}}$$

and on the right of (3.12) if and only if

$$x_{(1)} = \cdots = x_{(k)} = \bar{x} + s\sqrt{\tfrac{n-k}{k}} \quad \text{and} \quad x_{(k+1)} = \cdots = x_{(n)} = \bar{x} - s\sqrt{\tfrac{k}{n-k}}.$$

3.2.1. Wolkowicz & Styan (1979). Possibly the simplest proof of (3.12) is that presented in 1979 by Wolkowicz and Styan [135]. We use our Lemma 2.6.1 above, a version of the Cauchy-Schwarz inequality given by Wolkowicz and Styan (Lemma 2.1 in [137], p. 475):

$$(3.13) \qquad\qquad -s\sqrt{n}\,\mathbf{w'Cw} \le \mathbf{w'Cx} \le s\sqrt{n}\,\mathbf{w'Cw},$$

where $\mathbf{w}$ and $\mathbf{x}$ are real nonnull $n \times 1$ vectors and the centering matrix $\mathbf{C} = \mathbf{I} - \mathbf{ee'}/n$ as in (2.19), with $\mathbf{e}$ the $n \times 1$ vector of ones. Equality holds on the left (right) of (2.22) if and only if

$$(3.14) \qquad\qquad \mathbf{x} = c\mathbf{w} + d\mathbf{e}$$

for some scalars c and d with $c < 0$ ($c > 0$).

Now let $\mathbf{w} = \sum_{i=k}^{l} \mathbf{e}_i/(l-k+1)$ and $\mathbf{x} = \{x_{(i)}\}$, where $\mathbf{e}_i$ is defined as in (2.25) above and

$$(3.15) \qquad x_{\min} = x_{(n)} \le x_{(n-1)} \le \cdots \le x_{(2)} \le x_{(1)} = x_{\max}.$$

Then $\mathbf{w'Cx} = \bar{x}_{(k,l)} - \bar{x}$, where the "subsample mean"

$$(3.16) \qquad \bar{x}_{(k,l)} = \sum_{i=k}^{l} x_{(i)}/(l-k+1) \qquad \text{for} \quad 1 \le l \le k \le n.$$

Moreover, $\mathbf{w'Cw} = (l-k+1)^{-1} - n^{-1}$. Hence (3.13) implies

$$(3.17) \qquad \bar{x} - s\sqrt{\tfrac{k-1}{n-k+1}} \le \bar{x}_{(k,n)} \le \bar{x}_{(k,l)} \le \bar{x}_{(1,l)} \le \bar{x} + s\sqrt{\tfrac{n-l}{l}}$$

which, when $l = k$, reduces to

$$(3.18) \qquad \bar{x} - s\sqrt{\tfrac{k-1}{n-k+1}} \le x_{(k)} \le \bar{x} + s\sqrt{\tfrac{n-k}{k}}$$

as in (3.12). From (3.14) we note that equality holds in (3.12) if and only if $\mathbf{x} = c\mathbf{w} + d\mathbf{e}$ for some scalars c and d. The equality conditions for (3.12) follow at once. □

4. Some Matrix-theoretic Extensions Related to the Cauchy-Schwarz and Laguerre-Samuelson Inequalities

4.1. Bounds for Eigenvalues. When the real $n \times n$ matrix $\mathbf{A}$ has all its eigenvalues real, e.g., when $\mathbf{A}$ is symmetric, then the Laguerre-Samuelson, Brunk and Boyd-Hawkins inequalities provide bounds for the eigenvalues of $\mathbf{A}$ as observed by Wolkowicz and Styan [137], [138]; see also, e.g., Merikoski [66], Merikoski, Styan and Wolkowicz [68], Merikoski and Virtanen [69], Merikoski and Wolkowicz [70], and Tarazaga [129].

As Mirsky [75] and Brauer and Mewborn [19] pointed out, the mean and variance of the eigenvalues λ_i may be expressed in terms of the trace of $\mathbf{A}$ and the trace of $\mathbf{A}^2$:

$$(4.1) \qquad m = \frac{1}{n} \sum_{i=1}^{n} \lambda_i = \frac{1}{n} \text{tr} \mathbf{A}$$

and

$$(4.2) \qquad s^2 = \frac{1}{n} \sum_{i=1}^{n} \lambda_i^2 - \left(\frac{1}{n} \sum_{i=1}^{n} \lambda_i \right)^2 = \frac{1}{n} \text{tr} \mathbf{A}^2 - \left(\frac{1}{n} \text{tr} \mathbf{A} \right)^2.$$

Then from (1.13) and (1.14) we obtain:

$$(4.3) \qquad m + \frac{s}{\sqrt{n-1}} \le \lambda_{\max} = \lambda_1 \le m + s\sqrt{n-1}$$

and

$$(4.4) \qquad m - s\sqrt{n-1} \le \lambda_{\min} = \lambda_n \le m - \frac{s}{\sqrt{n-1}},$$

while from (1.15):

$$(4.5) \qquad m - s\sqrt{\frac{k-1}{n-k+1}} \le \lambda_k \le m + s\sqrt{\frac{n-k}{k}} \qquad \text{for } k = 2, \ldots, n-1,$$

where λ_k is the k-th largest eigenvalue of $\mathbf{A}, k = 2, \ldots, n-1$.

4.2. Some Matrix Inequalities Related to the Cauchy-Schwarz and Laguerre-Samuelson Inequalities. Two of our eight proofs of the Laguerre-Samuelson inequality were based explicitly on the Cauchy-Schwarz inequality which, as we noted at the end of Section 1, "may be perceived to be lurking in the background of many of the proofs"[17] of the Laguerre-Samuelson inequality, cf. §2.4, §2.6, and Lemma 2.6.1. Moreover, the discussion in [10], [24], [25], [91], and [131] of the proof given in §2.8 is all centered around the Cauchy-Schwarz inequality.

In their 1996 paper Pečarić, Puntanen and Styan [103] presented the following matrix-theoretic extension of the Cauchy-Schwarz inequality; here a g-inverse (generalized inverse) $\mathbf{X}^-$ is any matrix $\mathbf{X}^-$ such that $\mathbf{X}\mathbf{X}^-\mathbf{X} = \mathbf{X}$.

Theorem 4.1. *Let $\mathbf{A}$ be an $n \times n$ symmetric and nonnegative definite matrix with $\mathbf{A}^{\{p\}}$ defined as*

$$\mathbf{A}^{\{p\}} \;=\; \mathbf{A}^p; \quad p = 1, 2, \ldots,$$

$$=\; \mathbf{P_A} = \mathbf{A}(\mathbf{A}'\mathbf{A})^-\mathbf{A}'; \quad p = 0,$$

$$=\; (\mathbf{A}^+)^{|p|}; \quad p = -1, -2, \ldots,$$

where $\mathbf{A}^+$ is the (unique) Moore-Penrose inverse of $\mathbf{A}$, and $\mathbf{P_A}$ denotes the orthogonal projector onto the column space $\mathcal{C}(\mathbf{A})$ of $\mathbf{A}$. Let $\mathbf{t}$ and $\mathbf{u}$ be $n \times 1$ vectors, and let h and k be integers. Then

$$(4.6) \qquad (\mathbf{t}'\mathbf{A}^{\{(h+k)/2\}}\mathbf{u})^2 \leq \mathbf{t}'\mathbf{A}^{\{h\}}\mathbf{t} \cdot \mathbf{u}'\mathbf{A}^{\{k\}}\mathbf{u}$$

for $h, k = \ldots, -1, 0, 1, 2, \ldots$, with equality if and only if

$$(4.7) \qquad \mathbf{A}\mathbf{t} \propto \mathbf{A}^{\{1+(k-h)/2\}}\mathbf{u}.$$

Several extensions of the Theorem 5.1.1 and some statistical applications are also given in Pečarić, Puntanen and Styan [103].

When $h = 1$ and $k = -1$, then the inequality (4.6) becomes

$$(4.8) \qquad (\mathbf{t}'\mathbf{P_A}\mathbf{u})^2 \leq \mathbf{t}'\mathbf{A}\mathbf{t} \cdot \mathbf{u}'\mathbf{A}^+\mathbf{u},$$

cf. Bancroft [10].

Equality holds in (4.8) if and only if

$$(4.9) \qquad \mathbf{A}\mathbf{t} \propto \mathbf{P_A}\mathbf{u}.$$

When $\mathbf{t} = \mathbf{w}$, $\mathbf{u} = \mathbf{x}$ and $\mathbf{A} = \mathbf{C}$, the centering matrix $\mathbf{I}_n - n^{-1}\mathbf{e}\mathbf{e}'$ as in (2.19), then $\mathbf{A}^+ = \mathbf{P_A} = \mathbf{C}$ and (4.8) becomes

$$(4.10) \qquad (\mathbf{w}'\mathbf{C}\mathbf{x})^2 \leq \mathbf{w}'\mathbf{C}\mathbf{w} \cdot \mathbf{x}'\mathbf{C}\mathbf{x},$$

---

[17] Arnold and Balakrishnan [4], p. 45.

172

which is equivalent to (2.22) in Lemma 2.6.1, and the equality condition (4.9) becomes

$$(4.11) \qquad\qquad \mathbf{Cw} \propto \mathbf{Cx},$$

which is equivalent to (2.23) in Lemma 2.6.1[18].

We may also express (4.8) as

$$(4.12) \qquad (\mathbf{t'u})^2 \le \mathbf{t'At} \cdot \mathbf{u'A^-u} \qquad \text{for all} \quad \mathbf{u} \in \mathcal{C}(\mathbf{A})$$

and for any, and hence for every g-inverse $\mathbf{A^-}$, cf. Neudecker and Liu [91]. The quadratic form $\mathbf{u'A^-u}$ in (4.12) is invariant with respect to the choice of g-inverse $\mathbf{A^-}$ when $\mathbf{u} \in \mathcal{C}(\mathbf{A})$, since then $\mathbf{u} = \mathbf{Av}$ for some $\mathbf{v}$ and so $\mathbf{u'A^-u} = \mathbf{v'AA^-Av} = \mathbf{v'Av} = \mathbf{v'AA^\sim Av}$ for any g-inverse $\mathbf{A^\sim}$. Equality holds in (4.12) if and only if

$$(4.13) \qquad\qquad \mathbf{At} \propto \mathbf{u}.$$

Chaganty [24] presents (4.12) with the Moore-Penrose inverse $\mathbf{A^+}$ instead of a g-inverse $\mathbf{A^-}$ and observes that equality holds in (4.12) when $\mathbf{t} = \mathbf{A^+u}$ which, since $\mathbf{u} \in \mathcal{C}(\mathbf{A})$, implies $\mathbf{At} = \mathbf{u}$, cf. (4.13).

Trenkler [131] observes that Baksalary and Kala [9] showed that

$$(4.14) \qquad (\mathbf{t'u})^2 \le \alpha \mathbf{t'At} \qquad \text{for all} \quad \mathbf{u} \in \mathcal{C}(\mathbf{A})$$

provided that then $\mathbf{u'A^-u} \le \alpha$ for any, and hence for every g-inverse $\mathbf{A^-}$.

If we now let $\mathbf{u} = \mathbf{t} \in \mathcal{C}(\mathbf{A})$, then (4.12) becomes

$$(4.15) \qquad (\mathbf{t't})^2 \le \mathbf{t'At} \cdot \mathbf{t'A^-t} \qquad \text{for all} \quad \mathbf{t} \in \mathcal{C}(\mathbf{A})$$

for any, and hence for every g-inverse $\mathbf{A^-}$; when $\mathbf{t} \ne \mathbf{0}$ then equality holds in (4.15) if and only if $\mathbf{t}$ is an eigenvector of $\mathbf{A}$, cf. Lemma 2.1 of Dey and Gupta [29].

Acknowledgements. This paper is part of the M. Sc. thesis by the first author [44]. Some of the results in this paper were presented at the "One-and-a-half Days Full of Matrices and Statistics" Seminar: McGill University, Montréal, July 2–3, 1998, and some are scheduled for presentation at the "Seventh International Workshop on Matrices and Statistics, in Celebration of T. W. Anderson's 80th Birthday": Fort Lauderdale, Florida, December 11–14, 1998. In a companion paper [45] we study the related von Szökefalvi Nagy-Popoviciu and Nair-Thomson inequalities for the range (spread) and standard deviation. Our first research associated with the Laguerre-Samuelson Inequality began with

---

[18]Since $\mathbf{Cw} = k\mathbf{Cx}$ is equivalent to $\mathbf{x} = (1/k)\mathbf{w} + (k\bar{x} - \bar{w})\mathbf{e}$, where $\bar{w} = \mathbf{w'e}/n$.

joint work by the second author with Henry Wolkowicz in 1978 and this was followed by an ongoing collaboration with Jorma Kaarlo Merikoski, and with Henry Wolkowicz. Our thanks go also to Josip E. Pečarić and Themistocles M. Rassias for drawing our attention to several references, and to Rajendra Bhatia, Chandler Davis, S. W. Drury, Simo Puntanen, Hans Joachim Werner, and Keith J. Worsley for helpful discussions. Much of the biographical information was obtained by visiting the excellent O'Connor-Robertson Internet website [95], while web access to the databases MathSciNet (for *Mathematical Reviews*) and MATH Database (for *Zentralblatt für Mathematik*) has been of great help in compiling our bibliography. This research was supported in part by a research grant from the Natural Sciences and Engineering Research Council of Canada (to the second author).

Bibliography

1. David Abbott, ed., *The Biographical Dictionary of Scientists: Mathematicians*, Blond Educational (an imprint of Muller, Blond & White Limited), London, England, 1985.

2. Manzoor Ahmad, *On polynomials with real zeros*, Canadian Mathematical Bulletin **11** (1968), 237–240 [MR 37:6420, Zbl 165.05501].

3. Barry C. Arnold, *Schwarz, regression, and extreme deviance*, The American Statistician **28** (1974), 22–23 [MR 49:8223, Zbl 365.62070].

4. Barry C. Arnold and N[arayanaswamy] Balakrishnan, *Relations, Bounds and Approximations for Order Statistics*. Lecture Notes in Statistics, vol. 53, Springer-Verlag, New York, 1989 [MR 90i:62061, Zbl 703.62064].

5. Barry C. Arnold and Richard A. Groeneveld, *Bounds for deviations between sample and population statistics*, Biometrika **61** (1974), 387–389 [MR 51:716, Zbl 281.62012].

6. ———, *Bounds on deviations of estimates arising in finite population regression models*, Communications in Statistics–A, Theory and Methods **7** (1978), 1173–1179 [MR 80a:62015, Zbl 398.62056].

7. ———, *Bounds on expectations of linear systematic statistics based on dependent samples*, The Annals of Statistics **7** (1979), 220–223 [MR 80c:62061, Zbl 398.62036]. (Correction: **8** (1980), 1401 [Zbl 465.62043].)

8. ———, *Maximal deviation between sample and population means in finite populations*, Journal of the American Statistical Association **76** (1981), 443–445 [MR 82h:62016, Zbl 462.62012].

9. Jerzy K. Baksalary and R. Kala, *Partial orderings between matrices one of which is of rank one*, Bulletin of the Polish Academy of Sciences, Mathematics (Warsaw) **31** (1983), 5–7 [MR 85j:15016, Zbl 535.15006].

10. Diccon Bancroft, *Comment* (Letter to the Editor about Chaganty [24] and Olkin [96]), The American Statistician **48** (1994), 351.

11. Vic Barnett and Toby Lewis, *Outliers in Statistical Data*, Third Edition, Wiley, Chichester, 1994 [Zbl 801.62001].

12. Edwin F. Beckenbach and Richard Bellman, *Inequalities* [in English], Fourth [Revised] Printing. Ergebnisse der Mathematik und ihre Grenzgebiete, neue Folge, Heft 30, Springer-Verlag, Berlin, 1983 [MR 33:236, Zbl 513.26003].

13. Paul R. Beesack, *On bounds for the range of ordered variates*, Publikacije Elektrotehničkog Fakulteta Univerziteta u Beogradu, Serija Matematika i Fizika (Belgrade) **428** (1973), 93–96 [MR 48:5257, Zbl 274.62031].

14. _____, *On bounds for the range of ordered variates: II*, Aequationes Mathematicae **14** (1976), 293–301 [MR 54:2902, Zbl 336.62035].

15. Margherita Benzi, *Un "probabilista neoclassico": Francesco Paolo Cantelli* [in Italian with English and German summaries], Historia Mathematica **15** (1988), 53–72 [MR 89k:01022, Zbl 642.01012].

16. J[onathan] M. Borwein, G. P. H. Styan and H. Wolkowicz, *Some inequalities involving statistical expressions: Solution to Problem 81-10* (posed by Foster [32]). SIAM Review **24** (1982), 340–342. (Reprinted in *Problems in Applied Mathematics: Selections from SIAM Review* (Murray S. Klamkin, ed.), SIAM, Philadelphia, 1990, pp. 373–375.)

17. V[iktor Yakovlevich] Bouniakowsky [Buniakovski, Bunyakovsky], *Sur quelques inégalités concernant les intégrales ordinaires et les intégrales aux différences finies* [in French], Mémoires de l'Académie Impériale des Sciences de St.-Pétersbourg, Septième Série **1**, no. 9 (1859), pp. 1–18. (Cf. pp. 3–4.)

18. A. V. Boyd, *Bounds for order statistics*, Publikacije Elektrotehničkog Fakulteta Univerziteta u Beogradu, Serija Matematika i Fizika (Belgrade) **365** (1971), 31–32 [MR 46:9256, Zbl 245.62052].

19. Alfred Brauer and A. C. Mewborn, *The greatest distance between two characteristic roots of a matrix*, Duke Mathematical Journal **26** (1959), 653–661 [MR 22:10997, Zbl 095.01202].

20. Claude Brezinski, *Edmond Nicolas Laguerre* [in French], Polynômes Orthogonaux et Applications: Proceedings of the Laguerre Symposium held at Bar-le-Duc, October 15–18, 1984 (C. Brezinski, A. Draux, A. P. Magnus, P. Maroni and A. Ronveaux, eds.), Lecture Notes in Mathematics, vol. 1171, Springer-Verlag, Berlin, 1985, pp. xxi–xxvi.

21. K. A. Brownlee, *Statistical Theory and Methodology in Science and Engineering*, Second Edition, Wiley, New York, 1965 [Zbl 136.39203].

22. H. D. Brunk, *Note on two papers of K. R. Nair*, Journal of the Indian Society of Agricultural Statistics **11** (1959), 186–189.

23. [Baron] Augustin-Louis Cauchy, *Note II: Sur les formules qui résultent de l'emploi du signe > ou <, et sur les moyennes entre plusieurs quantités* [in French], Cours d'Analyse de l'École Royale Polytechnique, Première Partie: Analyse Algébrique. L'Imprimerie Royale Chez Debure Frères, Libraires du Roi et de la Bibliothèque du Roi, Paris, 1821, pp. 360–377. (Cf. pp. 373–374. Reprinted in *Œuvres Complètes d'Augustin Cauchy, Publiées sous la direction scientifique de l'Académie des Sciences et sous les auspices de M. le Ministre de l'Instruction Publique, Seconde Série: Mémoires Divers et Ouvrages*, vol. 3 (Mémoires publiés en corps d'ouvrage), Gauthier-Villars, Paris, 1897, pp. 360–377.)

24. N. Rao Chaganty, *Comment* (Letter to the Editor about Olkin [96]), The American Statistician **47** (1993), 158.

25. N. Rao Chaganty and Akhil K. Vaish, *Response* (Letter to the Editor about Bancroft [10] & Neudecker and Liu [91]), The American Statistician **48** (1994), 351–352.

26. _____, *On inequalities for outlier detection in statistical data analysis*, Journal of Applied Statistical Science **6** (1997), 235–243 [Zbl 980.21632].

175

27. Herbert A. David, *Order Statistics*, Second Edition. Wiley, New York, 1981 [MR 82i:62073, Zbl 553.62046].

28. _____, *General bounds and inequalities in order statistics*, Communications in Statistics–Theory and Methods **17** (1988), 2119–2134 [MR 89j:62068, Zbl 639.62042].

29. A. Dey and S. C. Gupta, *Singular weighing designs and estimation of total weight*, Communications in Statistics–A, Theory and Methods **6** (1977), 289–295 [MR 55:9435, Zbl 362.62034].

30. Meyer Dwass, *The extreme deviations inequality* (Letter to the Editor about Quesenberry [107] and Samuelson [117]), The American Statistician **29** (1975), 108.

31. Nicholas R. Farnum, *An alternate proof of Samuelson's inequality and its extensions*, The American Statistician **43** (1989), 46–47.

32. L. V. Foster, *Some inequalities involving statistical expressions: Problem 81-10*, SIAM Review **23** (1981), 256. (Problem solved by Borwein, Styan and Wolkowicz [16].)

33. Florin Gonzacenco, *Inegalități probabiliste și aplicații în statistică* [in Romanian with English summary: *Some probabilistic inequalities and their statistical applications*], Studii și Cercetări Matematice (Bucharest) **38** (1986), 290–297 [MR 88d:60054, Zbl 639.60025].

34. F[lorin] Gonzacenco and E[ugen] Mărgăritescu, *Best bounds for order statistics* [in English], Bulletin Mathématique de la Société des Sciences Mathématiques de la République Socialiste de Roumanie, Nouvelle Série (Bucharest) **31**, no. 79 (1987), 303–311 [MR 90h:62116, Zbl 634.62043].

35. Florin Gonzacenco, Eugen Mărgăritescu and Viorel Gh. Vodă, *Statistical consequences of some old and new algebraic inequalities* [in English], Revue Roumaine de Mathématiques Pures et Appliquées (Bucharest) **37** (1992), 877–886 [MR 93k:60048, Zbl 777.62104].

36. Richard A. Groeneveld, *Best bounds for order statistics and their expectations in range and mean units with applications*, Communications in Statistics–Theory and Methods **11** (1982), 1809–1815 [MR 83j:62070, Zbl 506.62031].

37. Anna Hart, *Standard deviation*, Encyclopedia of Statistical Sciences, Volume 8: Regressograms–St. Petersburg Paradox (Samuel Kotz, Norman L. Johnson and Campbell B. Read, eds.), Wiley, New York, 1988, pp. 625–629.

38. H. O. Hartley and H. A. David, *Universal bounds for mean range and extreme observation*, The Annals of Mathematical Statistics **25** (1954), 85–99 [Zbl 055.12801].

39. Hiroshi Haruki and Themistocles M. Rassias, *New integral representations for Bernoulli and Euler polynomials*, Journal of Mathematical Analysis and Applications **175** (1993), 81–90 [MR 94e:39016, Zbl 776.11009].

40. David A. Harville, *Matrix Algebra from a Statistician's Perspective*, Springer-Verlag, New York, 1997 [Zbl 881.15001].

41. Douglas M. Hawkins, *On the bounds of the range of order statistics*, Journal of the American Statistical Association **66** (1971), 644–645 [Zbl 228.62030].

42. [Friedrich Robert] Helmert, *Die Genauigkeit der Formel von Peters zur Berechnung des wahrscheinlichen Beobachtungsfehlers directer Beobachtungen gleicher Genauigkeit* [in German], Astronomische Nachrichten **88** (1876), 115–132.

43. Constatin Ionescu-Țiu, Problemei 15356 [in Romanian], Gazeta Matematică, ca. 1983. (Problem solved by Lupaș [54].)

176

44. Shane Tyler Jensen. *The Laguerre-Samuelson Inequality, with Extensions and Applications in Statistics and Matrix Theory.* M. Sc. thesis, Dept. of Mathematics and Statistics, McGill University, Montréal, 1998, in preparation.

45. Shane T. Jensen and George P. H. Styan. *Some comments and a bibliography on the von Szökefalvi Nagy-Popoviciu and Nair-Thomson inequalities, with extensions and applications in statistics and matrix theory.* Dept. of Mathematics and Statistics, McGill University, in preparation for presentation at the "Seventh International Workshop on Matrices and Statistics, in Celebration of T. W. Anderson's 80th Birthday": Fort Lauderdale, Florida, December 11–14, 1998.

46. D. G. Kabe, *On extensions of Samuelson's inequality* (Letter to the Editor about Wolkowicz and Styan [135]), The American Statistician **34** (1980), 249.

47. Oscar Kempthorne, *Personal communication*, 1973. (Cf. Arnold [3] and Arnold and Balakrishnan [4], pp. 45, 158.)

48. W. Kirby, *Algebraic boundedness of sample statistics*, Water Resources Research **10** (1974), 220–222.

49. Ludvik Kraus, *Poznámka k rovnicím, jež mají pouze realné kořeny* [in Czech: With a "Dodatek" by Ed. Weyr], Časopis pro Pěestování Matematiky a Fysiky: Část Vědecká (Prague) **15** (1886), 63–64.

50. [Edmond Nicolas] Laguerre, *Sur une méthode pour obtenir par approximation les racines d'une équation algébrique qui a toutes ses racines réelles* [in French]. Nouvelles Annales de Mathématiques (Paris), 2e Série **19** (1880), 161–171 & 193–202 [JFM 12:71]. (Reprinted in *Œuvres de Laguerre* (publiées sous les auspices de l'Académie des Sciences par MM. Ch. Hermite, H. Poincaré et E. Rouché), Gauthier-Villars, Paris, **1** (1898), 87–103.)

51. H. O. Lancaster, *The Helmert matrices*, The American Mathematical Monthly **72** (1965), 4–12 [MR 30:1134, Zbl 124.01102].

52. Alexandru Lupaş, *Problem 246* [in English], Matematicki Vesnik (Belgrade) **8**, no. 23 (1971), 333.

53. _____, *Inequalities for the roots of a class of polynomials* [in English], Publikacije Elektrotehničkog Fakulteta Univerziteta u Beogradu, Serija Matematika i Fizika (Belgrade) **594** (1977), 79–85 [Zbl 371.26009].

54. _____, *Asupra problemei 15356* (posed by Ionescu-Țiu [43]) [in Romanian], Gazeta Matematică **5**, no. 1/2 (1984), 56–60.

55. B. S. Madhava Rao and B. S. Sastry, *On the limits for the roots of a polynomial equation*, Journal of the Mysore University, Section B **1** (1940), 5–8 [MR 2:241d].

56. C[olin] L. Mallows and Donald Richter, *Inequalities of Chebyshev type involving conditional expectations*, The Annals of Mathematical Statistics **40** (1979), 1922–1932 [MR 40:6617, Zbl 187.14903].

57. E. Malo, *Note sur les équations algébriques dont toutes les racines sont réelles* [in French], Journal de Mathématiques Spéciales, 4e Série **4** (1894), 7–10 [JFM 26:120].

58. Eugen Mărgăritescu, *Bornes strictes pour les statistiques d'ordre et inégalités de type Samuelson* [in French], Revue Roumaine de Mathématiques Pures et Appliquées (Bucharest) **32** (1987), 343–349 [MR 88m:60043].

59. Eugen Mărgăritescu and Viorel Gh. Vodă, *O inegalitate algebrică și cîteva aplicații în statistică* [in Romanian with English summary: *An algebraic inequality and some*

applications in statistics], Studii şi Cercetări Matematice (Bucharest) **35** (1983), 376–387 [MR 85g:62125, Zbl 518.62059].

60. Thomas Mathew and Kenneth Nordström, *An inequality for a measure of deviation in linear models*, The American Statistician **51** (1997), 344–349 [MR 98g:62131].

61. A. T. McKay, *The distribution of the difference between the extreme observation and the sample mean in samples of n from a normal universe*, Biometrika **27** (1935), 466–471 [Zbl 013.03002].

62. A. T. McKay and E. S. Pearson, *A note on the distribution of range in samples of n*, Biometrika **25** (1933), 415–420.

63. A. Meir and A. Sharma, *Span of derivatives of polynomials*, The American Mathematical Monthly **74** (1967), 527–531 [MR 35:5561].

64. _____, *On zeros of derivatives of polynomials*, Canadian Mathematical Bulletin **11** (1968), 443–445 [MR 38:2285, Zbl 164.06405]. (Addendum: **11** (1968), 611 [Zbl 167.33101].)

65. Eugenio Melilli, *Cantelli, Francesco Paolo*, Leading Personalities in Statistical Sciences: From the Seventeenth Century to the Present (Norman L. Johnson and Samuel Kotz, eds.), Wiley, New York, 1997, pp. 228–232.

66. Jorma Kaarlo Merikoski, *On the trace and the sum of elements of a matrix*, Linear Algebra and Its Applications **60** (1984), 177–185 [MR 85h:15012, Zbl 559.15004].

67. Jorma Kaarlo Merikoski, Humberto Sarria and Pablo Tarazaga, *Bounds for singular values using traces*, Linear Algebra and Its Applications **210** (1994), 227–254 [MR 95h:15030, Zbl 813.15016].

68. Jorma Kaarlo Merikoski, George P. H. Styan and Henry Wolkowicz, *Bounds for ratios of eigenvalues using traces*, Linear Algebra and Its Applications **55** (1983), 105–124 [MR 85a:15019, Zbl 522:15008].

69. Jorma Kaarlo Merikoski and Ari Virtanen, *Bounds for eigenvalues using the trace and determinant*, Linear Algebra and Its Applications **264** (1997), 101–108 [Zbl 885.15011].

70. Jorma Kaarlo Merikoski and Henry Wolkowicz, *Improving eigenvalue bounds using extra bounds*, Linear Algebra and Its Applications **68** (1985), 93–113 [MR 87e:15038, Zbl 574.15010].

71. N[icolae] Mihăileanu. *Istoria Matematicii* [in Romanian: *History of Mathematics*], Editura Ştiinţifică şi Enciclopedică, Bucharest, 1981.

72. G[radimir] V. Milovanović, D[ragoslav] S. Mitrinović and Th[emistocles] M. Rassias, *Topics in Polynomials: Extremal Problems, Inequalities, Zeros*, World Scientific, Singapore, 1994 [MR 95m:30009, Zbl 848.26001].

73. G[radimir] V. Milovanović and Th[emistocles] M. Rassias, *Inequalities connected with trigonometric sums*, Constantin Carathéodory: An International Tribute (Themistocles M. Rassias, ed.), vol. 2, World Scientific, Singapore, 1991, pp. 875–941 [MR 93g:42002, Zbl 754.26008].

74. _____, *On the Markov-Duffin-Schaeffer inequalities*, Journal of Natural Geometry **5** (1994), 29–41 [MR 94h:26017, Zbl 788.26011].

75. L[eon] Mirsky, *The spread of a matrix*, Mathematika (London) **3** (1956), 127–130 [MR 18:460c, Zbl 073.00903].

76. _____, *Inequalities for normal and Hermitian matrices*, Duke Mathematical Journal **24** (1957), 591–599 [MR 19:832c, Zbl 081.25101].

77. Dragoslav S. Mitrinović, *Inégalités impliquées par le système des égalités* $a + b + c = p, bc + ca + ab = q$ [in French], Publikacije Elektrotehničkog Fakulteta Univerziteta u Beogradu, Serija Matematika i Fizika (Belgrade) **143–155** (1965), 5–7 [Zbl 184:00085].

78. ———, *Analytic Inequalities*, In cooperation with Petar M. Vasić, Die Grundlehren der mathematischen Wissenschaften in Einzeldarstellungen mit besonderer Berücksichtigung der Anwendungsgebiete, vol. 165, Springer-Verlag, New York, 1970 [MR 43:448, Zbl 199.38101]. (Cf. pp. 210–211. Translated into Serbo-Croatian and updated: Mitrinović [79]. Addenda by Mitrinović and Vasić [82] and by Mitrinović Lacković and Stanković [80].)

79. ———, *Analitičke Nejednakosti* [in Serbo-Croatian: *Analytic Inequalities*]. In cooperation with Petar M. Vasić, Translation of Mitrinović [78] from the English and updated, University of Belgrade Monographs, vol. 2, Građevinska Knjiga, Belgrade, 1970.

80. Dragoslav S. Mitrinović, Ivan B. Lacković and Miomir S. Stanković, *Addenda to the monograph* Analytic Inequalities *by Mitrinović* [78] *–II: On some convex sequences connected with N. Ozeki's results* [in English], Publikacije Elektrotehničkog Fakulteta Univerziteta u Beogradu, Serija Matematika i Fizika (Belgrade) **634–677** (1979), 3–24 [MR 82c:26011, Zbl 452.26007].

81. Dragoslav S. Mitrinović, Josip E. Pečarić and Arlington M. Fink, *Classical and New Inequalities in Analysis*, Mathematics and Its Applications: East European Series, vol. 61, Kluwer, Dordrecht, 1993 [MR 94c:00004, Zbl 771.26009].

82. Dragoslav S. Mitrinović and Petar M. Vasić, *Addenda to the monograph* Analytic Inequalities *by Mitrinović* [78]*–I* [in English], Publikacije Elektrotehničkog Fakulteta Univerziteta u Beogradu, Serija Matematika i Fizika (Belgrade) **577–598** (1977), 3–10 [MR 56:5818, Zbl 369.26010].

83. Vedula N. Murty, *A note on Samuelson's inequality* (Letter to the Editor about Farnum [31],) The American Statistician **44** (1990), 64–65.

84. Keshavan Raghavan Nair, Ph. D. thesis, University College London, 1947. (Parts published as Nair [85] and [86].)

85. ———, *Certain symmetrical properties of unbiased estimates of variance and covariance*, Journal of the Indian Society of Agricultural Statistics **1** (1948), 162–172 [MR 11:448a]. (Part of the thesis [84].)

86. ———, *The distribution of the extreme deviate from the sample mean and its Studentized form*, Biometrika **35** (1948), 118–144 [MR 9:602a]. (Part of the thesis [84].)

87. ———, *A note on the estimation of mean rate of change*, Journal of the Indian Society of Agricultural Statistics **8** (1956), 122–124.

88. ———, *A tail-piece to Brunk's paper* [22], Journal of the Indian Society of Agricultural Statistics **11** (1959), 189–190.

89. ———, *On extensions of Samuelson's inequality* (Letter to the Editor about Wolkowicz and Styan [135]), The American Statistician **34** (1980), 249–250.

90. ———, *In statistics by design*, Glimpses of India's Statistical Heritage (J. K. Ghosh, S. K. Mitra and K. R. Parthasarathy, eds.), Wiley Eastern, New Delhi, 1992, pp. 101–150 [MR 94c:62001, Zbl 832.01016].

91. Heinz Neudecker and Shuangzhe Liu, *Comment* (Letter to the Editor about Chaganty [24] and Olkin [96]), The American Statistician **48** (1994), 351.

92. Constantin Nicolau, *Recherches nouvelles sur les équations algébriques* [in French], Annales Scientifiques de l'Université de Jassy (Iaşi) **18** (1933), 15–69 [Zbl 007.39100].

93. ———, *Un théorème sur les équations algébriques* [in French], Comptes Rendus Hebdomadaires des Séances de l'Académie des Sciences (Paris) **208** (1939), 1958–1960 [Zbl 021.19703].

94. B. Niewenglowski, [*Problem No.*] *244* [in French], L'Intermédiaire des Mathématiciens (Paris) **1** (1894), 132–133.

95. John J. O'Connor and Edmund F. Robertson, *MacTutor History of Mathematics Archive:* http://www-history.mcs.st-andrews.ac.uk/history/BiogIndex.html.

96. Ingram Olkin, *A matrix formulation on how deviant an observation can be*, The American Statistician **46** (1992), 205–209.

97. Federico J. O'Reilly, *On a criterion for extrapolation in normal regression*, The Annals of Mathematical Statistics **3** (1975), 219–222 [Zbl 305.62040].

98. ———, *The extreme deviations inequality* (Letter to the Editor about Arnold [3], Dwass [30] and Quesenberry [107]), The American Statistician **30** (1976), 103.

99. Jagdish K. Patel, C. H. Kapadia and D[onald] B. Owen, *Handbook of Statistical Distributions*, Statistics: Textbooks and Monographs, vol. 20, Marcel Dekker, New York, 1976 [MR 56:9771, Zbl 367.62014].

100. E. S. Pearson, *"Student" as statistician*. Biometrika **30** (1938), 210–250 [Zbl 020.04005].

101. E. S. Pearson, assisted by C. Chandra Sekar, *The efficiency of statistical tools and a criterion for the rejection of outlying observations*, Biometrika **28** (1936), 308–320 [Zbl 015.26205]. (Includes Appendix by Scott [122].)

102. Josip E. Pečarić, *O nekim nejednakostima za korjene jedne klase polinoma* [in Serbo-Croatian with English summary: *On some inequalities for the roots of a class of polynomials*], Zbornik Fakulteta za Pomorstvo u Kotoru (Kotor) **9/10** (1983/1984), 123–129.

103. Josip E. Pečarić, Simo Puntanen and George P. H. Styan, *Some further matrix extensions of the Cauchy-Schwarz and Kantorovich inequalities, with some statistical applications*, Linear Algebra and Its Applications **237/238** (1996), 455–476 [MR 97c:15035, Zbl 860.15021].

104. Tiberiu Popoviciu, *Sur les équations algébriques ayant toutes leurs racines réelles* [in French], Mathematica (Cluj) **9** (1935), 129–145 [Zbl 014.10003].

105. P. Prescott, Letter quoted in *"Editor's Note"* (about Letters to the Editor by Kabe [46], Nair [89], and Wolkowicz and Styan [136]), The American Statistician **34** (1980), 251.

106. Simo Puntanen, *Matriiseja Tilastotieteilijälle* [in Finnish: *Matrices for the Statistician*], Preliminary Edition, Report B47, Dept. of Mathematical Sciences, University of Tampere, Tampere, Finland, 1998.

107. C. P. Quesenberry, *The extreme deviations inequality* (Letter to the Editor about Samuelson [117] and Arnold [3]), The American Statistician **28** (1974), 112.

108. C. P. Quesenberry and H[erbert] A. David, *Some tests for outliers*, Biometrika **48** (1961), 379–390 [Zbl 108.15604].

109. Themistocles M. Rassias, *A new inequality for complex-valued polynomial functions*, Proceedings of the American Mathematical Society **97** (1986), 296–298 [MR 87h:30005, Zbl 593.30004].

180

110. _____, *On certain properties of polynomials and their derivative*, Topics in Mathematical Analysis: A Volume Dedicated to the Memory of A. L. Cauchy (Themistocles M. Rassias, ed.), Series in Pure Mathematics, vol. 11, World Scientific, Singapore, 1989, pp. 758–802 [MR 92m:30009, Zbl 732.30005].

111. _____, *On polynomial inequalities and extremal problems*, General Inequalities: 6 (Oberwolfach 1990), International Series in Numerical Mathematics, vol. 103, Birkhäuser, Basel, 1992, pp. 161–174. [MR 94i:26008, Zbl 778.30009].

112. Th[emistocles] M. Rassias and Jaromír Šimša, *Finite Sums Decompositions in Mathematical Analysis*, Wiley, Chichester, 1995 [MR 96k:26006, Zbl 859.26005].

113. Th[emistocles] M. Rassias, S. N. Singh and H[ari] M. Srivastava, *Some q-generating functions associated with basic multiple hypergeometric series*, Computers & Mathematics with Applications **27** (1994), 33–39 [MR 95m:33018, Zbl 792.33013].

114. Th[emistocles] M. Rassias and H[ari] M. Srivastava, *Some general families of generating functions for the Laguerre polynomials*, Journal of Mathematical Analysis and Applications **174** (1993), 528–538 [MR 94g:33004, Zbl 787.33006].

115. Th[emistocles] M. Rassias, H[ari] M. Srivastava, and A. Yanushauskas (eds), *Topics in Polynomials of One and Several Variables and their Applications: Dedicated to the Memory of P. L. Chebyshev (1821–1894)*, World Scientific, Singapore, 1993 [MR 94m:00020, Zbl 849.00029].

116. Campbell B. Read. *Helmert, Friedrich Robert*, Leading Personalities in Statistical Sciences: From the Seventeenth Century to the Present (Norman L. Johnson and Samuel Kotz, eds.), Wiley, New York, 1997, pp. 42–43.

117. Paul A. Samuelson, *How deviant can you be?* Journal of the American Statistical Association **63** (1968), 1522–1525. (Reprinted in *The Collected Scientific Papers of Paul A. Samuelson*, vol. 3 (Robert C. Merton, ed.), MIT Press, Cambridge, Mass., 1972, pp. 340–343.)

118. _____, Unpublished letter to R. B. Murphy, 5 February 1980 [mentions "an Australian anticipation"]. (Cited by Wolkowicz and Styan [136].)

119. _____, *Economics in my time*, Lives of the Laureates: Thirteen Nobel Economists (William Breit and Roger W. Spencer, eds.), Third Edition, MIT Press, Cambridge, Mass., 1995, pp. 59–77.

120. Issai Schur, *Zwei Sätze über algebraische Gleichungen mit lauter reellen Wurzeln* [in German], Journal für die reine und angwandte Mathematik **144** (1914), 75–88 [JFM 45:169]. (Reprinted in *Issai Schur: Gesammelte Arbeiten*, vol. 2 (Alfred Brauer and Hans Rohrbach, eds.), Springer-Verlag, Berlin, 1973, pp. 56–69. Original version has author's name given as J. Schur.)

121. [Karl] H[ermann] A[mandus] Schwarz, *Ueber ein die Flächen kleinsten Flächeninhalts betreffendes Problem der Variationsrechnung: Festschrift zum Jubelgeburtstage des Herrn Karl Weierstrass* [in German], Acta Societatis Scientiarum Fennicæ (Helsinki) **15** (1888), 315–362. (Cf. pp. 343–345. Preface dated 31 October 1885. Reprinted in *Gesammelte Mathematische Abhandlungen von H. A. Schwarz*, vol. 1, Julius Springer, Berlin, 1890, pp. 223–270; cf. pp. 251–253.)

122. J. M. C. Scott, *Appendix*, In Pearson and Chandra Sekar [101], 1936, pp. 319–320.

123. William P. Smith. Letter quoted in *"Editor's Note"* (about Letters to the Editor by Kabe [46], Nair [89], and Wolkowicz and Styan [136]), The American Statistician **34** (1980), 251. (Cf. [4], pp. 46, 161. See also Prescott [105].)

124. Stephen M. Stigler, *The History of Statistics: The Measurement of Uncertainty before 1900*, The Belknap Press of Harvard University Press, Cambridge, Mass., and London, England, 1986.

125. Alan Stuart and J. Keith Ord, *Kendall's Advanced Theory of Statistics, Volume 1: Distribution Theory*, Sixth Edition, Edward Arnold, London, and Halsted Press, Wiley, New York, 1994.

126. Julius von Szökefalvi Nagy [Gyula Szökefalvi-Nagy], *Über algebraische Gleichungen mit lauter reellen Wurzeln* [in German], Jahresbericht der Deutschen Mathematiker-Vereinigung **27** (1918), 37–43 [JFM 46:125].

127. _____, Review of Nicolau [92], Zentralblatt für Mathematik und ihre Grenzgebiete **7** (1934), 391.

128. _____, Review of Nicolau [93], Zentralblatt für Mathematik und ihre Grenzgebiete **21** (1939), 197.

129. Pablo Tarazaga, *Eigenvalue estimates for symmetric matrices*, Linear Algebra and Its Applications **135** (1990), 171–179 [MR 91d:15036, Zbl 701.15012].

130. William R. Thompson, *On a criterion for the rejection of observations and the distribution of the ratio of deviation to sample standard deviation*, The Annals of Mathematical Statistics **6** (1935), 214–219 [Zbl 012.41102].

131. G[ötz] Trenkler, *Comment* (Letter to the Editor about Chaganty [24] and Olkin [96]), The American Statistician **48** (1994), 60.

132. Rodica-Cristina Vodă, *Margini pentru amplitudinea Studentizată de selecţie* [in Romanian with English summary: *Bounds for the sample Studentized range*]. Studii şi Cercetări Matematice (Bucharest) **35** (1983), 544–548 [MR 85c:62039, Zbl 549.62020].

133. Geoffrey S. Watson, Gülhan Alpargu and George P. H. Styan, *Some comments on six inequalities associated with the inefficiency of ordinary least squares with one regressor*, Linear Algebra and Its Applications **264** (1997), 13–53 [MR 98i:15023].

134. Heinrich Weber, *Die Sätze von Laguerre für Gleichungen mit nur reellen Wurzeln* [in German], Section 115 in *Lehrbuch der Algebra, Erster Band* , Third Edition, Chelsea, New York, 1961, pp. 364–371. (Original version: 1895/1896 and Second Edition: 1898, both pub. Friedrich Vieweg und Sohn, Braunschweig.)

135. Henry Wolkowicz and George P. H. Styan, *Extensions of Samuelson's inequality*, The American Statistician **33** (1979), 143–144 [MR 80h:62038].

136. _____, *Reply* (to Letters to the Editor by Kabe [46] and Nair [89] about Wolkowicz and Styan [135]), The American Statistician **34** (1980), 250–251.

137. _____, *Bounds for eigenvalues using traces*, Linear Algebra and Its Applications **29** (1980), 471–506 [MR 81k:15015, Zbl 435:15015 *sic*].

138. _____, *More bounds for eigenvalues using traces*, Linear Algebra and Its Applications **31** (1980), 1–17 [MR 81k:15016, Zbl 434:15003].

139. _____, *Samuelson-Nair inequality*, Encyclopedia of Statistical Sciences, Volume 8: Regressograms–St. Petersburg Paradox (Samuel Kotz, Norman L. Johnson and Campbell B. Read, eds.), Wiley, New York, 1988, pp. 258–259 [Zbl 706.62001].

140. *Zentralblatt für Mathematik / Mathematics Abstracts: MATH Database*, European Mathematical Society, FIZ Karlsruhe & Springer-Verlag, Berlin. (Three free hits are currently available at http://www.emis.de/cgi-bin/MATH.)

FRACTIONAL ORDER INEQUALITIES OF HARDY TYPE

ALOIS KUFNER
Mathematical Institute, Czech Academy of Sciences
Žitná 25, 115 67 Praha 1, Czech Republic

Abstract. Some necessary and sufficient conditions on the weight functions are given which allow to estimate the weighted norm of the fractional order λ, $0 < \lambda < 1$, by the weighted norm of the first order derivative.

1. Introduction

The Hardy inequality

$$(1.1) \qquad \left(\int_a^b |f(x)|^q u(x)dx \right)^{1/q} \le C \left(\int_a^b |f'(x)|^p v(x)dx \right)^{1/p}$$

allows to estimate the function f (in terms of a weighted L^q-norm) by its *derivative* (of first order, in terms of a weighted L^p-norm). If we rewrite (1.1) in the form

$$(1.2) \qquad \|f\|_{q,u} \le C\|f'\|_{p,v}$$

where $\|g\|_{r,w}$ denotes the norm in the weighted Lebesgue space $L^r(a, b; w) = L^r(w)$:

$$\|g\|_{r,w} = \left(\int_a^b |g(x)|^r w(x)dx \right)^{1/r},$$

the following natural question arises: Whether it possible to estimate (i) the function f by a *fractional order derivative* $f^{(\lambda)}$ with $0 < \lambda < 1$, and (ii) the fractional order derivative $f^{(\lambda)}$ by the first order derivative f'. In other words: Under what conditions we can derive inequalities of the form

$$(1.3) \qquad \|f\|_{q,u} \le C\|f^{(\lambda)}\|_{p,v}$$

and

$$(1.4) \qquad \|f^{(\lambda)}\|_{q,u} \le C\|f'\|_{p,v} \ ?$$

1991 *Mathematics Subject Classification.* Primary 26D10; Secondary 26A33.
Key words and phrases. Hardy inequality; fractional order derivative; weighted difference inequality.

T.M. Rassias and H.M. Srivastava (eds.), Analytic and Geometric Inequalities and Applications, 183–189.
© 1999 Kluwer Academic Publishers.

Following the notation used in the theory of fractional order (non-weighted) Sobolev spaces, where the L^r-(semi)norm of a "fractional order derivative $f^{(\lambda)}$" is defined as

$$(1.5) \qquad \|f^{(\lambda)}\|_r = \Big(\int_a^b \int_a^b \frac{|f(x) - f(y)|^r}{|x - y|^{1+\lambda r}} \, dx \, dy \Big)^{1/r},$$

the symbol $\|f^{(\lambda)}\|_{r,w}$ will be understood as

$$(1.6) \qquad \|f^{(\lambda)}\|_{r,w} = \Big(\int_a^b \int_a^b \frac{|f(x) - f(y)|^r}{|x - y|^{1+\lambda r}} w(x,y) \, dx \, dy \Big)^{1/r},$$

$0 < \lambda < 1$.

Since fractional order Sobolev spaces can be investigated via the theory of interpolation of Banach spaces, it is quite natural to use this theory also for the case of weighted spaces. Attempts to use interpolation theory for the investigation of inequalities of the form (1.3) and (1.4) have been made by Kufner and Triebel [6] and by Kufner and Persson [5], but the results are still not satisfactory enough.

Inequality (1.3) is dealt with in detail e.g. in the paper Heinig, Kufner and Persson [3]. Here, we will concentrate on inequality (1.4), more precisely, on the more general inequality

$$(1.7) \qquad \Big(\int_a^b \int_a^b |f(x) - f(y)|^q W(x,y) \, dy \, dx \Big)^{1/q} \le C \Big(\int_a^b |f'(x)|^p w(x) \, dx \Big)^{1/q}$$

For the sake of completeness, let us collect the assumptions: It is $-\infty \le a < b \le \infty$, $0 < q < \infty$, $1 < p < \infty$, w and W are weight functions, i.e. functions measurable and positive a.e. in (a,b) and in $(a,b) \times (a,b)$, respectively.

Remark 1. A special form of inequality (1.7) was derived in [4], namely

$$\int_0^\infty \int_0^\infty \frac{|f(x) - f(y)|^p}{|x - y|^{1+\lambda p}} \, dx \, dy \le C \int_0^\infty |f'(x)|^p x^{(1-\lambda)p} \, dx,$$

$0 < \lambda < 1$. This inequality is a counterpart to the Jakovlev-Grisvard inequality

$$\int_0^\infty |f(x)|^p x^{-\lambda p} \, dx \le C \int_0^\infty \int_0^\infty \frac{|f(x) - f(y)|^p}{|x - y|^{1+\lambda p}} \, dx \, dy$$

which was derived independently by Jakovlev [2] and Grisvard [1] for $0 < \lambda < 1$, $\lambda \ne 1/p$, and $f \in C_0^\infty(0, \infty)$.

Inequality (1.7) can be rewritten in the following form:

$$(1.8) \qquad \Big(\int_a^b \int_x^b \Big| \int_x^y F(t) \, dt \Big|^q \widetilde{W}(x,y) \, dy \, dx \Big)^{1/q} \le C \Big(\int_a^b F^p(x) w(x) \, dx \Big)^{1/q}$$

with $\widetilde{W}(x,y) = W(x,y) + W(y,x)$ and $F \ge 0$.

Indeed: If we denote $F(x) = f'(x)$, then $f(y) - f(x) = \int_x^y F(t)dt$ and Fubini's theorem yields

$$\int_a^b \int_a^b H(x,y)dydx = \int_a^b \int_a^x H(x,y)dydx + \int_a^b \int_x^b H(x,y)dydx =$$

$$= \int_a^b \int_y^b H(x,y)dxdy + \int_a^b \int_x^b H(x,y)dydx =$$

$$= \int_a^b \int_x^b [H(y,x) + H(x,y)]dydx$$

with $H(x,y) = |f(x) - f(y)|^q W(x,y) = |\int_x^y F(t)dt|^q W(x,y)$.

Remark 2. Obviously, inequality (1.7) can be rewritten also in the form

$$(1.8^*) \qquad \left(\int_a^b \int_a^x \Big| \int_x^y F(t)dt \Big|^q \widetilde{W}(x,y)dydx \right)^{1/q} \leq C \left(\int_a^b F^p(x)w(x)dx \right)^{1/p}.$$

In Section 2, we will formulate some *sufficient* conditions of the validity of inequality (1.7) for both cases $p \leq q$ and $p > q$. The proof follows the ideas of the proof of the corresponding inequality for $a = 0$, $b = \infty$ in Persson and Kufner [7] (where the equivalent inequality (1.8*) is used) and is therefore omitted.

In Section 3, we will give some *necessary and sufficient* condition of the validity of (1.7) using its equivalent form (1.8), for the case $1 < p \leq q < \infty$. The proof follows some ideas of H. Heinig which is here gratefully acknowledged.

2. Some sufficient conditions

Theorem 1. *(i) Let $1 < p \leq q < \infty$ and suppose that there is a weight function $w(x,y)$ on $(a,b) \times (a,b)$ such that*

$$B(x) := \sup_{x < t < b} \left(\int_t^b \widetilde{W}(x,y)dy \right)^{1/q} \left(\int_x^t w^{1-p'}(x,y)dy \right)^{1/p'} < \infty$$

for a.e. $x \in (a,b)$ with $p' = \frac{p}{p-1}$. Then inequality (1.8) - and consequently, inequality (1.7) - holds with

$$w(x) = \left(\int_a^x B^q(t)w^{q/p}(t,x)dt \right)^{p/q}.$$

(ii) Let $0 < q < p < \infty$, $1 < p < \infty$ and suppose that

$$\widetilde{B}(x) := \left(\int_x^b \left(\int_t^x \widetilde{W}(x,y)dy \right)^{r/p} \left(\int_x^t w^{1-p'}(x,y)dy \right)^{r/p'} \widetilde{W}(x,t)dt \right)^{1/r} < \infty$$

for a.e. $x \in (a,b)$ with $\frac{1}{r} = \frac{1}{q} - \frac{1}{p}$, and that

$$\left(\int_a^b \widetilde{B}^r(x)dx \right)^{1/r} < \infty.$$

Then inequality (1.8) holds with

$$w(x) = \left(\int_a^x w(t,x)dt \right).$$

Another simple sufficient condition, which does not depend on the mutual position of the parameters p and q, reads as follows:

Theorem 2. *Let* $1 < p, q < \infty$. *Denote*

$$V(x) = \int_a^x w^{1-p'}(t)dt.$$

and suppose that

$$B := \int_a^b \int_a^b |V(x) - V(y)|^{q/p'} W(x,y)dxdy < \infty.$$

Then inequality (1.7) holds with $C = B^{1/q}$.

3. A necessary and sufficient condition

3.1. Necessity. Suppose that (1.8) holds for every function $F \geq 0$ and choose for F the function

$$\widetilde{F}(x) = w^{1-p'}(x)\chi_{(\alpha,\beta)}(x) \quad \text{with} \quad a < \alpha < \beta < b.$$

For the right-hand side in (1.8) we then have

$$\left(\int_a^b \widetilde{F}^p(x)w(x)dx\right)^{1/p} = \left(\int_\alpha^\beta w^{p(1-p')}(x)w(x)dx\right)^{1/p} = \left(\int_\alpha^\beta w^{1-p'}(x)dx\right)^{1/p}$$

while for the left-hand side we have the (lower) estimate

$$\left(\int_a^b \int_x^b \left|\int_x^y \widetilde{F}(t)dt\right|^q \widetilde{W}(x,y)dydx\right)^{1/q} \geq$$

$$\geq \left(\int_a^\alpha \int_\beta^b \left|\int_x^y \widetilde{F}(t)dt\right|^q \widetilde{W}(x,y)dydx\right)^{1/q} =$$

$$= \left(\int_a^\alpha \int_\beta^b \left|\int_\alpha^\beta w^{1-p'}(t)dt\right|^q \widetilde{W}(x,y)dydx\right)^{1/q} =$$

$$= \left(\int_\alpha^\beta w^{1-p'}(t)dt\right)\left(\int_a^\alpha \int_\beta^b \widetilde{W}(x,y)dydx\right)^{1/q}.$$

Thus, it follows from inequality (1.8) that

$$\left(\int_\alpha^\beta w^{1-p'}(t)dt\right)\left(\int_a^\alpha \int_\beta^b \widetilde{W}(x,y)dydx\right)^{1/q} \leq C\left(\int_\alpha^\beta w^{1-p'}(x)dx\right)^{1/p}$$

and this yields the *necessary condition*

$$(3.1) \qquad \sup_{(\alpha,\beta)\subset(a,b)} \left(\int_a^\alpha \int_\beta^b \widetilde{W}(x,y)dydx\right)^{1/q}\left(\int_\alpha^\beta w^{1-p'}(x)dx\right)^{1/p'} \leq C.$$

Remark 3. Notice that condition (3.1) is necessary for (1.7) to hold independently on the mutual position of the parameters p and q.

Theorem 3. *Let $1 < p \leq q < \infty$ and suppose that the weight function $W(x, y)$ satisfies*

$$(3.2) \quad \sup_{a < t < b} \int_a^t \int_t^b \widetilde{W}(x, y) \left(\int_a^x \int_t^b \widetilde{W}(\sigma, \tau) d\tau d\sigma \right)^{-1/q}$$

$$\cdot \left(\int_a^x \int_y^b \widetilde{W}(\sigma, \tau) d\tau d\sigma \right)^{-1/q'} dy dx < \infty.$$

Then inequality (1.8) - and, consequently, inequality (1.7) - holds if and only if the weight functions W and w satisfy (3.1).

Proof. We have only to show that condition (3.1) is *sufficient* if (3.2) is satisfied. Denote

$$h(s, t) = \left(\int_s^t w^{1-p'}(\tau) d\tau \right)^{1/(p'q)}, \quad a < s < t < b.$$

Then we have for $a < x < y < b$ by Hölder's inequality

$$\int_x^y F(t) dt = \int_x^y F(t) w^{1/p}(t) h(x, t) h^{-1}(x, t) w^{-1/p}(t) dt \leq$$

$$\leq \left(\int_x^y F^p(t) w(t) h^p(x, t) dt \right)^{1/p} \left(\int_x^y h^{-p'}(x, t) w^{1-p'}(t) dt \right)^{1/p'}$$

and (3.1) implies that

$$\int_x^y h^{-p'}(x, t) w^{1-p'}(t) dt = \int_x^y \left[\int_x^t w^{1-p'}(\tau) d\tau \right]^{-1/q} w^{1-p'}(t) dt =$$

$$= \int_x^y q' \frac{d}{dt} \left[\int_x^t w^{1-p'}(\tau) d\tau \right]^{1/q'} dt = q' \left[\int_x^y w^{1-p'}(\tau) d\tau \right]^{1/q'} \leq$$

$$\leq q' C^{p'/q'} \left(\int_a^x \int_y^b \widetilde{W}(\sigma, \tau) d\tau d\sigma \right)^{-p'/(qq')}$$

and thus

$$\left| \int_x^y F(t) dt \right|^q \leq (q')^{q/p'} C^{q-1} \left(\int_x^y F^p(t) w(t) h^p(x, t) dt \right)^{q/p} G^{-1/q'}(x, y)$$

where we used the notation

$$G(x, y) = \int_a^x \int_y^b \widetilde{W}(\sigma, \tau) d\tau d\sigma.$$

This estimate and Fubini's theorem yield

$$(3.3) \quad \int_a^b \int_x^b \left| \int_x^y F(t) dt \right|^q \widetilde{W}(x, y) dy dx \leq$$

$$\leq C_1 \int_a^b \int_x^b \widetilde{W}(x, y) G^{-1/q'}(x, y) \left(\int_x^y F^p(t) w(t) h^p(x, t) dt \right)^{q/p} dy dx$$

$$= C_1 \int_a^b \int_a^y \widetilde{W}(x, y) G^{-1/q'}(x, y) \left(\int_x^y F^p(t) w(t) h^p(x, t) dt \right)^{q/p} dx dy.$$

188

An application of Minkowski's integral inequality yields

$$\int_a^y \widetilde{W}(x,y)G^{-1/q'}(x,y)\Big(\int_x^y F^p(t)w(t)h^p(x,t)dt\Big)^{q/p}dx =$$

$$=\int_a^y \Big(\int_x^y \widetilde{W}^{p/q}(x,y)G^{-p/(qq')}(x,y)F^p(t)w(t)h^p(x,t)dt\Big)^{q/p}dx \le$$

$$\le\Big(\int_a^y \Big(\int_a^t \widetilde{W}(x,y)G^{-1/q'}(x,y)F^q(t)w^{q/p}(t)h^q(x,t)dx\Big)^{p/q}dt\Big)^{q/p} =$$

$$=\Big(\int_a^y F^p(t)w(t)\Big(\int_a^t \widetilde{W}(x,y)G^{-1/q'}(x,y)h^q(x,t)dx\Big)^{p/q}dt\Big)^{q/p}.$$

If we denote

$$H(t,y) = \Big(\int_a^t \widetilde{W}(x,y)G^{-1/q'}(x,y)h^q(x,t)dx\Big)^{p/q},$$

use the last estimate in (3.3) a apply again Minkowski's integral inequality, we obtain that

$$\int_a^b \int_x^\infty \Big|\int_x^y F(t)dt\Big|^q \widetilde{W}(x,y)dydx \le$$

$$\le C_1 \int_a^b \Big(\int_a^y F^p(t)w(t)H(t,y)dt\Big)^{q/p}dy \le$$

$$\le C_1 \int_a^b \Big(\int_t^b F^q(t)w^{q/p}(t)H^{q/p}(t,y)dy\Big)^{p/q}dt\Big)^{q/p} =$$

$$= C_1\Big(\int_a^b F^p(t)w(t)\Big(\int_t^b H^{q/p}(t,y)dy\Big)^{p/q}dt\Big)^{q/p}.$$

This inequality yields (1.8) provided we can show that

$$\int_t^b H^{q/p}(t,y)dy \le \text{const} \quad \text{for every } t \in (a,b).$$

But the definition of H, h, G, Fubini's theorem and condition (3.1) (for $\alpha = x$, $\beta = t$) imply that

$$\int_t^b H^{q/p}(t,y)dy = \int_t^b \Big(\int_a^t \widetilde{W}(x,y)G^{-1/q'}(x,y)h^q(x,t)dx\Big)dy =$$

$$= \int_a^t h^q(x,t)\Big(\int_t^b \widetilde{W}(x,y)G^{-1/q'}(x,y)dy\Big)dx =$$

$$= \int_a^t \Big(\int_x^t w^{1-p'}(\tau)d\tau\Big)^{1/p'}\Big(\int_t^b \widetilde{W}(x,y)G^{-1/q'}(x,y)dy\Big)dx$$

$$\le C \int_a^t G^{-1/q}(x,t)\int_t^b \widetilde{W}(x,y)G^{-1/q'}(x,y)dydx$$

$$= C \int_a^t \int_t^b \widetilde{W}(x,y)G^{-1/q}(x,t)G^{-1/q'}(x,y)dydx$$

which is bounded due to condition (3.2). So, the proof of Theorem 3 is finished.

Acknowledgement. This work was supported by the Grant Agency of Czech Republic, grant No. 201/97/0744.

References

[1] P. Grisvard: *Espaces intermédiaires entre espaces de Sobolev avec poids.* Ann. Scuola Norm. Sup. Pisa 23 (1969), 373-386.

[2] G. N. Jakovlev: *Boundary properties of functions from the space $W_p^{(l)}$ on domains with angular points* (Russian), Dokl. Akad. Nauk SSSR 140 (1961), 73-76.

[3] H. P. Heinig, A. Kufner, L. E. Persson: *On some fractional order Hardy inequalities.* J. of Inequal. and Appl. 1 (1997), No. 1, 25-46.

[4] A. Kufner: *Hardy's inequality and related topics.* In: Function Spaces, Differential Operators and Nonlinear Analysis (J. Rákosník, ed.). Prometheus, Prague 1996, 89-99.

[5] A. Kufner, L. E. Persson: *Hardy inequalities of fractional order via interpolation.* In: Inequalities and applications (R. P. Agarwal, ed.), World Scientific, Singapore 1994, 417-430.

[6] A. Kufner, H. Triebel: *Generalizations of Hardy's inequality.* Conf. Sem. Mat. Univ. Bari 156 (1978), 21pp.

[7] L. E. Persson, A. Kufner: *Some difference inequalities with weights and interpolation.* Math. Inequalities and Appl. 1 (1998), No. 3, 437-444.

Theory of Differential and Integral Inequalities with Initial Time Difference and Applications

V.LAKSHMIKANTHAM

Florida Institute of technology,Applied Math Program, Melbourne, FL 32901

A.S.VATSALA

University of Southwestern Louisiana,Department of Mathematics, Lafayette, LA 70504

Key words: Initial value problems, variable initial times.

1. Introduction

Recently [4], an investigation of initial value problems of differential equations where the initial time changes with each solution in addition to the change of space (or dependent) variable is initiated. Since it is impossible not to make errors in the starting time, it is natural and important to vary the initial time as well. This creates problems in comparing any two solutions and there could be more than one choice of measuring the difference between any two solutions starting at different initial times. In [4], some preliminary results are attempted to understand the possible intricacies in dealing with such situations.

In this paper, we shall continue this study and investigate necessary differential and integral inequalities in this new framework and then apply the theory of such inequalities to discuss the method of upper

T.M. Rassias and H.M. Srivastava (eds.), Analytic and Geometric Inequalities and Applications, 191–203.
© 1999 *Kluwer Academic Publishers.*

and lower solutions, monotone iterative technique, global existence and some simple stability criteria. This study is new and awaits further advancement since it is closer to the real world situations than before.

2. Differential and integral inequalities

Let us begin with the following comparison result.

Theorem 2.1. *Assume that*

(i) $m \in C^1[R_+, R_+]$, $g \in C[R_+^2, R]$ *and* $m'(t) \leq g(t, m(t))$ *for* $t \geq t_0 \geq 0$;

(ii) *the maximal solution* $r(t) = r(t, t_0, w_0)$ *of*

$$w' = g(t, w), w(\tau_0) = w_0 \geq m(t_0), \tau_0 \geq 0, \tag{2.1}$$

exists on $[\tau_0, \infty)$;

(iii) $t_0 < \tau_0$ *and* $g(t, w)$ *is nondecreasing in* t *for each* w.

Then (a) $m(t) \leq r(t + \eta)$, $t \geq t_0$. *and* (b) $m(t - \eta) \leq r(t)$, $t \geq \tau_0$.

Proof. It is known [2] that if $w(t, \varepsilon)$ is any solution of

$$w' = g(t, w) + \epsilon, \ w(\tau_0) = w_0 + \epsilon, \tag{2.2}$$

for sufficiently small $\epsilon > 0$, then $\lim_{\epsilon \to 0} w(t, \epsilon) = r(t, \tau_0, w_0)$ on every compact interval $[\tau_0, \tau_0 + T]$. Hence setting $w_0(t, \epsilon) = w(\tau + \eta, \epsilon)$, we have

$$w_0'(t, \epsilon) = g(t + \eta, w_0(t, \epsilon)) + \epsilon > g(t + \eta, w_0(t, \epsilon)), t \geq t_0$$

and $w_0(t, \epsilon) = w_0(t_0 + \eta, \epsilon) = w(\tau_0, \epsilon) = w_0 + \epsilon$. It is therefore enough to show $m(t) \leq w_0(t, \epsilon)$, $t \geq t_0$ in order to prove (a). If this is not true, then there would exist a $t_1 > t_0$ such that

$$m(t_1) = w_0(t_1, \epsilon) \text{ and } m(t) < w_0(t, \epsilon), \ t_0 \leq t < t_1. \tag{2.3}$$

This implies that $m'(t_1) \geq w_0'(t_1, \epsilon)$, which yields

$$g(t_1, m(t_1)) \geq g(t_1 + \eta, w_0(t_1, \epsilon)) + \epsilon.$$

Since $\eta > 0$, the nondecreasing nature of g in t, then gives

$$g(t_1, m(t_1)) > g(t_1, w_0(t_1, \epsilon)),$$

which is a contradiction because of (2.3). Hence the proof of (a) is complete.

In order to prove (b), we set $m_0(t) = m(t - \eta)$, so that $m_0(\tau_0) = m(\tau_0 - \eta) = m(t_0)$. Now it is enough to show that

$$m_0(t) < w(t, \epsilon), t \geq \tau_0,$$

the proof of which is very similar to the foregoing proof. Hence The proof of Theorem 2.1 is complete.

Based on Theorem 2.1, one can prove the following result on integral inequalities.

Theorem 2.2. *Assume that*

(i) $m \in C[R_+, R_+]$, $g \in C[R_+^2, R]$ *and*

$$m(t) \leq m(t_0) + \int_{t_0}^t g(s, m(s))ds, t \geq t_0 \geq 0;$$

(ii) *the maximal solution* $r(t) = r(t, \tau_0, w_0)$ *of (2.1) exists for* $t \geq \tau_0$ *for* $m(t_0) \leq w_0$;

(iii) $t_0 < \tau_0$ *and* $g(t, w)$ *is nondecreasing in* t *and* w.

Then (a) $m(t) \leq r(t + \eta)$, $t \geq t_0$ and $m(t - \eta) \leq r(t)$, $t \geq \tau_0$.

Proof. we set $v(t) = m(t_0) + \int_{t_0}^t g(s, m(s))ds$, $t \geq t_0$, so that we have

$$v'(t) = g(t, m(t)) \leq g(t, v(t)), t \geq t_0.$$

using the nondecreasing character of $g(t, w)$ in w. Now by Theorem 2.1, we immediately obtain that

(a) $v(t) \leq r(t + \eta), t \geq t_0$ *or* (b) $v(t - \eta) \leq r(t), t \geq \tau_0$.

Since

$$m(t) \leq v(t), t \geq t_0 \text{ or } m(t - \eta) \leq v(t - \eta), t \geq \tau_0,$$

the conclusion of Theorem 2.2 follows right away and the proof is complete.

The special linear case of Theorem 2.2 is the Gronwall inequality in this frame work, which we state as a corollary.

Corollary 2.1. *Let* $m, \lambda \in C[R_+, R_+]$, *and*

$$m(t) \leq m(t_0) + \int_{t_0}^t \lambda(s, m(s))ds, t \geq t_0 \geq 0;$$

suppose that $t_0 < \tau_0$ *and* $\lambda(t)$ *is nondecreasing in* t. *Then*

$$m(t) \leq m(t_0) \exp[\int_{t_0}^t \lambda(s + \eta)ds], \ t \geq t_0 \text{ or}$$

$$m(t - \eta) \le w_0 \exp[\int_{t_0}^{t} \lambda(s)ds], \ t \ge \tau_0.$$

3. Method of lower and upper solutions

If we know the existence of lower and upper solutions $\alpha, \beta \in C^1[[0, T], R]$ such that $\alpha(t) \le \beta(t)$ on $[0, T]$, then we know that there exists a solution $x(t)$ of

$$x' = f(t, x) \tag{3.1}$$

satisfying $\alpha(t) \le x(t) \le \beta(t)$ on $0 \le t \le T$. See [1] for details. However if the lower and upper solutions do not start from the same point, such an existence result would take a different form as follows.

Theorem 3.1. *Assume that*

(i) $\alpha \in C^1[[t_0, t_0 + T], R], \ t_0, T > 0, \ \beta \in C^1[[\tau_0, \tau_0 + T], R], \tau_0 > 0,$ $f \in C[R_+ \times R, R]$ *and* $\alpha' \le f(t, \alpha(t)), \ t_0 \le t \le t_0 + T, \ \beta' \ge$ $f(t, \beta(t)), \ \tau_0 \le t \le \tau_0 + T$ *with* $\alpha(t_0) \le \beta(\tau_0);$

(ii) $t_0 < \tau_0, \ f(t, x)$ *is nondecreasing in t for each x and* $\alpha(t) \le$ $\beta(t + \eta), \ t_0 \le t \le t_0 + T, \ \eta = \tau_0 - t_0.$

Then there exists a solution $x(t)$ of (3.1) with $x(t_0) = x_0$, satisfying $\alpha(t) \le x(t) \le \beta(t + \eta), \ t_0 \le t \le t_0 + T.$

Proof. Set $\beta_0(t) = \beta(t + \eta)$ for $[t_0, t_0 + T]$ so that

$$\beta_0(t_0) = \beta(\tau_0) \ge \alpha(t_0) \text{ and } \beta_0'(t) \ge f(t + \eta, \beta_0(t)).$$

Suppose that $\alpha(t_0) \le x_0 \le \beta_0(t_0)$. Define

$$P : [t_0, t_0 + T] \times R \to R \text{ such that} \tag{1}$$

$$P(t, x) = \max[\alpha(t), \ \min(x, \beta_0(t))]. \qquad (3.2)$$

Then $f(t, P(t, x))$ defines a continuous extension of f to $[t_0, t_0 + T] \times R$ which is bounded. Since f is bounded on $\Omega = [(t, x) : t_0 \leq t \leq t_0 + T,$ $\alpha(t) \leq x \leq \beta_0(t)]$. Hence there exists a solution $x(t)$ of

$$x' = f(t, p(t, x)), x(t_0) = x_0,$$

on $[t_0, t_0 + T]$. For $\epsilon > 0$, we consider

$$\alpha_\epsilon(t) = \alpha(t) - \epsilon(1 + t) \text{ and } \beta_{0\epsilon} = \beta_0(t) + \epsilon(1 + t).$$

Clearly $\alpha_\epsilon(t_0) < x_0 < \beta_{0\epsilon}(t_0)$. We shall show that $\alpha_\epsilon(t) < x(t) < \beta_{0\epsilon}(t)$ for $t_0 \leq t \leq t_0 + T$. Suppose that $t_1 \in (t_0, t_0 + T]$ is such that

$$\alpha_\epsilon(t) < x(t) < \beta_{0\epsilon}(t) \text{ on } [t_0, t_1) \text{ and } x(t_1) = \beta_{0\epsilon}(t_1). \qquad (3.3)$$

Then $x(t_1) > \beta_0(t_1)$ and hence $P(t_1, x(t_1)) = \beta_0(t_1)$. Also, we have $\alpha(t_1) \leq P(t_1, x(t_1)) \leq \beta_0(t_1)$. Therefore, we get, using the nondecreasing nature of f in t and (3.2), (3.3),

$$\begin{aligned}
\beta_0'(t_1) &= \beta'(t_1 + \eta) \geq f(t_1 + \eta, \beta(t_1 + \eta)) \\
&= f(t_1 + \eta, P(t_1, x(t_1))) \geq f(t_1, P(t_1, x(t_1))) = x'(t_1).
\end{aligned}$$

Since $\beta_{0\epsilon}'(t_1) > \beta_0'(t_1)$, we see that $\beta_{0\epsilon}'(t_1) > x'(t_1)$. This contradicts $x(t) < \beta_{0\epsilon}(t)$ for $t \in [t_0, t_1)$. Consequently, $\alpha_\epsilon(t) < x(t) < \beta_{0\epsilon}(t)$ for $t_0 \leq t \leq t_0 + T$. Letting $\epsilon \to 0$, we get $\alpha(t) \leq x(t) \leq \beta_0(t)$ on $[t_0, t_0 + T]$, and the proof is complete.

Corollary 3.1. *Under the assumption of Theorem 3.1, if we assume that*

$\alpha(t-\eta) \leq \beta(t)$ *for* $\tau_0 \leq t \leq \tau_0+T$, *then there exists a solution* $x(t)$ *of* (3.1) *with* $x(\tau_0) = x_0$ *satisfying* $\alpha(t-\eta) \leq x(t) \leq \beta(t)$, $\tau_0 \leq t \leq \tau_0+T$.

For the proof, we proceed with $\alpha_0(t) = \alpha(t-\eta)$ for $\tau_0 \leq t \leq \tau_0+T$, noting that $\alpha_0(\tau_0) = \alpha(t_0)$ and $\alpha_0'(t) \leq f(t-\eta, \alpha_0(t))$, instead of β. The rest of the proof is similar.

4. Monotone iterative technique.

We shall develop monotone iterative technique in the present frame work.

Theorem 4.1. *Assume that condition (i), (ii) of Theorem 3.1 hold. Suppose further that*

$$f(t,x) - f(t,y) \geq -M(x-y), \ M \geq 0, \tag{4.1}$$

whenever $\alpha(t) \leq y \leq x \leq \beta(t+\eta)$, $t_0 \leq t \leq t_0 + T$. *Then there exist monotone sequences* $\{\alpha_n(t)\}, \{\tilde{\beta}_n(t)\}$ *such that* $\alpha_n(t) \to \rho(t)$ *and* $\tilde{\beta}_n(t) \to \tilde{r}(t)$ *as* $n \to \infty$, *uniformly and monotonically on* $[t_0, t_0 + T]$ *where* $\tilde{\beta}_0(t) = \beta(t+\eta)$, $\alpha_0(t) = \alpha(t)$. *Moreover,* $\rho, \tilde{r}$ *are minimal and maximal solutions of* $x' = f(t,x)$, $x(\tau_0) = x_0$ *on* $[\tau_0, \tau_0 + T]$ *and* $x' = f(t,x)$, $x(t_0) = x_0$ *on* $[t_0, t_0 + T]$ *in the sector* $\left[\alpha, \tilde{\beta}\right]$

Proof. We recall that $\tilde{\beta}_0(t) = \beta(t+\eta)$ implies that $\tilde{\beta}_0(t_0) = \beta(\tau_0) \geq \alpha(t_0)$ and $\tilde{\beta}_0'(t) \geq f(t+\eta, \tilde{\beta}_0(t))$, $t \in [t_0, t_0 + T]$. We define the iterates as follows:

$$\alpha'_{n+1}(t) = f(t, \alpha_n(t)) - M(\alpha_{n+1}(t) - \alpha_n(t)), \alpha_{n+1}(t_0) = x_0$$

$$\beta'_{n+1}(t) = f(t+\eta, \tilde{\beta}_n(t)) - M(\tilde{\beta}_{n+1}(t) - \tilde{\beta}_n(t)), \tilde{\beta}_{n+1}(t_0) = x_0, t \in [t_0, t_0 + T].$$

Let $p = \tilde{\beta}_1 - \tilde{\beta}_0$ so that $p(t_0) \leq 0$ and

$$p' \leq f(t+\eta, \tilde{\beta}_0) - M(\tilde{\beta}_1 - \tilde{\beta}_0) - f(t+\eta, \tilde{\beta}_0) = -Mp.$$

This shows that $\tilde{\beta}_1 \leq \tilde{\beta}_0$ for $t \in [t_0, t_0 + T]$. It is easy to see that $\alpha_0 \leq \alpha_1, t \in [t_0, t_0 + T]$. To show $\alpha_1 \leq \tilde{\beta}_1, t \in [t_0, t_0 + T]$, consider

$$p = \alpha_1 - \tilde{\beta}_1, \text{ so that } p(t_0) = 0 \text{ and}$$

$$p' \leq f(t, \alpha_0) - M(\alpha_1 - \alpha_0) - [f(t+\eta, \tilde{\beta}_0) - -M(\tilde{\beta}_1 - \tilde{\beta}_0)]$$

$$\leq f(t+\eta, \alpha_0) - f(t+\eta, \tilde{\beta}_0) - -M[\alpha_1 - \alpha_0 + \tilde{\beta}_1 - \tilde{\beta}_0]$$

$$\leq M[\tilde{\beta}_0 - \alpha_0] - -M[\alpha_1 - \alpha_0 + \tilde{\beta}_1 - \tilde{\beta}_0] = -Mp,$$

using (4.1) and nondecreasing nature of $f(t, x)$ in t. This yields $\alpha_1 \leq \tilde{\beta}_1$, $t \in [t_0, t_0 + T]$. Thus we have

$$\alpha_0 \leq \alpha_1 \leq \tilde{\beta}_1 \leq \tilde{\beta}_0, t \in [t_0, t_0 + T].$$

Following similar arguments, it is now easy to prove by induction that

$$\alpha_0 \leq \alpha_1 \leq \alpha_2 \leq \cdots \leq \alpha_n \leq \tilde{\beta}_n \leq \cdots \leq \tilde{\beta}_2 \leq \tilde{\beta}_1 \leq \tilde{\beta}_0, t \in [t_0, t_0 + T].$$

Then the standard proof shows that $\alpha_n \to \rho$ and $\tilde{\beta}_n \to \tilde{r}$ as $n \to \infty$ uniformly and monotonically on $[t_0, t_0 + T]$. See [4] for details. Further

$\rho, \tilde{r}$ satisfy

$$\rho'(t) = f(t, \rho), \rho(t_0) = x_0, \ \tilde{r}'(t) = f(t + \eta, \tilde{r}(t)), \ \tilde{r}(t_0) = x_0. \qquad (4.2)$$

Let $x(t)$ be any solution of (a) $x' = f(t, x)$, $x(t_0) = x_0$ such that $\alpha_0(t) \le x(t) \le \tilde{\beta}_0(t), t \in [t_0, t_0 + T]$. Then we know by the standard proof [1] that $\rho \le x$, $t \in [t_0, t_0 + T]$. But $x' = f(t, x) \le f(t + \eta, x)$, $x(t_0) = x_0$ in view of the nondecreasing character of f in t. Since $\alpha_0 \le x \le \tilde{\beta}_0$, $t \in [t_0, t_0 + T]$, one can prove by induction that $\alpha_n \le x \le \tilde{\beta}_0$, $t \in [t_0, t_0 + T]$, and this proves that $x \le \tilde{r}$, $t \in [t_0, t_0 + T]$ completing the proof.

Let $x(t)$ be any solution of $x' = f(t, x)$, $x(\tau_0) = x_0$ on $[\tau_0, \tau_0 + T]$. Then setting $\tilde{x}(t) = x(t + \eta)$, we see that $\tilde{x}'(t) = f(t + \eta, \tilde{x})$, $\tilde{x}(t_0) = x_0$, $t \in [t_0, t_0 + T]$. Suppose that $\alpha_0 \le \tilde{x} \le \tilde{\beta}_0, t \in [t_0, t_0 + T]$. Then we can show as before by induction that $\alpha_n \le \tilde{x} \le \tilde{\beta}_n, t \in [t_0, t_0 + T]$ which proves that $\rho \le \tilde{x} \le \tilde{r}$ on $[t_0, t_0 + T]$. Thus we see that $\rho, \tilde{r}$ are extremal solutions of IVPS (a) and (b). The proof is complete.

5. Global existence.

Consider the differential system

$$x' = f(t, x), x(t_0) = x_0, \ t_0 \ge 0, \qquad (5.1)$$

where $f \in C[R_+ \times R^n, R^n]$. The following global existence result utilizes the integral inequalities result Theorem 2.2 which is in the present set up.

Theorem 5.1. *Assume that* $f \in C[R_+ \times R^n, R^n]$ *and*

$$|f(t, x)| \leq g(t, |x|), \ (t, x) \in R_+ \times R^n, \tag{5.2}$$

where $g \in C[R_+^2, R_+]$, $g(t, w)$ *is nondecreasing in* t *and* w. *Suppose further that the maximal solution of*

$$w' = g(t, w), w(\tau_0) = w_0 \geq 0, \ \tau_0 > t_0, \tag{5.3}$$

exists on $[\tau_0, \infty)$. *Then the solutions* $x(t, t_0, x_0)$ *of (5.1) such that* $|x_0| \leq w_0$ *exists on* $[t_0, \infty)$.

Proof. Let $x(t, t_0, x_0)$ be any solution of (5.1) with $|x_0| \leq w_0$, which exists on $[t_0, \beta)$, $t_0 < \beta < \infty$ and that the value of β cannot be increased. Setting $m(t) = |x(t, t_0, x_0)|$, we see that for $t_0 \leq t < \beta$,

$$m(t) \leq m(t_0) + \int_{t_0}^t |f(s, x(s, t_0, x_0))| \, ds \leq m(t_0) + \int_{t_0}^t g(s, m(s)) ds,$$

in view of assumption (5.2). Theorem (2.2) then yields

$$m(t) = |x(t, t_0, x_0)| \leq r(t + \eta, \tau_0, w_0), \ t_0 \leq t < \beta \tag{5.4}$$

where $r(t, \tau_0, w_0)$ is the maximal solution of (5.3).

For any $t_1, t_2 \in [t_0, \beta)$ such that $t_1 < t_2$, we have

$$|x(t_1, t_0, x_0) - x(t_2, t_0, x_0)| \leq \int_{t_1}^{t_2} g(s, |x(s, t_0, x_0)|) ds$$

$$\leq \int_{t_1}^{t_2} g(s, r(s + \eta, \tau_0, w_0)) ds = r(t_0 + \eta, \tau_0, w_0) - r(t_2 + \eta, \tau_0, w_0).$$

Here we have employed (5.2), (5.4) and the monotonic nature of $g(t, w)$ in w. Since $\lim_{t+\eta \to \beta^-} r(t + \eta, \tau_0, w_0)$ exists and is finite, it follows, using

Cauchy criterion, that as $t_1, t_2 \to \beta$, we have $\lim\limits_{t \to \beta^-} x(t, t_0, x_0)$ exists. We then define $x(\beta, t_0, x_0) = \lim\limits_{t \to \beta^-} x(t, t_0, x_0)$ and consider the IVP

$$x' = f(t, x), \quad x(\beta) = x(\beta, t_0, x_0).$$

By local existence, we find that $x(t, t_0, x_0)$ can be continued beyond β, contradicting our assumptions. Hence every solution $x(t, t_0, x_0)$ with $|x_0| \le w_0$ of (5.1) exists on $[t_0, \infty)$ and the proof is complete.

6. Stability criteria

Let $x(t, \tau_0, x_0), x(t, t_0, y_0)$ be the solutions of (5.1) through (τ_0, x_0), (t_0, y_0) respectively, Define

$$v(t) = x(t, \tau_0, x_0) - x(t - \eta, t_0, y_0) \equiv x(t) - \tilde{x}(t), \qquad (6.1)$$

$\eta = \tau_0 - t_0 > 0$. Then $v(\tau_0) = x_0 - y_0$ and

$$
\begin{aligned}
v'(t) &= f(t, x(t)) - f(t - \eta, \tilde{x}(t)) \\
&= f(t, v(t) + \tilde{x}(t)) - f(t - \eta, \tilde{x}(t)) \equiv \tilde{f}(t, v(t); \eta) \text{ for } t \ge \tau_0.
\end{aligned}
$$

Note that $\tilde{f}(t, 0; \eta) \ne 0$ so that there is no trivial solution for the differential equation

$$v' = \tilde{f}(t, v; \eta), \quad v(\tau_0) = x_0 - y_0. \qquad (6.2)$$

Assume that

$$\lim\limits_{h \to 0+} Sup \frac{1}{h} [|v + h\tilde{f}(t, v; \eta)| - |v|] \le g(t, |v|, |\eta|). \qquad (6.3)$$

where $g \in C[R_+^3, R]$. Then if we set $m(t) = |v(t)|$, we arrive at the differential inequality

$$D^+ m(t) \leq g(t, m(t), |\eta|), \quad m(\tau_0) = |x_0 - y_0|.$$

Consequently, it follows that

$$m(t) \leq \tau(t, \tau_0, m(\tau_0); |\eta|), \quad t \geq \tau_0, \tag{6.4}$$

where $r(t, \tau_0, w_0; |\eta|)$ is the maximal solution of

$$w' = g(t, w, |\eta|), \quad w(\tau_0) = m(\tau_0). \tag{6.5}$$

Let us suppose that any solution $w(t, \tau_0, w_0; |\eta|)$ satisfies the following definition.

Definition: We say that the differential equation (6.5) is practically stable, if given $0 < \lambda < A$, there exists a $\sigma = \sigma(\lambda, A) > 0$ such that

$$w_0 < \lambda \text{ and } |\eta| < \sigma \text{ implies } w(t, \tau_0, w_0; |\eta|) < A, \ t \geq \tau_0.$$

Then we get from (6.1) and (6.4) the practical stability of the differential system (5.1), namely,

$$|x_0 - y_0| < \lambda \text{ and } |\eta| < \sigma \text{ implies } |x(t, \tau_0, x_0) - x(t - \eta, t_0, y_0)| < A, \ t \geq \tau_0$$

We have proved the following stability result.

Theorem 6.1. *Assume that (6.3) holds where $\tilde{f}$ is defined in terms of the function f in (5.1). Then practical stability of (6.5) implies the practical stability of the system (5.1).*

Other concepts of practical stability may be proved on the basis of Theorem 6.1. See [3] for details of practical stability. As an example of the comparison function, consider $g(t, w, |\eta|) = -\alpha w + \beta |\eta|$, $\alpha, \beta > 0$. Then

$$w(t, \tau_0, w_0; |\eta|) \leq w(\tau_0)e^{-\alpha(t-\tau_0)} + \frac{\beta}{2}|\eta|, t \geq \tau_0.$$

Consequently, choosing $\sigma = (A - \lambda)\frac{\alpha}{\beta}$, we see that $w_0 < \lambda$ and $|\eta| < \sigma$ implies $w(t, \tau_0, w_0; |\eta|) < A$, $t \geq \tau_0$.

References

1.Ladde, G.S., Lakshmikantham, V., and Vatsala, A.S.: Monotone Iterative Technique for nonlinear differential Equations, Pitman, Boston 1985.

2.Lakshmikantham, V., and Leela, S.: Differential and Integral Inequalities, Vol I, Academic Press, New York 1969.

3.Lakshmikantham, V., Leela, S., and Martynyuk: Stability analysis of non-linear systems, Marcel Dekker, Inc., New York 1989.

4.Lakshmikantham, V., and Vatsala, A.S.: Differential Inequalities with Initial Time Difference and Applications, to appear in Journal of Inequalities and Applications.

NUMERICAL RADII OF SOME COMPANION MATRICES AND BOUNDS FOR THE ZEROS OF POLYNOMIALS

HANSJÖRG LINDEN
Fachbereich Mathematik, Fernuniversität-Gesamthochschule,
Postfach 940, Lützowstr. 125, D-58084 Hagen, Germany.

Abstract. Let a monic polynomial $P_n(x) := x^n - a_1 x^{n-1} - \cdots - a_n$, $a_j \in \mathbb{C}$, $j = 1, 2, \ldots, n$, be given. Bounds for the the numerical radii of some special companion matrices of P_n are computed. From these estimates bounds for the zeros of P_n are derived.

1. Introduction

Let

$$P_n(x) := x^n - a_1 x^{n-1} - \cdots - a_n, \ a_j \in \mathbb{C}, \ j = 1, 2, \ldots, n, \ a_n \neq 0, \tag{1}$$

be a monic polynomial of degree $n \geq 3$. In [1, 4, 5] using results on the numerical range and the numerical radius of the Frobenius companion matrix of P_n bounds for the zeros of P_n were given: In [1] (see also [5], Section 5.3) an estimate of the numerical radius of a matrix is applied to the Frobenius companion matrix, and in [4] a formula for the numerical radius of a matrix of rank one is applied to the Frobenius companion matrix which was decomposed in the sum of a matrix of rank one and a right shift matrix.

In this paper we extend these methods and give further estimates for the zeros of P_n using properties of the numerical ranges and the numerical radii of some other types of companion matrices of P_n. In [9] we used these types of (generalized) companion matrices, which are based on special multiplicative decompositions of the coefficients of the polynomial, to obtain estimates for the zeros of the polynomial P_n mainly by the application of Gersgorin's theorem to the companion matrices or by computing the singular values of the companion matrices and using majorization relations of H. Weyl between the eigenvalues and singular values of a matrix. Further informations on bounds for zeros of polynomials can be found in [6, 11, 12, 13, 14].

2. A Companion Matrix of Frobenius Type

In this section we consider a companion matrix of P_n which can be obtained by a

1991 *Mathematics Subject Classification.* Primary 15A60, 26C10, 30C15.
Key words and phrases. Numerical radius; Companion matrix; Bounds for zeros of polynomials.

T.M. Rassias and H.M. Srivastava (eds.), Analytic and Geometric Inequalities and Applications, 205–229.
© 1999 *Kluwer Academic Publishers.*

similarity transformation of the Frobenius companion matrix of P_n.

Proposition 1 (cf. [9]). *Let P_n be the monic polynomial of degree $n \geq 3$ given by (1). Let there exist complex numbers $c_1, c_2, \ldots, c_n \in \mathbb{C}$, $0 \neq b_1, b_2, \ldots, b_{n-1} \in \mathbb{C}$ such that*

$$
\begin{aligned}
a_1 &:= c_1, \\
a_2 &:= c_2 b_1, \\
&\;\;\vdots \\
a_n &:= c_n b_{n-1} \cdots b_2 b_1.
\end{aligned}
\tag{2}
$$

Let A be the n-by-n-matrix

$$
A := \begin{pmatrix}
0 & b_{n-1} & 0 & \cdots & 0 \\
\vdots & \ddots & \ddots & & \\
0 & \cdots & & 0 & b_1 \\
c_n & c_{n-1} & \cdots & c_2 & c_1
\end{pmatrix}.
\tag{3}
$$

Then

$$
\det(x E_n - A) = P_n(x),
$$

where E_n is the n-by-n identity matrix.

From Proposition 1 follows that the eigenvalues of A are equal to the zeros of P_n.

The matrix A is normal only in some special cases (cf. [9] for a discussion of the normality conditions for A).

Decompositions of type (2) of the coefficients of P_n are always possible. The simpliest one is $c_k := a_k$, $k = 1, \ldots, n$, $b_1 = \cdots = b_{n-1} := 1$. In this case (3) is the Frobenius companion matrix. In [9] we proposed several different special choices for the decompositions (2).

Let M be a n-by-n-matrix. We denote by $(.,.)$ the usual inner product in $\mathbb{C}^n$ and by $\|.\| := \sqrt{(.,.)}$ the corresponding norm. Then the numerical range (or the field of values) $F(M)$ of M is defined by $F(M) := \{(Mx, x) : x \in \mathbb{C}^n, \|x\| = 1\}$, and the numerical radius $r(M)$ of M is defined by $r(M) := \max\{|z| : z \in F(M)\}$ (cf. [5], [7], Chapter 1). Since we have $\sigma(M) \subset F(M)$, where $\sigma(M)$ denotes the spectrum of M, estimates for $r(M)$ give estimates for the eigenvalues of M. Thus we derive in the following upper bounds for the numerical radius of the companion matrix A, the inverse matrix A^{-1}, and some other companion matrices. In the proofs we use the norm properties of the numerical radius and results on the numerical radii of some special matrices. These are a result from [4] on the numerical radii of matrices of rank one (see [4], Theorem 1), a result from [2] on the numerical radii of tridiagonal matrices (see [2], Theorem 3), results on certain zero-one matrices from [3, 5, 10],

and a result on the numerical radius of a matrix from [1] (see Theorem 4 of [1]).
For convenience we state the result of [1].

Theorem A (cf. [1]). *Let $M = (a_{jk})$ be a n-by-n-matrix, let $m \in \{1, \ldots, n\}$, and let M_m be the $(n-1)$-by-$(n-1)$-matrix, which is obtained from M by omitting the mth row and the mth column (the mth projection of M), and let $d_m \geq r(M_m)$ be an arbitrary constant. Then*

$$r(M) \leq \frac{1}{2}(|a_{mm}| + d_m)$$

$$+ \frac{1}{2}\left[(|a_{mm}| - d_m)^2 + \left(\left(\sum_{\substack{k=1 \\ k \neq m}}^{n} |a_{mk}|^2 \right)^{\frac{1}{2}} + \left(\sum_{\substack{k=1 \\ k \neq m}}^{n} |a_{km}|^2 \right)^{\frac{1}{2}} \right)^2 \right]^{\frac{1}{2}}.$$

In the next proposition estimates of the numerical radius of A are given.

Proposition 2. *Let A be as given by (3), and let*

$$\beta_1 := \min\left\{ \cos\frac{\pi}{n+1} \max_{k=1,\ldots,n-1} |b_k|, \ \frac{1}{2} \max_{k=1,\ldots,n-2} (|b_k| + |b_{k+1}|) \right\},$$

$$\beta_2 := \min\left\{ \cos\frac{\pi}{n} \max_{k=2,\ldots,n-1} |b_k|, \ \frac{1}{2} \max_{k=2,\ldots,n-2} (|b_k| + |b_{k+1}|) \right\}.$$

Then we have

$$r(A) \leq \beta_1 + \frac{1}{2}\left(|c_1| + \sqrt{\sum_{k=1}^{n} |c_k|^2} \right), \tag{4}$$

and

$$r(A) \leq \frac{1}{2}(|c_1| + \beta_2) + \frac{1}{2}\left[(|c_1| - \beta_2)^2 + \left(|b_1| + \sqrt{\sum_{k=2}^{n} |c_k|^2} \right)^2 \right]^{\frac{1}{2}}. \tag{5}$$

If $b_1 = b_2 = \cdots = b_{n-1} =: b$, then

$$r(A) \leq |b| \cos\frac{\pi}{n+1} + \frac{1}{2}\left(|c_1| + \sqrt{\sum_{k=1}^{n} |c_k|^2} \right), \tag{6}$$

and

$$r(A) \leq \frac{1}{2}\left(|c_1| + |b| \cos\frac{\pi}{n} \right)$$

$$+ \frac{1}{2}\left[\left(|c_1| - |b| \cos\frac{\pi}{n} \right)^2 + \left(|b| + \sqrt{\sum_{k=2}^{n} |c_k|^2} \right)^2 \right]^{\frac{1}{2}}. \tag{7}$$

Proof. First we prove (4). We set $A := B + C$ with

$$B := \begin{pmatrix} 0 & b_{n-1} & 0 & \cdots & 0 \\ \vdots & \ddots & \ddots & & \\ 0 & \cdots & & 0 & b_1 \\ 0 & 0 & \cdots & 0 & 0 \end{pmatrix}, \quad C := \begin{pmatrix} 0 & 0 & 0 & \cdots & 0 \\ \vdots & \ddots & \ddots & & \\ 0 & \cdots & & 0 & 0 \\ c_n & c_{n-1} & \cdots & c_2 & c_1 \end{pmatrix}.$$

Then $r(A) \leq r(B) + r(C)$. Since

$$C x = (0, \ldots, 0, \sum_{k=1}^{n} c_{n-k+1} x_k)^T = (x, \bar{c}) e_n,$$

where $c = (\bar{c}_n, \bar{c}_{n-1}, \ldots, \bar{c}_1)^T$, $e_n = (0, \ldots, 0, 1)^T$, we have

$$r(C) = \frac{1}{2}(\|\bar{c}\|\|e_n\| + |(\bar{c}, e_n)|) = \frac{1}{2}\left(\sqrt{\sum_{k=1}^{n} |c_k|^2} + |c_1|\right)$$

(cf. [4], Theorem 1). Furthermore it follows from Theorem 3 in [2] that $F(B)$ is a circular disk centered at the origin with

$$r(B) \leq \min\left\{ \cos\frac{\pi}{n+1} \max_{k=1,\ldots,n-1} |b_k|, \; \frac{1}{2} \max_{k=1,\ldots,n-2}(|b_k| + |b_{k+1}|)\right\}.$$

From this the estimate (4) follows.

For the proof of (5) we apply Theorem A for $m = n$. The nth projection B_n of A is given by

$$B_n := \begin{pmatrix} 0 & b_{n-1} & 0 & \cdots & 0 \\ \vdots & \ddots & & \ddots & \\ 0 & \cdots & & 0 & b_2 \\ 0 & 0 & \cdots & 0 & 0 \end{pmatrix},$$

and from Theorem 3 in [2] it follows that

$$r(B_n) \leq \min\left\{ \cos\frac{\pi}{n} \max_{k=2,\ldots,n-1} |b_k|, \; \frac{1}{2} \max_{k=2,\ldots,n-2}(|b_k| + |b_{k+1}|)\right\}.$$

Now (5) follows by a short computation from Theorem A for $m = n$.

If $b_1 = \ldots = b_{n-1} =: b$, then $r(B) = |b| \cos\frac{\pi}{n+1}$ and $r(B_n) = |b| \cos\frac{\pi}{n}$ (cf. [3], Proposition 1). From this (6) and (7) follow. $\square$

Remark 1. We have given both estimates in (4), (5) and in (6), (7), respectively, (and we will do this also in the following), since the bound in (4) ((6)) is not always better than the bound in (5) ((7)), and vice versa. For example: *If $b = 1$ and $a_1 = 0$ (and thus also if $a_1 \neq 0$ and if $|a_1|$ is sufficiently small) the estimate (6) is better than the estimate (7). But if $a_1 \neq 0$ and $|a_1|$ is sufficiently large the estimate (7) is better than the estimate (6).*

Proof.   We denote the right sides of (6), (7) by B_1, B_2, respectively. Let $a_1 = 0$. Then the inequality $B_1 < B_2$ is equivalent to the inequality

$$\left(\cos\frac{\pi}{n+1}\left(\cos\frac{\pi}{n+1} - \cos\frac{\pi}{n}\right) - \frac{1}{4}\right) < \left(\frac{1}{2} - \cos\frac{\pi}{n+1} + \frac{1}{2}\cos\frac{\pi}{n}\right)\sqrt{\sum_{k=2}^{n}|a_k|^2}.$$

We now have

$$\cos\frac{\pi}{n+1}\left(\cos\frac{\pi}{n+1} - \cos\frac{\pi}{n}\right) \; - \; \frac{1}{4}$$

$$= \; 2\sin\frac{(2n+1)\pi}{2n(n+1)}\sin\frac{\pi}{2n(n+1)}\cos\frac{\pi}{n+1} - \frac{1}{4}$$

$$\leq \; 2\sin\frac{7\pi}{24}\sin\frac{\pi}{24} - \frac{1}{4} < 0$$

for $n \geq 3$. Furthermore we have

$$\frac{1}{2} - \cos\frac{\pi}{n+1} + \frac{1}{2}\cos\frac{\pi}{n} > \frac{1}{2} + \frac{1}{2}\left(1 - \frac{\pi^2}{2n^2}\right) - \left(1 - \frac{\pi^2}{2(n+1)^2} + \frac{\pi^4}{24(n+1)^4}\right)$$

$$= \; \pi^2\frac{n^4 - (4 + \frac{1}{6}\pi^2)n^2 - 4n - 1}{4n^2(n+1)^4} > 0$$

for $n \geq 3$. Thus $B_1 < B_2$ is valid.

Let $a_1 \neq 0$. Then the inequality $B_2 < B_1$ is equivalent to

$$\frac{1}{2} + \sqrt{\sum_{k=2}^{n}|a_k|^2} \; < \; |a_1|\cos\frac{\pi}{n} + 2\cos\frac{\pi}{n+1}\left(\cos\frac{\pi}{n+1} - \cos\frac{\pi}{n}\right)$$

$$+ \left(2\cos\frac{\pi}{n+1} - \cos\frac{\pi}{n}\right)\sqrt{\sum_{k=1}^{n}|a_k|^2}.$$

Therefore, for $|a_1|$ sufficiently large the right side of the inequality above becomes larger than the left side, and $B_2 < B_1$ becomes valid in this case. $\square$

We apply now similar methods to the inverse matrix A^{-1} of A.

Proposition 3.   *Let A be as given by (3), and let*

$$\tilde{\beta}_1 \; := \; \min\left\{\cos\frac{\pi}{n+1}\max_{k=1,\dots,n-1}\frac{1}{|b_k|}, \; \frac{1}{2}\max_{k=1,\dots,n-2}\left(\frac{1}{|b_k|} + \frac{1}{|b_{k+1}|}\right)\right\},$$

$$\tilde{\beta}_2 \; := \; \min\left\{\cos\frac{\pi}{n}\max_{k=1,\dots,n-2}\frac{1}{|b_k|}, \; \frac{1}{2}\max_{k=1,\dots,n-3}\left(\frac{1}{|b_k|} + \frac{1}{|b_{k+1}|}\right)\right\}.$$

Then we have

$$r(A^{-1}) \leq \tilde{\beta}_1 + \frac{1}{2|c_n|} \left(\left| \frac{c_{n-1}}{b_{n-1}} \right| + \sqrt{1 + \sum_{k=1}^{n-1} \left| \frac{c_k}{b_k} \right|^2} \right), \tag{8}$$

and

$$r(A^{-1}) \leq \frac{1}{2} \left(\frac{|c_{n-1}|}{|c_n b_{n-1}|} + \tilde{\beta}_2 \right)$$

$$+ \frac{1}{2} \left[\left(\frac{|c_{n-1}|}{|c_n b_{n-1}|} - \tilde{\beta}_2 \right)^2 + \left(\frac{1}{|b_{n-1}|} + \frac{1}{|c_n|} \sqrt{1 + \sum_{k=1}^{n-2} \left| \frac{c_k}{b_k} \right|^2} \right)^2 \right]^{\frac{1}{2}} \tag{9}$$

If $b_1 = b_2 = \cdots = b_{n-1} =: b$, then

$$r(A^{-1}) \leq \frac{1}{2|bc_n|} \left(2|c_n| \cos \frac{\pi}{n+1} + |c_{n-1}| + \sqrt{|b|^2 + \sum_{k=1}^{n-1} |c_k|^2} \right), \tag{10}$$

and

$$r(A^{-1}) \leq \frac{1}{2|bc_n|} \left(|c_n| \cos \frac{\pi}{n} + |c_{n-1}| + \left[\left(|c_{n-1}| - |c_n| \cos \frac{\pi}{n} \right)^2 \right. \right.$$

$$\left. \left. + \left(|c_n| + \sqrt{|b|^2 + \sum_{k=1}^{n-2} |c_k|^2} \right)^2 \right]^{\frac{1}{2}} \right). \tag{11}$$

Proof. Since the inverse matrix A^{-1} of A is given by

$$A^{-1} = \begin{pmatrix} -\frac{c_{n-1}}{c_n b_{n-1}} & -\frac{c_{n-2}}{c_n b_{n-2}} & \cdots & -\frac{c_1}{c_n b_1} & \frac{1}{c_n} \\ \frac{1}{b_{n-1}} & & & & \\ & \frac{1}{b_{n-2}} & & & \\ & & \ddots & & \\ & & & \frac{1}{b_1} & 0 \end{pmatrix}, \tag{12}$$

we proceed analogously as in the proof of Proposition 2.

For the proof of (8) we set $A^{-1} := \tilde{B} + \tilde{C}$ with

$$\tilde{B} := \begin{pmatrix} 0 & 0 & \cdots & 0 & 0 \\ \frac{1}{b_{n-1}} & 0 & & \cdots & 0 \\ & \frac{1}{b_{n-2}} & \ddots & & \\ & & \ddots & \ddots & \\ 0 & \cdots & 0 & \frac{1}{b_1} & 0 \end{pmatrix},$$

$$
\tilde{C} := \begin{pmatrix}
-\frac{c_{n-1}}{c_n \bar{b}_{n-1}} & -\frac{c_{n-2}}{c_n \bar{b}_{n-2}} & \cdots & -\frac{c_1}{c_n \bar{b}_1} & \frac{1}{c_n} \\
0 & 0 & & \cdots & 0 \\
& 0 & & & \\
& & \ddots & & \\
0 & \cdots & 0 & 0 & 0
\end{pmatrix}.
$$

Then $r(A^{-1}) \le r(\tilde{B}) + r(\tilde{C})$. Since

$$
\tilde{C}x = \left(-\sum_{k=1}^{n-1} \frac{c_{n-k}}{c_n \bar{b}_{n-k}} x_k + \frac{1}{c_n} x_n, 0, \ldots, 0 \right)^T = (x, \bar{c})e_1,
$$

where

$$
\widehat{c} = \left(-\frac{\bar{c}_{n-1}}{\bar{c}_n \bar{b}_{n-1}}, -\frac{\bar{c}_{n-2}}{\bar{c}_n \bar{b}_{n-2}}, \ldots, -\frac{\bar{c}_1}{\bar{c}_n \bar{b}_1}, \frac{1}{\bar{c}_n} \right)^T, \quad e_1 = (1, 0, \ldots, 0)^T,
$$

the assertion now follows by analogous computations of $r(\tilde{B})$ and $r(\tilde{C})$ as in the proof of Proposition 2.

To prove (9) we now apply Theorem A for $m = 1$. The first projection $\tilde{B}_1$ of A^{-1} is given by

$$
\tilde{B}_1 := \begin{pmatrix}
0 & 0 & \cdots & 0 & 0 \\
\frac{1}{b_{n-2}} & 0 & & \cdots & 0 \\
& \frac{1}{b_{n-3}} & \ddots & & \\
& & \ddots & & \\
0 & \cdots & 0 & \frac{1}{b_1} & 0
\end{pmatrix},
$$

and again from Theorem 3 in [2] it follows that

$$
r(\tilde{B}_1) \le \min \left\{ \cos \frac{\pi}{n} \max_{k=1,\ldots,n-2} \frac{1}{|b_k|}, \frac{1}{2} \max_{k=1,\ldots,n-3} \left(\frac{1}{|b_k|} + \frac{1}{|b_{k+1}|} \right) \right\}.
$$

Now (9) follows from Theorem A for $j = 1$ by a short computation.

If $b_1 = \ldots = b_{n-1} =: b$, then $r(\tilde{B}) = \frac{1}{|b|} \cos \frac{\pi}{n+1}$ and $r(\tilde{B}_1) = \frac{1}{|b|} \cos \frac{\pi}{n}$ (cf. [3], Proposition 1). From this (10) and (11) follow. $\square$

From Propositions 2 and 3 we can determine annuli for the zeros of P_n.

Theorem 1. *Let the monic polynomial P_n satisfy the assumptions of Proposition 1. Then all zeros of P_n lie in the intersection of the annuli given by*

$$
\left(\tilde{\beta}_1 + \frac{1}{2|c_n|} \left(\left| \frac{c_{n-1}}{b_{n-1}} \right| + \sqrt{1 + \sum_{k=1}^{n-1} \left| \frac{c_k}{b_k} \right|^2} \right) \right)^{-1} \le |x|
$$

$$
\le \beta_1 + \frac{1}{2} \left(|c_1| + \sqrt{\sum_{k=1}^{n} |c_k|^2} \right),
$$

and

$$\left(\frac{1}{2}\left(\frac{|c_{n-1}|}{|c_n b_{n-1}|} + \tilde{\beta}_2\right)\right.$$

$$\left. + \frac{1}{2}\left[\left(\frac{|c_{n-1}|}{|c_n b_{n-1}|} - \tilde{\beta}_2\right)^2 + \left(\frac{1}{|b_{n-1}|} + \frac{1}{|c_n|}\sqrt{1 + \sum_{k=1}^{n-2}\left|\frac{c_k}{b_k}\right|^2}\right)^2\right]^{\frac{1}{2}}\right)^{-1}$$

$$\le |x| \le \frac{1}{2}(|c_1| + \beta_2) + \frac{1}{2}\left[(|c_1| - \beta_2)^2 + \left(|b_1| + \sqrt{\sum_{k=2}^{n}|c_k|^2}\right)^2\right]^{\frac{1}{2}}.$$

If $b_1 = \cdots = b_{n-1} =: b$, then the annuli are given by

$$2|bc_n|\left(2|c_n|\cos\frac{\pi}{n+1} + |c_{n-1}| + \sqrt{|b|^2 + \sum_{k=1}^{n-1}|c_k|^2}\right)^{-1}$$

$$\le |x| \le |b|\cos\frac{\pi}{n+1} + \frac{1}{2}\left(|c_1| + \sqrt{\sum_{k=1}^{n}|c_k|^2}\right), \tag{13}$$

and

$$2|bc_n|\left(|c_n|\cos\frac{\pi}{n} + |c_{n-1}|\right.$$

$$\left. + \left[\left(|c_{n-1}| - |c_n|\cos\frac{\pi}{n}\right)^2 + \left(|c_n| + \sqrt{|b|^2 + \sum_{k=1}^{n-2}|c_k|^2}\right)^2\right]^{\frac{1}{2}}\right)^{-1} \le |x|$$

$$\le \frac{1}{2}\left(|c_1| + |b|\cos\frac{\pi}{n}\right) + \frac{1}{2}\left[\left(|c_1| - |b|\cos\frac{\pi}{n}\right)^2 + \left(|b| + \sqrt{\sum_{k=2}^{n}|c_k|^2}\right)^2\right]^{\frac{1}{2}} \tag{14}$$

Proof. The assertions follow from Propositions 2 and 3. $\square$

To make a comparison between some of the bounds of Theorem 1 and two known bounds of Carmichael-Mason (cf. [11], p. 125) and Parodi (cf. [14], p. 131, see also [8]), we consider the special cases of (13) and (14) for $b := 1$ and thus $c_k = a_k, k = 1, \ldots, n$. In this case (13) means, that all zeros of P_n lie in the circle given by

$$|x| \le \cos\frac{\pi}{n+1} + \frac{1}{2}\left(|a_1| + \sqrt{\sum_{k=1}^{n}|a_k|^2}\right) =: B_{FK}.$$

This bound is given in [4] and compared with the Carmichael-Mason bound

$$|x| \leq \sqrt{1 + \sum_{k=1}^{n} |a_k|^2} =: B_{CM}$$

with the result that the bound B_{FK} is not always better than B_{CM} and vice versa. But if $a_1 = 0$ and $\sqrt{\sum_{k=2}^{n} |a_k|^2}$ is fairly large, then the bound B_{FK} is better than B_{CM}. The same assertion is valid for the bound

$$|x| \leq \left[\frac{1}{2} \left(1 + \sum_{k=1}^{n} |a_k|^2 + \sqrt{\left(1 + \sum_{k=1}^{n} |a_k|^2 \right)^2 - 4|a_n|^2} \right) \right]^{\frac{1}{2}} =: B_P$$

of Parodi, although the bound B_P is always better than B_{CM}.

We now consider (14). In the special case $b := 1$ the formula (14) means that

$$|x| \leq \frac{1}{2} \left(|a_1| + \cos\frac{\pi}{n} \right) + \frac{1}{2} \left[\left(|a_1| - \cos\frac{\pi}{n} \right)^2 + \left(1 + \sqrt{\sum_{k=2}^{n} |a_k|^2} \right)^2 \right]^{\frac{1}{2}} =: B_A.$$

This bound is given in [1]. In Remark 1 we have compared the bound B_A with the bound B_{FK}. Furthermore, the bound B_A is not always better than B_{CM} and vice versa. But if $a_1 = 0$ and $\sqrt{\sum_{k=2}^{n} |a_k|^2}$ is fairly large, then the bound B_A is better than B_{CM}. The same assertion is valid for B_P.

Now we apply the method of proof of Proposition 2 to the companion matrix A_α of the monic polynomial

$$Q_{n+1}(x) := (x - \alpha)P_n(x),$$

where $\alpha \in \mathbb{C}$. Q_{n+1} has the same zeros as P_n and in addition the zero α.

Proposition 4. *For $0 \neq \alpha \in \mathbb{C}$ let A_α be given by*

$$A_\alpha := \begin{pmatrix} 0 & \alpha & 0 & \cdots & & 0 \\ & & b_{n-1} & & & \\ & \ddots & & & \ddots & \\ 0 & \cdots & & 0 & & b_1 \\ -c_n & c_n - \alpha\frac{c_{n-1}}{b_{n-1}} & \cdots & & c_2 - \alpha\frac{c_1}{b_1} & c_1 + \alpha \end{pmatrix}, \tag{15}$$

and let

$$\beta_{\alpha,1} := \min\left\{ \cos\frac{\pi}{n+2} \max\{|\alpha|, |b_k|, k = 1, \ldots, n-1\}, \right.$$

$$\left. \frac{1}{2} \max\{|\alpha| + |b_{n-1}|, |b_k| + |b_{k+1}|, k = 1, \ldots, n-2\} \right\},$$

$$\beta_{\alpha,2} \quad := \quad \min\left\{\cos\frac{\pi}{n+1}\max\{|\alpha|, |b_k|, k=2,\ldots,n-1\},\right.$$
$$\left.\frac{1}{2}\max\{|\alpha|+|b_{n-1}|, |b_k|+|b_{k+1}|, k=2,\ldots,n-2\}\right\}.$$

Then

$$r(A_\alpha) \le \beta_{\alpha,1} + \frac{1}{2}\left(|c_1+\alpha| + \sqrt{|c_n|^2 + |c_1+\alpha|^2 + \sum_{k=1}^{n-1}\left|c_{k+1}-\alpha\frac{c_k}{b_k}\right|^2}\right),$$

and

$$r(A_\alpha) \le \frac{1}{2}(|c_1+\alpha| + \beta_{\alpha,2})$$
$$+\frac{1}{2}\left[(|c_1+\alpha| - \beta_{\alpha,2})^2 + \left(|b_1| + \sqrt{|c_n|^2 + \sum_{k=1}^{n-1}\left|c_{k+1}-\alpha\frac{c_k}{b_k}\right|^2}\right)^2\right]^{\frac{1}{2}}.$$

If $b_1 = \cdots = b_{n-1} =: b := \alpha$, then

$$r(A_b) \le |b|\cos\frac{\pi}{n+2} + \frac{1}{2}\left(|c_1+b| + \sqrt{|c_n|^2 + |c_1+b|^2 + \sum_{k=1}^{n-1}|c_{k+1}-c_k|^2}\right),$$

and

$$r(A_b) \le \frac{1}{2}\left(|c_1+b| + |b|\cos\frac{\pi}{n+1}\right) + \frac{1}{2}\left[\left(|c_1+b| - |b|\cos\frac{\pi}{n+1}\right)^2\right.$$
$$\left. + \left(|b| + \sqrt{|c_n|^2 + \sum_{k=1}^{n-1}|c_{k+1}-c_k|^2}\right)^2\right]^{\frac{1}{2}}.$$

If furthermore $c_1 \ne 0$ and $b_1 = \cdots = b_{n-1} = \alpha := -c_1 = -a_1$, then

$$r(A_{-a_1}) \le |a_1|\cos\frac{\pi}{n+2} + \frac{1}{2}\sqrt{|a_n|^2|a_1|^{-2(n-1)} + \sum_{k=1}^{n-1}|a_{k+1}+a_1 a_k|^2|a_1|^{-2k}},$$

and

$$r(A_{-a_1}) \le \frac{1}{2}|a_1|\cos\frac{\pi}{n+1} + \frac{1}{2}\left[\left(|a_1|\cos\frac{\pi}{n+1}\right)^2\right.$$
$$\left. + \left(|a_1| + \sqrt{|a_n|^2|a_1|^{-2(n-1)} + \sum_{k=1}^{n-1}|a_{k+1}+a_1 a_k|^2|a_1|^{-2k}}\right)^2\right]^{\frac{1}{2}}.$$

Proof. The proof is analogous to the proof of Proposition 2. □

Theorem 2. *(a) Let the monic polynomial P_n satisfy the assumptions of Proposition 1, and let $0 \neq \alpha \in \mathbb{C}$. Then all zeros of P_n lie in the intersection of the circles given by*

$$|x| \leq \beta_{\alpha,1} + \frac{1}{2}\left(|c_1 + \alpha| + \sqrt{|c_n|^2 + |c_1 + \alpha|^2 + \sum_{k=1}^{n-1}\left|c_{k+1} - \alpha\frac{c_k}{b_k}\right|^2}\right)$$

and

$$|x| \leq \frac{1}{2}(|c_1 + \alpha| + \beta_{\alpha,2})$$

$$+ \frac{1}{2}\left[(|c_1 + \alpha| - \beta_{\alpha,2})^2 + \left(|b_1| + \sqrt{|c_n|^2 + \sum_{k=1}^{n-1}\left|c_{k+1} - \alpha\frac{c_k}{b_k}\right|^2}\right)^2\right]^{\frac{1}{2}}.$$

(b) All zeros of the monic polynomial P_n given by (1) lie in the intersection of the circles given by

$$|x| \leq \cos\frac{\pi}{n+2} + \frac{1}{2}\left(|a_1 + 1| + \sqrt{|a_n|^2 + |a_1 + 1|^2 + \sum_{k=1}^{n-1}|a_{k+1} - a_k|^2}\right)$$

and

$$|x| \leq \frac{1}{2}\left(|a_1 + 1| + \cos\frac{\pi}{n+1}\right)$$

$$+ \frac{1}{2}\left[\left(|a_1 + 1| - \cos\frac{\pi}{n+1}\right)^2 + \left(1 + \sqrt{|a_n|^2 + \sum_{k=1}^{n-1}|a_{k+1} - a_k|^2}\right)^2\right]^{\frac{1}{2}}.$$

Proof. (a) This follows immediately from the first and the second assertion of Proposition 4.

(b) This follows from the third and the fourth assertion of Proposition 4 by the special choices $b_1 = \cdots = b_{n-1} = 1 = \alpha$. □

Now we choose some special decompositions of the coefficients of the polynomial P_n. Then we get:

Proposition 5. *(a) Let A be as given by (3) with $b_1 = \cdots = b_{n-1} =: b$, where b is a nth root of a_n. Then*

$$r(A) \leq \sqrt[n]{|a_n|} + \frac{1}{2}\left(|a_1| + \sqrt{\sum_{k=1}^{n-1}\frac{|a_k|^2}{\left(\sqrt[n]{|a_n|}\right)^{2(k-1)}}}\right),$$

and

$$r(A^{-1}) \le \frac{1}{|a_n|}\left(\left(\sqrt[n]{|a_n|}\right)^{n-1} + \frac{1}{2}\left(|a_{n-1}| + \sqrt{\sum_{k=1}^{n-1}|a_k|^2\left(\sqrt[n]{|a_n|}\right)^{2(n-k-1)}}\right)\right).$$

(b) (i) Let $j \in \{2,\ldots,n-1\}$ such that $a_j \neq 0$, and let A be as given by (3) with $b_1 = \cdots = b_{n-1} =: b$, where b is a jth root of a_j. Then

$$r(A) \le \frac{5}{4}\sqrt[j]{|a_j|} + \frac{1}{2}|a_1| + \frac{1}{2}\left[\left(|a_1| - \frac{1}{2}\sqrt[j]{|a_j|}\right)^2 + \sum_{\substack{k=2\\k\neq j}}^{n}\frac{|a_k|^2}{\left(\sqrt[j]{|a_j|}\right)^{2(k-1)}}\right]^{\frac{1}{2}}.$$

(ii) Let A be as given by (3) with $b_1 = \cdots = b_{n-1} =: b$. Assume that there exists $j \in \{1,\ldots,n-2\}$ such that $c_n = -c_j$. Then

$$r(A^{-1}) \le \frac{1}{2|bc_n|}\left(|c_{n-1}| + \frac{5}{2}|c_n| + \left[\left(|c_{n-1}| - \frac{1}{2}|c_n|\right)^2 + |b|^2 + \sum_{\substack{k=1\\k\neq j}}^{n-2}|c_k|^2\right]^{\frac{1}{2}}\right).$$

(iii) Let A be as given by (3) with $b_1 = \cdots = b_{n-1} =: b$, let $j \in \{1,\ldots,n-2\}$ such that $a_j \neq 0$, and let b be an $(n-j)$th root of $-\frac{a_n}{a_j}$, then

$$r(A^{-1}) \le \frac{1}{2|a_n|}\left(|a_{n-1}| + \frac{5}{2}|a_n|\left|\frac{a_j}{a_n}\right|^{\frac{1}{n-j}} + \left[\left(|a_{n-1}| - \frac{1}{2}|a_n|\left|\frac{a_j}{a_n}\right|^{\frac{1}{n-j}}\right)^2\right.\right.$$

$$\left.\left. + \left|\frac{a_n}{a_j}\right|^{\frac{2(n-1)}{n-j}} + \sum_{\substack{k=1\\k\neq j}}^{n-2}|a_k|^2\left|\frac{a_n}{a_j}\right|^{\frac{2(n-k-1)}{n-j}}\right]^{\frac{1}{2}}\right).$$

(c) (i) Let $a_1 \neq 0$, and let A be as given by (3) with $b_1 = \cdots = b_{n-1} =: b := a_1$. Then

$$r(A) \le \frac{3}{2}|a_1| + \frac{1}{2}\sqrt{\sum_{k=2}^{n}\frac{|a_k|^2}{|a_1|^{2(k-1)}}}.$$

(ii) Let A be as given by (3) with $b_1 = \cdots = b_{n-1} =: b$, and assume that $c_n = -c_{n-1}$. Then

$$r(A^{-1}) \le \frac{1}{2|b|}\left(3 + \frac{1}{|c_n|}\sqrt{|b|^2 + \sum_{k=1}^{n-2}|c_k|^2}\right).$$

(iii) Let A be as given by (3) with $b_1 = \cdots = b_{n-1} =: b$, let $a_{n-1} \neq 0$, and let $b := -\frac{a_n}{a_{n-1}}$, then

$$r(A^{-1}) \leq \frac{|a_{n-1}|}{2|a_n|} \left(3 + \frac{|a_n|^{n-2}}{|a_{n-1}|^{n-1}} \sqrt{\left|\frac{a_n}{a_{n-1}}\right|^2 + \sum_{k=1}^{n-2} |a_k|^2 \left|\frac{a_{n-1}}{a_n}\right|^{2(k-1)}} \right).$$

Proof. (a) Now we have

$$A := \begin{pmatrix} 0 & b & 0 & \cdots & 0 \\ \vdots & \ddots & & \ddots & \\ 0 & \cdots & & 0 & b \\ b & c_{n-1} & \cdots & c_2 & c_1 \end{pmatrix}$$

and $A = B_n + C_n$ with

$$B_n := \begin{pmatrix} 0 & b & 0 & \cdots & 0 \\ \vdots & \ddots & & \ddots & \\ 0 & \cdots & & 0 & b \\ b & 0 & \cdots & 0 & 0 \end{pmatrix}, \quad C_n := \begin{pmatrix} 0 & 0 & 0 & \cdots & 0 \\ \vdots & \ddots & & \ddots & \\ 0 & \cdots & & 0 & 0 \\ 0 & c_{n-1} & \cdots & c_2 & c_1 \end{pmatrix}.$$

Then $r(A) \leq r(B_n) + r(C_n)$. Since $r(B_n) = |b|$ (cf. [5], Theorem 5.3-1) and

$$r(C_n) = \frac{1}{2} \left(\sqrt{\sum_{k=1}^{n-1} |c_k|^2} + |c_1| \right),$$

the estimate for $r(A)$ follows.

For A^{-1} now we have

$$A^{-1} = \begin{pmatrix} -\frac{c_{n-1}}{b^2} & -\frac{c_{n-2}}{b^2} & \cdots & -\frac{c_1}{b^2} & \frac{1}{b} \\ \frac{1}{b} & & & & \\ & \frac{1}{b} & & & \\ & & \ddots & & \\ & & & \frac{1}{b} & 0 \end{pmatrix}.$$

and $A^{-1} = \widehat{B}_n + \widehat{C}_n$ with

$$\widehat{B}_n := \begin{pmatrix} 0 & 0 & \cdots & 0 & \frac{1}{b} \\ \frac{1}{b} & 0 & & \cdots & 0 \\ & \frac{1}{b} & \ddots & & \\ & & \ddots & & \\ 0 & \cdots & 0 & \frac{1}{b} & 0 \end{pmatrix}, \quad \widehat{C}_n := \begin{pmatrix} -\frac{c_{n-1}}{b^2} & -\frac{c_{n-2}}{b^2} & \cdots & -\frac{c_1}{b^2} & 0 \\ 0 & 0 & & \cdots & 0 \\ & 0 & & & \\ & & \ddots & & \\ 0 & \cdots & 0 & 0 & 0 \end{pmatrix}.$$

Then again $r(A^{-1}) \leq r(\widehat{B}_n) + r(\widehat{C}_n)$. Since $r(\widehat{B}_n) = \frac{1}{|b|}$ and

$$r(\widehat{C}_n) = \frac{1}{2|b|^2}\left(|c_{n-1}| + \sqrt{\sum_{k=1}^{n-1}|c_k|^2}\right)$$

the assertion follows now by a short computation.

(b)(i) We now have

$$A := \begin{pmatrix} 0 & b & 0 & & \cdots & & & 0 \\ & \ddots & \ddots & & & & & \\ \vdots & & & & b & & & \\ & & & & & \ddots & \ddots & \\ 0 & & \cdots & & & & 0 & b \\ c_n & c_{n-1} & \cdots & c_{j+1} & b & c_{j-1} & \cdots & c_2 & c_1 \end{pmatrix}$$

and $A = B_j + C_j$ with

$$B_j := \begin{pmatrix} 0 & b & 0 & & \cdots & & & 0 \\ & \ddots & \ddots & & & & & \\ & & & b & & & & \\ \vdots & & & 0 & & & & \\ & & & & b & & & \\ & & & & & \ddots & \ddots & \\ 0 & & \cdots & & & & 0 & b \\ 0 & 0 & \cdots & 0 & b & 0 & \cdots & 0 & 0 \end{pmatrix},$$

$$C_j := \begin{pmatrix} 0 & 0 & 0 & & \cdots & & & 0 \\ & \ddots & \ddots & & & & & \\ \vdots & & & & b & & & \\ & & & & & \ddots & \ddots & \\ 0 & & \cdots & & & & 0 & 0 \\ c_n & c_{n-1} & \cdots & c_{j+1} & 0 & c_{j-1} & \cdots & c_2 & c_1 \end{pmatrix}.$$

Then again $r(A) \leq r(B_j) + r(C_j)$. We have $r(B_j) = |b|$ (cf. [5], Theorem 5.3-1), and the application of Theorem A to the matrix C_j for $m = n$ gives

$$r(C_j) \leq \frac{1}{2}\left(|c_1| + \frac{1}{2}|b|\right) + \frac{1}{2}\left[\left(|c_1| - \frac{1}{2}|b|\right)^2 + \sum_{\substack{k=2 \\ k \neq j}}^{n} |c_k|^2\right]^{\frac{1}{2}}.$$

From this the assertion follows.

(b)(ii) The proof for the estimate of $r(A^{-1})$ is similar to the proof for the estimate of $r(A)$ in (i) by the application of an analogous method to the inverse matrix A^{-1} of A, and is therefore omitted.

(b)(iii) follows from (b)(ii), since the assumptions of (b)(iii) imply $c_n = -c_j$.

(c)(i) We now have

$$A := \begin{pmatrix} 0 & b & 0 & \cdots & 0 \\ \vdots & \ddots & \ddots & & \\ 0 & \cdots & & 0 & b \\ c_n & c_{n-1} & \cdots & c_2 & b \end{pmatrix}$$

and $A = B_1 + C_1$ with

$$B_1 := \begin{pmatrix} 0 & b & 0 & \cdots & 0 \\ \vdots & \ddots & \ddots & & \vdots \\ & & & b & 0 \\ 0 & \cdots & & 0 & 0 \\ 0 & \cdots & & 0 & b \end{pmatrix}, \quad C_1 := \begin{pmatrix} 0 & 0 & 0 & \cdots & 0 \\ \vdots & \ddots & \ddots & & \vdots \\ 0 & \cdots & & 0 & 0 \\ 0 & \cdots & & 0 & b \\ c_n & c_{n-1} & \cdots & c_2 & 0 \end{pmatrix}.$$

Then again $r(A) \leq r(B_1) + r(C_1)$. Since $r(B_1) = |b|$ (cf. [5], Theorem 5.3-1) and $r(C_1) \leq \frac{1}{2}|b| + \frac{1}{2}\sqrt{\sum_{k=2}^{n}|c_k|^2}$, the assertion follows.

(c)(ii) The assumptions imply that now we have

$$A^{-1} = \begin{pmatrix} \frac{1}{b} & -\frac{c_{n-2}}{bc_n} & \cdots & -\frac{c_1}{bc_n} & \frac{1}{c_n} \\ \frac{1}{b} & & & & \\ & \frac{1}{b} & & & \\ & & \ddots & & \\ & & & \frac{1}{b} & 0 \end{pmatrix},$$

and $A^{-1} = \widehat{B}_1 + \widehat{C}_1$ with

$$\widehat{B}_1 = \begin{pmatrix} \frac{1}{b} & 0 & \cdots & & 0 \\ 0 & 0 & \cdots & & 0 \\ 0 & \frac{1}{b} & 0 & \cdots & 0 \\ \vdots & \ddots & \ddots & \ddots & \vdots \\ 0 & \cdots & 0 & \frac{1}{b} & 0 \end{pmatrix}, \quad \widehat{C}_1 = \begin{pmatrix} 0 & -\frac{c_{n-2}}{bc_n} & \cdots & -\frac{c_1}{bc_n} & \frac{1}{c_n} \\ \frac{1}{b} & 0 & \cdots & & 0 \\ 0 & 0 & & \cdots & 0 \\ \vdots & \ddots & \ddots & \ddots & \vdots \\ 0 & \cdots & 0 & 0 & 0 \end{pmatrix}.$$

Thus the estimate for $r(A^{-1})$ follows analogously to the estimate for $r(A)$ in (i).

(c)(iii) follows from (c)(ii), since the assumptions of (iii) imply $c_n = -c_{n-1}$. $\square$

Remark 2. If we apply Theorem A for $m = n - j + 1$ to the matrix C_j from the proof of Proposition 5(b)(i) we have the estimate

$$r(C_j) \leq \frac{1}{4} \left(|c_1| + \sqrt{\sum_{\substack{k=1 \\ k \neq j}}^{n} |c_k|^2} \right) + \frac{1}{2} \left[|b|^2 + \frac{1}{4} \left(|c_1| + \sqrt{\sum_{\substack{k=1 \\ k \neq j}}^{n} |c_k|^2} \right)^2 \right]^{\frac{1}{2}}.$$

Using this now we have the estimate

$$r(A) \leq \sqrt[j]{|a_j|} + \frac{1}{4} \left(|a_1| + \sqrt{\sum_{\substack{k=1 \\ k \neq j}}^{n} \frac{|a_k|^2}{\left(\sqrt[j]{|a_j|}\right)^{2(k-1)}}} \right)$$

$$+ \frac{1}{2} \left[\sqrt[j]{|a_j|^2} + \frac{1}{4} \left(|a_1| + \sqrt{\sum_{\substack{k=1 \\ k \neq j}}^{n} \frac{|a_k|^2}{\left(\sqrt[j]{|a_j|}\right)^{2(k-1)}}} \right)^2 \right]^{\frac{1}{2}}.$$

This bound can be better than the bound of Proposition 5(b)(i) and vice versa. Furthermore, the application of the same method to the matrix A^{-1} in Proposition 5(b)(ii) now gives

$$r(A^{-1}) \leq \frac{1}{4|bc_n|} \left(4|c_n| + |c_{n-1}| + \sqrt{|b|^2 + \sum_{\substack{k=1 \\ k \neq j}}^{n-1} |c_k|^2} \right.$$

$$+ \left[4|c_n|^2 + \left(|c_{n-1}| + \sqrt{|b|^2 + \sum_{\substack{k=1 \\ k \neq j}}^{n-1} |c_k|^2} \right)^2 \right]^{\frac{1}{2}} \right),$$

and a corresponding estimate in the case of Proposition 5(b)(iii).

Theorem 3. *Let P_n be the monic polynomial of degree $n \geq 3$ given by (1). Then all zeros of P_n lie in the intersection of the annuli given by*

$$|a_n| \left(\left(\sqrt[n]{|a_n|}\right)^{n-1} + \frac{1}{2} \left(|a_{n-1}| + \sqrt{\sum_{k=1}^{n-1} |a_k|^2 \left(\sqrt[n]{|a_n|}\right)^{2(n-k+1)}} \right) \right)^{-1}$$

$$\leq |x| \leq \sqrt[n]{|a_n|} + \frac{1}{2} \left(|a_1| + \sqrt{\sum_{k=1}^{n-1} \frac{|a_k|^2}{\left(\sqrt[n]{|a_n|}\right)^{2(k-1)}}} \right),$$

and

$$2|a_n|\left(\left(|a_{n-1}|+\frac{5}{2}|a_n|\left|\frac{a_j}{a_n}\right|^{\frac{1}{n-j}}\right.\right.+$$

$$\left.\left[\left(|a_{n-1}|-\frac{1}{2}|a_n|\left|\frac{a_j}{a_n}\right|^{\frac{1}{n-j}}\right)^2+\left|\frac{a_n}{a_j}\right|^{\frac{2(n-1)}{n-j}}+\sum_{\substack{k=1\\k\neq j}}^{n-2}|a_k|^2\left|\frac{a_n}{a_j}\right|^{\frac{2(n-k-1)}{n-j}}\right]^{\frac{1}{2}}\right)^{-1}$$

$$\leq|x|\leq\frac{3}{2}\sqrt[n]{|a_j|}+\frac{1}{2}\left(|a_1|+\sqrt{\sum_{\substack{k=1\\k\neq j}}^{n}\frac{|a_k|^2}{\left(\sqrt[n]{|a_j|}\right)^{2(k-1)}}}\right)$$

for $j\in\{2,\ldots,n-1\}$ with $a_j\neq 0$, and

$$\frac{2|a_n|}{|a_{n-1}|}\left(3+\frac{|a_n|^{n-2}}{|a_{n-1}|^{n-1}}\sqrt{\left|\frac{a_n}{a_{n-1}}\right|^2+\sum_{k=1}^{n-2}|a_k|^2\left|\frac{a_{n-1}}{a_n}\right|^{k-1}}\right)^{-1}$$

$$\leq|x|\leq\frac{3}{2}|a_1|+\frac{1}{2}\sqrt{\sum_{k=2}^{n}\frac{|a_k|^2}{|a_1|^{2(k-1)}}}$$

for $a_1,a_{n-1}\neq 0$.

Proof. The assertion follows from Proposition 5(a), (b)(i) and (ii), (c)(i) and (iii). □

In the next theorem we extend the method of proof of Proposition 5 to matrices of type given in (15).

Theorem 4. *Let P_n be the monic polynomial of degree $n\geq 3$ given by (1), and let α_n denote any nth root of a_n. Then all zeros of P_n lie in the intersection of the circles given by*

$$|x|\leq\sqrt[n]{|a_n|}+\frac{1}{2}\left(|a_1-\alpha_n|+\sqrt{|a_1-\alpha_n|^2+\sum_{k=1}^{n-1}|a_{k+1}+a_k\alpha_n|^2|a_n|^{-\frac{2k}{n}}}\right),$$

and

$$|x|\leq\frac{3}{4}|a_1|+\frac{1}{2}\sqrt{\frac{|a_n|^22^{2(n-1)}}{|a_1|^{2(n-1)}}+\sum_{k=1}^{n-1}\frac{2^{2k}}{|a_1|^{2k}}\left|a_{k+1}-\frac{1}{2}a_1a_k\right|^2},$$

for $a_1\neq 0$.

Proof. We consider the companion matrix $\widehat{A}_{-\alpha_n}$ of the polynomial $Q_{n+1}(x):=$

$(x + \alpha_n)P_n(x)$ given by

$$
\widehat{A}_{-\alpha_n} := \begin{pmatrix}
0 & \alpha_n & 0 & \cdots & & 0 \\
& & \alpha_n & & & \\
& \ddots & & & \ddots & \\
0 & \cdots & & 0 & & \alpha_n \\
\alpha_n & c_n + c_{n-1} & \cdots & c_2 + c_1 & & c_1 - \alpha_n
\end{pmatrix}.
$$

Then a similar computation of $r(\widehat{A}_{-\alpha_n})$ to the computation of $r(A)$ in Proposition 5(a) gives the first estimate.

For the proof of the second estimate we consider the companion matrix $\widehat{A}_{-a_1/2}$ of the polynomial $Q_{n+1}(x) := (x + \frac{1}{2}a_1)P_n(x)$ given by

$$
\widehat{A}_{-a_1/2} := \begin{pmatrix}
0 & \frac{1}{2}a_1 & 0 & \cdots & & 0 \\
& & \frac{1}{2}a_1 & & & \\
& \ddots & & & \ddots & \\
0 & \cdots & & 0 & & \frac{1}{2}a_1 \\
c_n & c_n + c_{n-1} & \cdots & c_2 + c_1 & & \frac{1}{2}a_1
\end{pmatrix}.
$$

Then a similar computation of $r(\widehat{A}_{-a_1/2})$ to the computation of $r(A)$ in Proposition 5(c)i gives the second estimate. $\square$

3. A Companion Matrix of Generalized Frobenius Type

In this section we consider a generalized companion matrix of the polynomial P_n in the sense that the characteristic polynomial of this matrix has the same zeros as P_n and additionally the zero 0 with a certain multiplicity. This type of matrix has been also used in [9] to get bounds for the zeros of P_n from eigenvalues and singular values. The elements of the matrix are based on multiplicative decompositions of the coefficients of the polynomial, where in contrast to the decomposition (2) there is no dependence between the coefficients. Throughout this section we assume that $n \geq 3$.

Proposition 6 (cf. [9]). *Let P_n be the monic polynomial of degree $n \geq 1$ given by (1). Let there exist complex numbers $a_1^{(1)}, a_2^{(1)}, a_2^{(2)}, \ldots, a_n^{(1)}, a_n^{(2)}, \ldots, a_n^{(n)} \in \mathbb{C}$ such that*

$$
\begin{aligned}
a_1 &:= a_1^{(1)}, \\
a_2 &:= a_2^{(1)} a_2^{(2)}, \\
&\ \ \vdots \\
a_n &:= a_n^{(1)} a_n^{(2)} \cdots a_n^{(n)}.
\end{aligned}
\tag{16}
$$

Let $\widehat{A}$ be the $\widehat{n}$-by-$\widehat{n}$-matrix given by

$$
\begin{pmatrix}
0 & a_n^{(2)} & 0 & \cdots & 0 & 0 & \cdots & \cdots & \cdots & 0 & 0 \\
\vdots & \ddots & \ddots & & \vdots & \vdots & & & & \vdots & \vdots \\
\vdots & & \ddots & \ddots & 0 & \vdots & & & & \vdots & \vdots \\
\vdots & & & \ddots & a_n^{(n-1)} & 0 & \cdots & \cdots & \cdots & 0 & 0 \\
0 & \cdots & \cdots & \cdots & 0 & 0 & \cdots & \cdots & \cdots & 0 & a_n^{(n)} \\
\vdots & & & \vdots & & \ddots & & & & & \vdots \\
\vdots & & & \vdots & & & \ddots & & & & \vdots \\
\vdots & & & \vdots & & & & 0 & a_3^{(2)} & 0 & 0 \\
\vdots & & & \vdots & & & & 0 & 0 & 0 & a_3^{(3)} \\
0 & & & \vdots & & & & 0 & 0 & 0 & a_2^{(2)} \\
a_n^{(1)} & 0 & \cdots & \cdots & 0 & \cdots & \cdots & a_3^{(1)} & 0 & a_2^{(1)} & a_1^{(1)}
\end{pmatrix}, \quad (17)
$$

where $\widehat{n} := 1 + n(n-1)/2$. Then

$$
\det(x E_{\widehat{n}} - \widehat{A}) = x^{(n-1)(n-2)/2} P_n(x),
$$

where $E_{\widehat{n}}$ is the $\widehat{n}$-by-$\widehat{n}$ identity matrix.

The matrix $\widehat{A}$ is normal only in some special cases (cf. [9] for a discussion of the normality conditions for $\widehat{A}$).

Decompositions of type (16) of the coefficients of P_n are always possible. The simpliest one is if we choose $a_k^{(1)} := a_k$, $k = 1, 2, \ldots, n$, $a_k^{(j)} := 1$, $j = 2, \ldots, k$, $k = 2, \ldots, n$. In this case the matrix $\widehat{A}$ is a (generalized) Frobenius companion matrix. In [9] we proposed several special choices of the decomposition coefficients in (16), which give special estimates for the zeros of P_n.

Now we give estimates for the numerical radius of $\widehat{A}$.

Proposition 7. *Let $\widehat{A}$ be as given by (17), and let*

$$
\begin{aligned}
\widehat{\beta}_1 \ := \ \min\Big\{ & \cos\frac{\pi}{\widehat{n}+1} \max_{k=3,\ldots,n} \max_{j=2,\ldots,k-1} |a_k^{(j)}|, \\
& \frac{1}{2} \max_{k=3,\ldots,n} \max_{j=2,\ldots n-2} \left(|a_k^{(j)}| + |a_k^{(j+1)}| \right) \Big\}, \\
\widehat{\beta}_2 \ := \ \min\Big\{ & \cos\frac{\pi}{\widehat{n}} \max_{k=3,\ldots,n} \max_{j=2,\ldots,k-1} |a_k^{(j)}|, \\
& \frac{1}{2} \max_{k=3,\ldots,n} \max_{j=2,\ldots n-2} \left(|a_k^{(j)}| + |a_k^{(j+1)}| \right) \Big\}.
\end{aligned}
$$

224

Then we have

$$r(\widehat{A}) \le \widehat{\beta}_1 + \frac{1}{2}\left(|a_1^{(1)}| + \sqrt{\sum_{k=1}^{n}|a_k^{(1)}|^2} + \sqrt{\sum_{k=2}^{n}|a_k^{(k)}|^2}\right) \tag{18}$$

and

$$r(\widehat{A}) \le \frac{1}{2}\left(|a_1^{(1)}| + \widehat{\beta}_2\right)$$

$$+ \frac{1}{2}\left[\left(|a_1^{(1)}| - \widehat{\beta}_2\right)^2 + \left(\left(\sum_{k=2}^{n}|a_k^{(1)}|^2\right)^{\frac{1}{2}} + \left(\sum_{k=2}^{n}|a_k^{(k)}|^2\right)^{\frac{1}{2}}\right)^2\right]^{\frac{1}{2}}. \tag{19}$$

If $a_k^{(j)} = b$ for $k = 3,\ldots,n, j = 2,\ldots,k-1$, then

$$r(\widehat{A}) \le |b| \cos\frac{\pi}{n} + \frac{1}{2}\left(|a_1^{(1)}| + \sqrt{\sum_{k=1}^{n}|a_k^{(1)}|^2} + \sqrt{\sum_{k=2}^{n}|a_k^{(k)}|^2}\right) \tag{20}$$

and

$$r(\widehat{A}) \le \frac{1}{2}\left(|a_1^{(1)}| + |b|\cos\frac{\pi}{n}\right)$$

$$+ \frac{1}{2}\left[\left(|a_1^{(1)}| - |b|\cos\frac{\pi}{n}\right)^2 + \left(\left(\sum_{k=2}^{n}|a_k^{(1)}|^2\right)^{\frac{1}{2}} + \left(\sum_{k=2}^{n}|a_k^{(k)}|^2\right)^{\frac{1}{2}}\right)^2\right]^{\frac{1}{2}} \tag{21}$$

Proof. First we prove (18). We set $\widehat{A} := \widehat{B} + \widehat{C} + \widehat{D}$ with

$$\widehat{B} := \begin{pmatrix}
0 & a_n^{(2)} & 0 & \cdots & 0 & 0 & \cdots & \cdots & \cdots & 0 & 0 \\
\vdots & \ddots & \ddots & & \vdots & \vdots & & & & \vdots & \vdots \\
\vdots & & \ddots & \ddots & 0 & \vdots & & & & \vdots & \vdots \\
\vdots & & & \ddots & a_n^{(n-1)} & 0 & \cdots & \cdots & \cdots & 0 & 0 \\
0 & \cdots & \cdots & \cdots & 0 & 0 & \cdots & \cdots & \cdots & 0 & 0 \\
\vdots & & & & \vdots & \ddots & & & & \vdots \\
\vdots & & & & \vdots & & \ddots & & & \vdots \\
\vdots & & & & \vdots & & & 0 & a_3^{(2)} & 0 & 0 \\
\vdots & & & & \vdots & & & 0 & 0 & 0 & 0 \\
0 & & & & \vdots & & & 0 & 0 & 0 & 0 \\
0 & 0 & \cdots & \cdots & 0 & & \cdots & \cdots & 0 & 0 & 0
\end{pmatrix},$$

$$\widehat{C} := \begin{pmatrix} 0 & 0 & 0 & \cdots & 0 & 0 & \cdots & \cdots & \cdots & 0 & 0 \\ \vdots & \ddots & \ddots & & \vdots & \vdots & & & & \vdots & \vdots \\ \vdots & & \ddots & \ddots & 0 & \vdots & & & & \vdots & \vdots \\ \vdots & & & \ddots & 0 & 0 & \cdots & \cdots & \cdots & 0 & 0 \\ 0 & \cdots & \cdots & \cdots & 0 & 0 & \cdots & \cdots & \cdots & 0 & 0 \\ \vdots & & & & \vdots & & \ddots & & & & \vdots \\ \vdots & & & & \vdots & & & \ddots & & & \vdots \\ \vdots & & & & \vdots & & & & 0 & 0 & 0 & 0 \\ \vdots & & & & \vdots & & & & 0 & 0 & 0 & 0 \\ 0 & & & & \vdots & & & & 0 & 0 & 0 & 0 \\ a_n^{(1)} & 0 & \cdots & \cdots & 0 & \cdots & \cdots & a_3^{(1)} & 0 & a_2^{(1)} & a_1^{(1)} \end{pmatrix} ,$$

$$\widehat{D} := \begin{pmatrix} 0 & 0 & 0 & \cdots & 0 & 0 & \cdots & \cdots & \cdots & 0 & 0 \\ \vdots & \ddots & \ddots & & \vdots & \vdots & & & & \vdots & \vdots \\ \vdots & & \ddots & \ddots & 0 & \vdots & & & & \vdots & \vdots \\ \vdots & & & \ddots & 0 & 0 & \cdots & \cdots & \cdots & 0 & 0 \\ 0 & \cdots & \cdots & \cdots & 0 & 0 & \cdots & \cdots & \cdots & 0 & a_n^{(n)} \\ \vdots & & & & \vdots & & \ddots & & & & \vdots \\ \vdots & & & & \vdots & & & \ddots & & & \vdots \\ \vdots & & & & \vdots & & & & 0 & 0 & 0 & 0 \\ \vdots & & & & \vdots & & & & 0 & 0 & 0 & a_3^{(3)} \\ 0 & & & & \vdots & & & & 0 & 0 & 0 & a_2^{(2)} \\ 0 & 0 & \cdots & \cdots & 0 & \cdots & \cdots & 0 & 0 & 0 & 0 \end{pmatrix} .$$

Then $r(\widehat{A}) \le r(\widehat{B}) + r(\widehat{C}) + r(\widehat{D})$. Again from Theorem 3 in [2] it follows that

$$r(\widehat{B}) \le \min \left\{ \cos \frac{\pi}{\widehat{n}+1} \max_{k=3,\ldots,n} \max_{j=2,\ldots,k-1} |a_k^{(j)}|, \right.$$
$$\left. \frac{1}{2} \max_{k=3,\ldots,n} \max_{j=2,\ldots n-2} \left(|a_k^{(j)}| + |a_k^{(j+1)}| \right) \right\}.$$

Since $\widehat{C}x = (x, \widehat{c})e_{\widehat{n}}$, where $\widehat{c} = (a_n^{(1)}, 0, \ldots, 0, \ldots, 0, a_3^{(1)}, 0, a_2^{(1)}, a_1^{(1)})^T$, $e_{\widehat{n}} = (0, \ldots, 0, 1)^T$, we have from Theorem 1 in [4] that

$$r(\widehat{C}) = \frac{1}{2} \left(|a_1^{(1)}| + \sqrt{\sum_{k=1}^{n} |a_k^{(1)}|^2} \right).$$

Since $\widehat{D}x = (x, e_{\widehat{n}})\widehat{d}$, where $\widehat{d} = (0, \ldots, 0, a_n^{(n)}, \ldots, 0, a_3^{(3)}, a_2^{(2)}, 0)^T$, again from Theorem 1 in [4] we have that

$$r(\widehat{D}) = \frac{1}{2}\sqrt{\sum_{k=2}^{n} |a_k^{(k)}|^2}.$$

From this the estimate (18) follows. For the proof of (19) we apply Theorem A for $m = \widehat{n}$. The $\widehat{n}$th projection $B_{\widehat{n}}$ of the matrix $\widehat{A}$ fulfills $r(B_{\widehat{n}}) \leq \widehat{\beta}_2$ (Theorem 3 in [2]). From this (19) follows by a short computation. For the proof of (20) and (21) we note that the assumptions imply that now $r(\widehat{B}) = |b| \cos \frac{\pi}{n}$ and $r(B_{\widehat{n}}) = |b| \cos \frac{\pi}{n}$ (cf. [3], Proposition 1). From this (20) and (21) follow. $\square$

It turns out that in the case of the matrix $\widehat{A}$ the estimate (19) ((21)) is always better than the estimate (18) ((20)). Therefore in the following we only take the estimates (19) and (21) for further use.

Theorem 5. *(a) Let the monic polynomial P_n satisfy the assumptions of Proposition 6. Then all zeros of P_n lie in the circle given by*

$$|x| \leq \frac{1}{2}\left(\left|a_1^{(1)}\right| + \widehat{\beta}_2\right)$$
$$+\frac{1}{2}\left[\left(\left|a_1^{(1)}\right| - \widehat{\beta}_2\right)^2 + \left(\left(\sum_{k=2}^{n} |a_k^{(1)}|^2\right)^{\frac{1}{2}} + \left(\sum_{k=2}^{n} |a_k^{(k)}|^2\right)^{\frac{1}{2}}\right)^2\right]^{\frac{1}{2}}.$$

If $a_k^{(j)} = b$ for $k = 3, \ldots, n, j = 2, \ldots, k-1$, then the circle is given by

$$|x| \leq \frac{1}{2}\left(\left|a_1^{(1)}\right| + |b| \cos \frac{\pi}{n}\right)$$
$$+\frac{1}{2}\left[\left(\left|a_1^{(1)}\right| - |b| \cos \frac{\pi}{n}\right)^2 + \left(\left(\sum_{k=2}^{n} |a_k^{(1)}|^2\right)^{\frac{1}{2}} + \left(\sum_{k=2}^{n} |a_k^{(k)}|^2\right)^{\frac{1}{2}}\right)^2\right]^{\frac{1}{2}}.$$

(b) Let P_n be the monic polynomial of degree $n \geq 3$ given by (1).

(i) Let $\alpha > 0$ be a positive real number. Then all zeros of P_n lie in the circle given by

$$|x| \leq \frac{1}{2}\left(|a_1| + \alpha \cos \frac{\pi}{n}\right) + \frac{1}{2}\left[\left(|a_1| - \alpha \cos \frac{\pi}{n}\right)^2 + 4\sum_{k=2}^{n} \frac{|a_k|}{\alpha^{k-2}}\right]^{\frac{1}{2}}. \tag{22}$$

(ii) All zeros of P_n lie in the intersection of the circles given by

$$|x| \leq \frac{1}{2}\left(|a_1| + \cos \frac{\pi}{\widehat{n}} \max_{k=3,\ldots,n} \sqrt[k]{|a_k|}\right)$$

$$+\frac{1}{2}\left[\left(|a_1|-\cos\frac{\pi}{n}\max_{k=3,\ldots,n}\sqrt[k]{|a_k|}\right)^2+4\sum_{k=2}^{n}\sqrt[k]{|a_k|^2}\right]^{\frac{1}{2}},$$

$$|x|\ \leq\ \sum_{k=1}^{n}\sqrt[k]{|a_k|}.$$

Proof. (a) follows from the estimates (19) and (21) of Proposition 7. (b)(i) follows from estimate (21) of Proposition 7 by the choice $a_k^{(j)}=\alpha$ for $k=3,\ldots,n, j=2,\ldots,k-1$, and the choice $a_k^{(1)}=a_k^{(k)}$ to be equal to a squareroot of $\frac{a_k}{\alpha^{k-2}}$ for $k=2,\ldots,n$. (b)(ii) (the first estimate) follows from the first assertion of (a) by the choice $a_k^{(j)}$ to be the same kth root of a_k for $j=1,\ldots,k$, and for $k=2,\ldots,n$. To prove (b)(ii) (the second estimate) we decompose $\widehat{A}$ in the sum of matrices of type

where $a_k^{(j)}$ for $j=1,\ldots,k$ is equal to the same kth root of a_k. Then the numerical radius of each of these matrices is equal to $\sqrt[k]{|a_k|}$. From this the assertion follows. $\square$

In Theorem 5(b)(i) special choices of α give different bounds. For example the choices

$$\alpha:=1,$$

and

$$\alpha:=\max_{k=2,\ldots,n}|a_k|^{\frac{1}{k}}$$

are useful. The last estimate in Theorem 5(b)(ii) is a result of Walsh (cf. [11], p. 126).

For the case $a_1 = 0$ we compare the bound (22)

$$B_{L_1} := \frac{1}{2}\alpha\cos\frac{\pi}{n} + \frac{1}{2}\left[\alpha^2\cos^2\frac{\pi}{n} + 4\sum_{k=2}^{n}\frac{|a_k|}{\alpha^{k-2}}\right]^{\frac{1}{2}}$$

with a bound from [9]

$$B_{L_2} := \max\{\alpha, \left(\sum_{k=2}^{n}\frac{|a_k|}{\alpha^{k-2}}\right)^{\frac{1}{2}}.$$

If

$$\alpha \leq \left(\sum_{k=2}^{n}\frac{|a_k|}{\alpha^{k-2}}\right)^{\frac{1}{2}},$$

then the bound B_{L_2} is better than B_{L_1}. But if

$$\left(\sum_{k=2}^{n}\frac{|a_k|}{\alpha^{k-2}}\right)^{\frac{1}{2}} \leq \alpha\sqrt{1 - \cos\frac{\pi}{n}},$$

then B_{L_1} is better than B_{L_2}.

We also can combine the different types of estimates in Theorem 5. But since there are many different possibilities of combinations we give only one example.

Theorem 6. *Let P_n be the monic polynomial of degree $n \geq 3$ given by (1). Let $\emptyset \neq I_1, I_2 \subset \{1,\ldots,n\}, I_1 \cap I_2 = \emptyset, I_1 \cup I_2 = \{1,\ldots,n\}, \max I_1 \geq 3$, and let $\alpha > 0$ be a positive real number.*

(a) Let $1 \in I_1$. Then all zeros of P_n lie in the circle given by

$$|x| \leq \sum_{k\in I_2}\sqrt[k]{|a_k|} + \frac{1}{2}\left(|a_1| + \alpha\cos\frac{\pi}{\max I_1}\right)$$

$$+ \frac{1}{2}\left[\left(|a_1| - \alpha\cos\frac{\pi}{\max I_1}\right)^2 + 4\sum_{k\in I_1\backslash\{1\}}\frac{|a_k|}{\alpha^{k-2}}\right]^{\frac{1}{2}}.$$

(b) Let $1 \notin I_1$. Then all zeros of P_n lie in the circle given by

$$|x| \leq \sum_{k\in I_2}\sqrt[k]{|a_k|} + \frac{1}{2}\alpha\cos\frac{\pi}{\max I_1} + \frac{1}{2}\left[\alpha^2\cos^2\frac{\pi}{\max I_1} + 4\sum_{k\in I_1}\frac{|a_k|}{\alpha^{k-2}}\right]^{\frac{1}{2}}.$$

Proof. We decompose the matrix $\hat{A}$ in the sum of two matrices each of which consists of blocks of type defined in the proof of Theorem 5(b)(ii) (the second estimate). The first matrix corresponds to elements a_k for $k \in I_1$. To get an estimate of the numerical radius of this matrix we apply the method of proof of the second estimate for the numerical radius of $\hat{A}$ in Proposition 7 and additionally use the special choice of Theorem 5(b)(i). The second matrix corresponds to elements a_k for $k \in I_2$. To this matrix we apply the method of proof of the third estimate in Theorem 5(b). From this the assertions follow. $\square$

References

[1] A. A. Abdurakhmanov, *The geometry of the Hausdorff domain in localization problems for the spectrum of arbitrary matrices*, Math. USSR Sbornik **59** (1988), 39-51.

[2] M.-T. Chien, *On the numerical range of tridiagonal operators*, Lin. Alg. Appl. **246** (1996), 203-214.

[3] K. R. Davidson and J. A. R. Holbrook, *Numerical radii of zero-one matrices*, Michigan Math. J. **35** (1988), 261-267.

[4] M. Fujii and F. Kubo, *Buzano's inequality and bounds for roots of algebraic equations*, Proc. Am. Math. Soc. **117** (1993), 359-361.

[5] K. E. Gustafson and D. K. M. Rao, *Numerical Range*, Springer-Verlag, New York, 1997.

[6] R. A. Horn and C. R. Johnson, *Matrix Analysis*, Cambridge University Press, Cambridge, 1990.

[7] R. A. Horn and C. R. Johnson, *Topics in Matrix Analysis*, Cambridge University Press, Cambridge, 1991.

[8] F. Kittaneh, *Singular values of companion matrices and bounds on zeros of polynomials*, SIAM J. Matrix Anal. Appl. **16** (1995), 333-340.

[9] H. Linden, *Bounds for the zeros of polynomials from eigenvalues and singular values of some companion matrices*, Lin. Alg. Appl. **271** (1998), 41-82.

[10] M. Marcus and B. N. Shure, *The numerical range of certain* $0, 1 - Matrices$, Lin. Alg. Appl. **7** (1979), 111-120.

[11] M. Marden, *Geometry of Polynomials*, Amer. Math. Soc. Surveys, Vol. 3, 2nd ed., Providence, R. I., 1969.

[12] M. Mignotte, *Mathematics for Computer Algebra*, Springer, New York, 1992.

[13] G. V. Milovanović, D. S. Mitrinović, and Th. M. Rassias, *Topics in Polynomials: Extremal Problems, Inequalities, Zeros*, World Scientific, Singapore New Jersey London Hong Kong, 1994.

[14] M. Parodi, *La Localisation des Valeurs Caractéristiques des Matrices et ses Applications*, Gauthier-Villars, Paris, 1959.

BOUNDS ON ENTROPY MEASURES FOR MIXED POPULATIONS

M. MATIĆ
Dept. Math., FESB, R. Boškovića b.b., Split, Croatia

C. E. M. PEARCE
*Dept. Appl. Math. The University of Adelaide, Adelaide
SA 5005, Australia*

J. PEČARIĆ
Faculty of Textile Technology, Pierottijeva 6, Zagreb, Croatia

Abstract. We derive a number of bounds relating entropy measures for a mixed population to parameters of the n constituent populations. The results extend some recent studies of Dragomir and Goh for the two–population case and are tighter than those given by them for the case $n = 2$.

1. Introduction

Recently a number of improvements have been made to some of the fundamental inequalities for various entropy measures to do with discrete–valued random variables in information theory. We note in particular the work of Dragomir and Goh [1–3]. See also the present authors [4,5].

In this article we address the situation of n random variables X_i having the same finite range but different probability distributions. If we mix populations represented by these random variables with respective proportions α_i (with $\alpha_i > 0$ and $\sum_{i=1}^{n} \alpha_i = 1$), we derive a new population represented by the random variable X with

$$(1.1) \qquad p(x) = \sum_{i=1}^{n} \alpha_i p_i(x),$$

where $p_i(x) = P(X_i = x)$ gives the probability distribution of X_i ($i = 1, \ldots, n$) and $p(x)$ that of X. Such a situation can arise, in particular, in problems involving a number of sources being fed into a common channel.

We are interested in relations between information measures for X and those for the original random variables X_i. For example, for $b > 1$ the b–entropy of X is

1991 *Mathematics Subject Classification.* 94A17, 26D15
Keywords and phrases. Bounds, entropy measures, mixed populations

T.M. Rassias and H.M. Srivastava (eds.), Analytic and Geometric Inequalities and Applications, 231–244.

defined by

$$H(X) := \sum_x p(x) \log_b 1/p(x).$$

The entropy function H is a concave functional of the probability distribution, that is,

$$(1.2) \qquad \sum_{i=1}^n \alpha_i H(X_i) \leq H(X),$$

so that $H(X) - \sum_i \alpha_i H(X_i)$ is bounded below by zero. But what can be said about upper bounds for this quantity? We provide several answers to this question in Section 2, along with a simple derivation of (1.2).

In Section 3, 4 and 5 we give corresponding results for joint entropy, conditional entropy and mutual information respectively in the context of mixed populations. Our results both extend to general n known results for the case $n = 2$ derived in [3] and sharpen existing results for that case.

Our arguments make use of several basic results. One is the following lemma established as [4, Lemma 1.2]. The authors have found this to have a key role in the study of entropy measures.

Lemma A. *Let $\xi_k \in (0, \infty)$ $(1 \leq k \leq n)$ and $\rho := \max_{i,k} \xi_i/\xi_k$. Suppose $p_k \geq 0$ with $\sum_{k=1}^n p_k = 1$ and $b > 1$. Then*

$$0 \leq \log_b \left(\sum_{k=1}^n p_k \xi_k \right) - \sum_{k=1}^n p_k \log_b \xi_k \leq \frac{1}{4 \ln b} (\sqrt{\rho} - 1/\sqrt{\rho})^2.$$

If

$$\rho \leq \Phi(\varepsilon) := 1 + 2\varepsilon \ln b + 2\sqrt{\varepsilon \ln b(1 + \varepsilon \ln b)}$$

for $\varepsilon > 0$, then

$$0 \leq \log_b \left(\sum_{k=1}^n p_k \xi_k \right) - \sum_{k=1}^n p_k \log_b \xi_k \leq \varepsilon.$$

To prove inequalities for joint entropy, we shall need the following result of Dragomir and Goh [1, Proposition 4.1].

Lemma B. *Let $\xi_k \in (0, \infty)$, $p_k \geq 0$ with $\sum_{k=1}^n p_k = 1$ and $b > 1$. Then*

$$0 \leq \log_b \left(\sum_{k=1}^n p_k \xi_k \right) - \sum_{k=1}^n p_k \log_b \xi_k \leq \frac{1}{\ln b} \left[\sum_{k=1}^n \frac{p_k}{\xi_k} \sum_{j=1}^n p_j \xi_j - 1 \right].$$

We shall also need the following discrete version of the Grüss inequality (see [7, Chapter 10]).

Lemma C. *Let (a_i) and (b_i) be n-tuples of real numbers and $p_i \geq 0$ $(1 \leq i \leq n)$. If $\alpha \leq a_i \leq A$ and $\beta \leq b_i \leq B$, then*

$$\left| \sum_{i=1}^{n} p_i \sum_{j=1}^{n} p_j a_j b_j - \sum_{i=1}^{n} p_i a_i \sum_{j=1}^{n} p_j b_j \right| \leq \frac{1}{4}(A - \alpha)(B - \beta) \left(\sum_{i=1}^{n} p_i \right)^2.$$

2. Entropy for a mixed population

In this section we suppose as in the introduction that populations given by $p_i(\cdot)$ are mixed in accordance with probability weights α_i $(1 \leq i \leq n)$ to produce a population given by $p(\cdot)$. We derive several upper bounds for $H(X) - \sum_{i=1}^{n} \alpha_i H(X_i)$. We derive *en passant* that a lower bound is zero, that is to say, that H is concave.

Theorem 2.1. *With X_i and X as above, we have*

$$0 \leq H(X) - \sum_{i=1}^{n} \alpha_i H(X_i) \leq \frac{1}{\ln b} \left[\sum_x \sum_{i=1}^{n} \alpha_i p_i^2(x)/p(x) - 1 \right].$$

Proof. By (1.1) we have

$$H(X) - \sum_{i=1}^{n} \alpha_i H(X_i) = \sum_x p(x) \log_b 1/p(x) - \sum_{i=1}^{n} \alpha_i \sum_x p_i(x) \log_b 1/p_i(x)$$

(2.1)

$$= -\sum_{i=1}^{n} \alpha_i \sum_x p_i(x) \log_b \frac{p(x)}{p_i(x)}.$$

Using Lemma B with $p_k = p_i(x)$ and $\xi_k = p(x)/p_i(x)$, we get

$$0 \leq \log_b \left(\sum_x p_i(x) \frac{p(x)}{p_i(x)} \right) - \sum_x p_i(x) \log_b \frac{p(x)}{p_i(x)}$$

$$= -\sum_x p_i(x) \log_b \frac{p(x)}{p_i(x)}$$

(2.2)

$$\leq \frac{1}{\ln b} \left[\sum_x \frac{p_i^2(x)}{p(x)} \sum_y p_i(y) \frac{p(y)}{p_i(y)} - 1 \right]$$

$$= \frac{1}{\ln b} \left[\sum_x \frac{p_i^2(x)}{p(x)} - 1 \right] \quad (i = 1, 2, \ldots, n).$$

The desired result follows from (2.1) and (2.2). □

Theorem 2.2. *Define*

$$M_i := \max_x p_i(x)/p(x) \ , \ m_i := \min_x p_i(x)/p(x) \qquad (i = 1, 2, \ldots, n).$$

Then

(2.3)
$$0 \leq H(X) - \sum_{i=1}^{n} \alpha_i H(X_i) \leq \frac{1}{4 \ln b} \sum_{i=1}^{n} \alpha_i (M_i - m_i)^2.$$

Also

$$(2.4) \qquad 0 \le H(X) - \sum_{i=1}^{n} \alpha_i H(X_i) \le \frac{1}{4\ln b} \sum_{i=1}^{n} \alpha_i \left(\sqrt{\rho_i} - 1/\sqrt{\rho_i}\right)^2,$$

where $\rho_i = M_i/m_i$ $(i = 1, 2, \ldots, n)$.

Proof. By Lemma C with $p_i = p(x)$ and $a_i = b_i = p_j(x)/p(x)$, we have

$$(2.5) \qquad \sum_{x} \frac{p_j^2(x)}{p(x)} - 1 = \sum_{y} p(y) \sum_{x} p(x) \left(\frac{p_j(x)}{p(x)}\right)^2 - \left(\sum_{x} p(x) \frac{p_j(x)}{p(x)}\right)^2$$
$$\le \frac{1}{4}(M_j - m_j)^2 \quad (1 \le j \le n).$$

Inequality (2.3) follows from (2.1), (2.2) and (2.5). To prove (2.4), set $p_k = p_i(x)$ and $\xi_k = p(x)/p_i(x)$ in Lemma A. We get

$$(2.6) \qquad \begin{aligned} 0 &\le \log_b \left(\sum_{x} p_i(x) \frac{p(x)}{p_i(x)}\right) - \sum_{x} p_i(x) \log_b \frac{p(x)}{p_i(x)} \\ &= -\sum_{x} p_i(x) \log_b \frac{p(x)}{p_i(x)} \\ &\le \frac{1}{4\ln b} \left(\sqrt{\rho_i} - 1/\sqrt{\rho_i}\right)^2 \qquad (i = 1, 2, \ldots, n). \end{aligned}$$

Now (2.4) follows from (2.1) and (2.6). $\qquad\qquad\square$

Remark. Let $\max_i(M_i - m_i) = M - m$. Since

$$\max_{x,y} \left\{ \max_{i=1,2,\ldots,n} \left\{ \left| \frac{p_i(x)}{p(x)} - \frac{p_i(y)}{p(y)} \right| \right\} \right\} = \max\{M_i - m_i\} = M - m,$$

we get from (2.3) that

$$0 \le H(X) - \sum_{i=1}^{n} \alpha_i H(X_i) \le \frac{1}{4\ln b}(M - m)^2.$$

Furthermore, the function $f(x) = (\sqrt{x} - 1/\sqrt{x})^2$ is increasing for $x \ge 1$. Since $\rho_i = M_i/m_i \ge 1$ for $i = 1, 2, \ldots, n$, we derive from (2.4) that

$$0 \le H(X) - \sum_{i=1}^{n} \alpha_i H(X_i) \le \frac{1}{4\ln b} \left(\sqrt{\rho} - 1/\sqrt{\rho}\right)^2,$$

where $\rho = \max\{\rho_i\}$.

Corollary 2.3. *If*

$$\max_{x,y} \left\{ \max_{i=1,2,\ldots,n} \left\{ \left| \frac{p_i(x)}{p(x)} - \frac{p_i(y)}{p(y)} \right| \right\} \right\} \le 2\sqrt{\varepsilon \ln b}$$

for some $\varepsilon = \varepsilon(\alpha_1, \alpha_2, \ldots, \alpha_n) > 0$, then

$$0 \le H(X) - \sum_{i=1}^{n} \alpha_i H(X_i) \le \varepsilon.$$

Proof. The result follows from the preceding remark and (2.3). $\square$

Remark. In the case $n = 2$, this strengthens a result of Dragomir and Goh [3, Cor. 4.2], which gives the result with $\sqrt{2\varepsilon \ln b}$ in place of $2\sqrt{\varepsilon \ln b}$.

The upper bounds on $H(X) - \sum_{i=1}^{n} \alpha_i H(X_i)$ derived in Theorems 2.1 and 2.2 involve the quantities α_i. To conclude this section we derive one that does not. We employ the following result.

Lemma 2.4. *Suppose*

$$g(\alpha_1, \ldots, \alpha_n) := \frac{\sum_{i=1}^{n} \alpha_i t_i u_i^2}{\sum_{i=1}^{n} \alpha_i t_i u_i} \quad (0 \le \alpha_i \le 1),$$

where $0 < t_i, u_i \le 1$ $(1 \le i \le n)$ are fixed. Then

$$\min_i \{u_i\} \le g \le \max_i \{u_i\} \quad \forall \alpha_i \in [0, 1].$$

Proof. We may regard g as an average $g = \sum_{i=1}^{n} w_i u_i$ of the values u_i using the weights

$$w_i = \alpha_i t_i u_i \Big/ \sum_j \alpha_j t_j u_j.$$

Since $w_i \ge 0$ and $\sum_i w_i = 1$, the quantity g lies between the least and greatest of the values u_i. $\square$

Theorem 2.5. *We have*

$$0 \le H(X) - \sum_{i=1}^{n} \alpha_i H(X_i) \le \frac{1}{\ln b} \left[\sum_x \max_i \{p_i(x)\} - 1 \right].$$

Proof. By (1.1), we may set $t_i = 1$, $u_i = p_i(x)$ in Lemma 2.4 to obtain

$$\sum_{i=1}^{n} \alpha_i p_i(x)^2 / p(x) \le \max_i \{p_i(x)\}.$$

The desired result follows from Theorem 2.1. $\square$

3. Joint entropy for a mixed population

Now we shall give results for joint entropy analogous to those for entropy. For X, Y discrete random variables with finite ranges, the joint entropy of X and Y is defined by

$$H(X,Y) := \sum_{x,y} p(x,y) \log_b 1/p(x,y)$$

(see [6, p. 25]), where $p(x,y) := P\{X = x, Y = y\}$ and the summation is over those pairs (x,y) for which $p(x,y) > 0$.

Theorem 3.1. *Suppose* (X_i, Y_i) *($1 \le i \le n$) are pairs of discrete random variables with the same finite range and respective joint probability distributions* $p_i(x,y)$. *Suppose further that the pair* (X,Y) *of discrete random variables satisfies*

$$p(x,y) := \sum_{i=1}^{n} \alpha_i p_i(x,y), \quad \alpha_i > 0, \ \sum_{i=1}^{n} \alpha_i = 1.$$

Then

$$0 \le H(X,Y) - \sum_{i=1}^{n} \alpha_i H(X_i, Y_i) \le \frac{1}{\ln b} \left[\sum_{x,y} \sum_{i=1}^{n} \alpha_i p_i^2(x,y)/p(x,y) - 1 \right].$$

Proof. We argue as in Theorem 2.1 with $p(x,y), p_i(x,y)$ in place of $p(x), p_i(x)$ respectively. □

Theorem 3.2. *Let* (X_i, Y_i) *($1 \le i \le n$) and* (X,Y) *be as in the previous theorem. Define*

$$M_i := \max_{x,y} \frac{p_i(x,y)}{p(x,y)}, \quad m_i := \min_{x,y} \frac{p_i(x,y)}{p(x,y)} \quad (i = 1, 2, \ldots, n).$$

Then

$$0 \le H(X,Y) - \sum_{i=1}^{n} \alpha_i H(X_i, Y_i) \le \frac{1}{4 \ln b} \sum_{i=1}^{n} \alpha_i (M_i - m_i)^2.$$

Also we have

$$0 \le H(X,Y) - \sum_{i=1}^{n} \alpha_i H(X_i, Y_i) \le \frac{1}{4 \ln b} \sum_{i=1}^{n} \alpha_i \left(\sqrt{\rho_i} - 1/\sqrt{\rho_i} \right)^2,$$

where $\rho_i = M_i/m_i$, *($i = 1, 2, \ldots, n$).*

Proof. The argument follows that of Theorem 2.2. □

Again we may argue as in the remark following Theorem 2.2 to derive the following.

Corollary 3.3. *If*

$$\max_{(x,y),(u,v)} \left\{ \max_{i=1,2,\dots,n} \left\{ \left| \frac{p_i(x,y)}{p(x,y)} - \frac{p_i(u,v)}{p(u,v)} \right| \right\} \right\} \le 2\sqrt{\varepsilon \ln b}$$

for some $\varepsilon = \varepsilon(\alpha_1, \alpha_2, \dots, \alpha_n) > 0$, *then*

$$0 \le H(X,Y) - \sum_{i=1}^{n} \alpha_i H(X_i, Y_i) \le \varepsilon.$$

Finally we have a result analogous to Theorem 2.5.

Theorem 3.4. *Let* (X_i, Y_i) *and* (X,Y) *be as in Theorem 3.1. Then*

$$0 \le H(X,Y) - \sum_{i=1}^{n} \alpha_i H(X_i, Y_i) \le \frac{1}{\ln b} \left[\sum_{x,y} \max_i \{p_i(x,y)\} - 1 \right].$$

4. Conditional entropy for a mixed population

For a pair of discrete random variables X, Y with finite ranges, the conditional
b–entropy of X given Y is defined by

$$H(X|Y) := \sum_{x,y} p(x,y) \log_b 1/p(x|y),$$

where $p(x,y) := P\{X = x, Y = y\}$ and $p(x|y) := P\{X = x | Y = y\} = p(x,y)/p(y)$
(see [6, p. 22]).

In this section we prove that conditional entropy is a concave functional of the joint
probability distribution $p(x,y)$. Also we derive some upper bounds for the difference
$H(X|Y) - \sum_{i=1}^{n} \alpha_i H(X_i|Y_i)$.

Theorem 4.1. *With* (X_i, Y_i) $(1 \le i \le n)$ *and* (X,Y) *as in Theorem 3.1, we have*

$$(4.1) \quad 0 \le H(X|Y) - \sum_{i=1}^{n} \alpha_i H(X_i|Y_i) \le \frac{1}{\ln b} \left[\sum_{x,y} \sum_{i=1}^{n} \alpha_i p_i(y) p_i^2(x|y)/p(x|y) - 1 \right].$$

Proof. Since $p(x,y) = \sum_{i=1}^{n} \alpha_i p_i(x,y)$, we have by the definition of conditional
entropy that

$$H(X|Y) - \sum_{i=1}^{n} \alpha_i H(X_i|Y_i) = \sum_{x,y} p(x,y) \log_b 1/p(x|y)$$

$$(4.2) \qquad\qquad\qquad\qquad - \sum_{i=1}^{n} \alpha_i \sum_{x,y} p_i(x,y) \log_b 1/p_i(x|y)$$

$$= - \sum_{i=1}^{n} \alpha_i \sum_{x,y} p_i(x,y) \log_b \frac{p(x|y)}{p_i(x|y)}.$$

Now $\sum_{x,y} p_i(y)p(x|y) = 1$ for $i = 1, 2, \ldots, n$, so by Lemma B with $p_k = p_i(x,y)$ and $\xi_k = p(x|y)/p_i(x|y)$ we get that

$$
0 \le \log_b \left(\sum_{x,y} p_i(x,y) \frac{p(x|y)}{p_i(x|y)} \right) - \sum_{x,y} p_i(x,y) \log_b \frac{p(x|y)}{p_i(x|y)}
$$

(4.3)

$$
= - \sum_{x,y} p_i(x,y) \log_b \frac{p(x|y)}{p_i(x|y)}
$$

$$
\le \frac{1}{\ln b} \left[\sum_{x,y} p_i(x,y) \frac{p_i(x|y)}{p(x|y)} \sum_{u,v} p_i(u,v) \frac{p(u|v)}{p_i(u|v)} - 1 \right]
$$

$$
= \frac{1}{\ln b} \left[\sum_{x,y} \frac{p_i(y)p_i^2(x|y)}{p(x|y)} - 1 \right] \quad (i = 1, 2, \ldots, n).
$$

The desired result now follows from (4.2) and (4.3). $\qquad\square$

Theorem 4.2. *Let (X_i, Y_i) $(1 \le i \le n)$ and (X, Y) be as in Theorem 3.1. Define*

$$
M_i := \max_{x,y} \frac{p_i(x|y)}{p(x|y)} \;, \quad m_i := \min_{x,y} \frac{p_i(x|y)}{p(x|y)} \quad (i = 1, 2, \ldots, n).
$$

Then

(4.4)
$$
0 \le H(X|Y) - \sum_{i=1}^{n} \alpha_i H(X_i|Y_i) \le \frac{1}{4\ln b} \sum_{i=1}^{n} \alpha_i (M_i - m_i)^2.
$$

Also

(4.5)
$$
0 \le H(X|Y) - \sum_{i=1}^{n} \alpha_i H(X_i|Y_i) \le \frac{1}{4\ln b} \sum_{i=1}^{n} \alpha_i \left(\sqrt{\rho_i} - 1/\sqrt{\rho_i} \right)^2,
$$

where $\rho_i = M_i/m_i$, $(i = 1, 2, \ldots, n)$.

Proof. To prove (4.4), note that for $j = 1, 2, \ldots, n$ we have

$$
\sum_{x,y} \frac{p_j(y)p_j^2(x|y)}{p(x|y)} - 1
$$

$$
= \sum_{x,y} p_j(y)p(x|y) \sum_{u,v} p_j(v)p(u|v) \left(\frac{p_j(u|v)}{p(u|v)} \right)^2 - \left(\sum_{x,y} p_j(y)p(x|y) \frac{p_j(x|y)}{p(x|y)} \right)^2.
$$

Now apply Lemma C with $p_i = p_j(y)p(x|y)$ and $a_i = b_i = p_j(x|y)/p(x|y)$. We obtain

(4.6)
$$
\sum_{x,y} \frac{p_j(y)p_j^2(x|y)}{p(x|y)} - 1 \le \frac{1}{4}(M_j - m_j)^2 \quad (j = 1, 2, \ldots, n).
$$

Combining (4.1) and (4.6) gives (4.4). Furthermore, by Lemma A with $p_k = p_j(x,y)$ and $\xi_k = p(x|y)/p_j(x|y)$, we have for $j = 1, 2, \ldots, n$ that

$$0 \le \log_b \left(\sum_{x,y} p_j(x,y) \frac{p(x|y)}{p_j(x|y)} \right) - \sum_{x,y} p_j(x,y) \log_b \frac{p(x|y)}{p_j(x|y)}$$

(4.7)
$$= -\sum_{x,y} p_j(x,y) \log_b \frac{p(x|y)}{p_j(x|y)}$$

$$\le \frac{1}{4 \ln b} \left(\sqrt{\rho_j} - 1/\sqrt{\rho_j} \right)^2.$$

Relation (4.5) follows from (4.2) and (4.7). □

Remark. With the same argument as in the remark after Theorem 2.2, we have

$$\max_{(x,y),(u,v)} \left\{ \max_{i=1,2,\ldots,n} \left\{ \left| \frac{p_i(x|y)}{p(x|y)} - \frac{p_i(u|v)}{p(u|v)} \right| \right\} \right\} = \max\{M_i - m_i\} = M - m.$$

So

$$0 \le H(X|Y) - \sum_{i=1}^{n} \alpha_i H(X_i|Y_i) \le \frac{(M-m)^2}{4 \ln b}.$$

Also we have

$$0 \le H(X|Y) - \sum_{i=1}^{n} \alpha_i H(X_i|Y_i) \le \frac{1}{4 \ln b} \left(\sqrt{\rho} - 1/\sqrt{\rho} \right)^2,$$

where $\rho = \max\{\rho_i\}$. If $M - m \le 2\sqrt{\varepsilon \ln b}$, then

$$0 \le H(X|Y) - \sum_{i=1}^{n} \alpha_i H(X_i|Y_i) \le \varepsilon.$$

In the case $n = 2$ this strengthens [3, Cor. 4.6], where the same conclusion is given but with $\sqrt{2\varepsilon \ln b}$ in place of $2\sqrt{\varepsilon \ln b}$.

As with entropy, the upper bounds on $H(X|Y) - \sum_{i=1}^{n} \alpha_i H(X_i|Y_i)$ derived in Theorems 4.1 and 4.2 depend on the values α_i. We have the following result which does not.

Theorem 4.3. *Under the conditions of Theorem 3.1,*

$$0 \le H(X|Y) - \sum_{i=1}^{n} \alpha_i H(X_i|Y_i) \le \frac{1}{\ln b} \left[\sum_{x,y} \max_i\{p_i(y)\} \max_i\{p_i(x|y)\} - 1 \right].$$

Proof. We have

(4.8)
$$\sum_{i=1}^{n} \alpha_i p_i(y) \frac{p_i^2(x|y)}{p(x|y)} = p(y) \sum_{i=1}^{n} \alpha_i p_i(y) p_i^2(x|y) \Big/ \left(\sum_{i=1}^{n} \alpha_i p_i(x,y) \right)$$

$$= p(y) \times \sum_{i=1}^{n} \alpha_i p_i(y) p_i^2(x|y) \Big/ \left(\sum_{i=1}^{n} \alpha_i p_i(y) p_i(x|y) \right).$$

240

By Lemma 2.4 with $t_i = p_i(y)$, $u_i = p_i(x|y)$, the second factor on the right–hand side in the second line of (4.8) is less than or equal to $\max_i\{p_i(x|y)\}$. Also

$$p(y) = \sum_{i=1}^{n} \alpha_i p_i(y) \le \max_i\{p_i(y)\}.$$

The desired result follows from (4.1). $\qquad\qquad\qquad\qquad\qquad\qquad\qquad\square$

5. The mutual information of a mixed population

The mutual information of two discrete random variables X, Y is defined by

$$I(X,Y) := H(X) - H(X|Y) = \sum_{x,y} p(x,y)\log_b \frac{p(x,y)}{p(x)p(y)} = H(Y) - H(Y|X).$$

Suppose that (X_i, Y_i) $(1 \le i \le n)$ and (X,Y) are pairs of random variables such that

(5.1) $$p(x) = \sum_{i=1}^{n} \alpha_i p_i(x)$$

with $\alpha_i > 0$ and $\sum_{i=1}^{n} \alpha_i = 1$, and that

(5.2) $$p_i(y|x) = p(y|x) \qquad (i = 1, 2, \dots, n).$$

It is well–known that in the case $n = 2$ we have

$$\alpha I(X_1, Y_1) + (1 - \alpha)I(X_2, Y_2) \le I(X,Y),$$

that is, for fixed transition probabilities, the mutual information $I(X,Y)$ is a concave functional of the input probabilities (see McEliece [6, Theorem 1.6]).
By contrast, if the roles of the input and transition probabilities are reversed, that is, if

$$p_1(x) = p_2(x) = p(x)$$

and

$$p(y|x) = \alpha p_1(y|x) + (1 - \alpha)p_2(y|x),$$

then

$$\alpha I(X_1, Y_1) + (1 - \alpha)I(X_2, Y_2) \ge I(X,Y).$$

That is, for fixed input probabilities and $p(y|x)$ a convex linear combination of $p_1(y|x)$ and $p_2(y|x)$, we have that $I(X,Y)$ is a convex functional of the transition probabilities ([6, Theorem 1.7]).

In this section we shall see that the n–pair generalizations of the above two results follow from our results on entropy and conditional entropy. In the former case

$$I(X,Y) - \sum_{i=1}^{n} \alpha_i I(X_i, Y_i)$$

is nonnegative and in the latter

$$\sum_{i=1}^{n} \alpha_i I(X_i, Y_i) - I(X,Y)$$

is nonnegative.

Further, we derive upper bounds for these quantities in the respective cases. The upper bound in both cases is the same as that for

$$H(X,Y) - \sum_{i=1}^{n} \alpha_i H(X_i, Y_i)$$

derived earlier under the same conditions.

Theorem 5.1. *Let (X_i, Y_i) $(i = 1, 2, \ldots, n)$ and (X,Y) be pairs of discrete random variables satisfying (5.1) and (5.2). Then*

$$0 \le I(X,Y) - \sum_{i=1}^{n} \alpha_i I(X_i, Y_i) \le \frac{1}{\ln b} \left[\sum_y \sum_{i=1}^{n} \alpha_i p_i^2(y)/p(y) \ - 1 \right].$$

Proof. We have

(5.3)
$$I(X,Y) - \sum_{i=1}^{n} \alpha_i I(X_i, Y_i) = H(Y) - \sum_{i=1}^{n} \alpha_i H(Y_i)$$
$$- \left[H(Y|X) - \sum_{i=1}^{n} \alpha_i H(Y_i|X_i) \right].$$

Since (5.1) and (5.2) hold, we deduce that

$$p(x,y) = \left[\sum_{i=1}^{n} \alpha_i p_i(x) \right] p(y|x) = \sum_{i=1}^{n} \alpha_i p_i(x,y)$$

and $p(y) = \sum_{i=1}^{n} \alpha_i p(y_i)$. So the difference in brackets in (5.3) is

$$\sum_{x,y} p(x,y) \log_b 1/p(y|x) - \sum_{i=1}^{n} \alpha_i \sum_{x,y} p_i(x,y) \log_b 1/p(y|x) = 0,$$

that is,

(5.4)
$$I(X,Y) - \sum_{i=1}^{n} \alpha_i I(X_i, Y_i) = H(Y) - \sum_{i=1}^{n} \alpha_i H(Y_i).$$

Now we can apply Theorem 2.1 with Y in place of X to obtain desired result. $\square$

Theorem 5.2. *Under the assumptions of the previous theorem, we have*

$$(5.5) \qquad 0 \le I(X,Y) - \sum_{i=1}^{n} \alpha_i I(X_i, Y_i) \le \frac{1}{4\ln b} \sum_{i=1}^{n} \alpha_i (M_i - m_i)^2,$$

where

$$M_i := \max_y \frac{p_i(y)}{p(y)} \;, \quad m_i := \min_y \frac{p_i(y)}{p(y)} \qquad (i = 1, 2, \dots, n).$$

Also

$$(5.6) \qquad 0 \le I(X,Y) - \sum_{i=1}^{n} \alpha_i I(X_i, Y_i) \le \frac{1}{4\ln b} \sum_{i=1}^{n} \alpha_i \left(\sqrt{\rho_i} - 1/\sqrt{\rho_i}\right)^2,$$

where $\rho_i = M_i/m_i, \; (i = 1, 2, \dots, n)$.

Proof. Since (5.4) holds, we can apply Theorem 2.2 with Y in place of X. □

Remark. As in Section 2 we have

$$\max_{y,v} \left\{ \max_{i=1,2,\dots,n} \left\{ \left| \frac{p_i(y)}{p(y)} - \frac{p_i(v)}{p(v)} \right| \right\} \right\} = \max_i \{M_i - m_i\} = M - m.$$

Thus from (5.5) we get

$$0 \le I(X,Y) - \sum_{i=1}^{n} \alpha_i I(X_i, Y_i) \le \frac{1}{4\ln b}(M - m)^2.$$

Furthermore, from (5.6) we derive

$$0 \le I(X,Y) - \sum_{i=1}^{n} \alpha_i I(X_i, Y_i) \le \frac{1}{4\ln b} \left(\sqrt{\rho} - 1/\sqrt{\rho}\right)^2,$$

where $\rho = \max_i\{\rho_i\}$. Now if $M - m \le 2\sqrt{\varepsilon \ln b}$, then

$$0 \le I(X,Y) - \sum_{i=1}^{n} \alpha_i I(X_i, Y_i) \le \varepsilon.$$

In the case $n = 2$ this result subsumes [3, Cor. 5.8], which gives the result under the more demanding condition $M - m \le \sqrt{2\varepsilon \ln b}$.

Theorem 5.3. *Under the assumptions of Theorem 5.1, we have*

$$0 \le I(X,Y) - \sum_{i=1}^{n} \alpha_i I(X_i, Y_i) \le \frac{1}{\ln b} \left[\sum_y \max_i \{p_i(y)\} - 1 \right].$$

Proof. The result follows from Theorem 2.5. □

Theorem 5.4. *Suppose* (X_i, Y_i) $(1 \le i \le n)$ *and* (X, Y) *are pairs of discrete random variables such that*

(5.7)
$$p_i(x) = p(x) \qquad (1 \le i \le n)$$

and

(5.8)
$$p(y|x) = \sum_{i=1}^{n} \alpha_i p_i(y|x)$$

with $\alpha_i > 0$ *and* $\sum_{i=1}^{n} \alpha_i = 1$. *Then*

$$0 \le \sum_{i=1}^{n} \alpha_i I(X_i, Y_i) - I(X, Y) \le \frac{1}{\ln b} \left[\sum_{x,y} \sum_{i=1}^{n} \alpha_i p_i(y) p_i^2(x|y)/p(x|y) \ - 1 \right].$$

Proof. Since (5.7) and (5.8) hold, we have $H(X) = H(X_i)$ for $1 \le i \le n$. So

(5.9)
$$\sum_{i=1}^{n} \alpha_i I(X_i, Y_i) - I(X, Y) = H(X|Y) - \sum_{i=1}^{n} \alpha_i H(X_i|Y_i).$$

The result follows by Theorem 4.1. $\qquad\qquad\qquad\qquad\qquad\qquad\qquad$ □

Remark. It is clear from (5.9) that Theorems 4.2 and 4.3 apply with

$$\sum_{i=1}^{n} \alpha_i I(X_i, Y_i) - I(X, Y)$$

in place of

$$H(X|Y) - \sum_{i=1}^{n} \alpha_i H(X_i|Y_i).$$

References

1. S. S. Dragomir and C. J. Goh, 'A counterpart of Jensen's discrete inequality for differentiable convex mappings and applications in information theory', *Math. Comput. Modelling* **24 (2)** (1996), 1–11.

2. ——, 'Some bounds on entropy measures in information theory', *Appl. Math. Lett.* **10 (3)** (1997), 23–28.

3. ——, 'Further counterparts of some inequalities in information theory' submitted.

4. M. Matić, C. E. M. Pearce and J. Pečarić, 'Improvements of some bounds on entropy measures in information theory', *Math. Ineq. Applic.* **1** (1998), 295–304.

5. ——, 'Refinements of some bounds in information theory', *J. Austral. Math. Soc. Ser. B* to appear.

6. R. J. McEliece, *The theory of information and coding* (Addison-Wesley, 1977).

7. D. S. Mitrinović, J. E. Pečarić and A. M. Fink, *Classical and new inequalities in analysis* (Kluwer Acad. Publ., 1993).

EXTREMAL PROBLEMS AND INEQUALITIES OF MARKOV-BERNSTEIN TYPE FOR POLYNOMIALS

GRADIMIR V. MILOVANOVIĆ
Faculty of Electronic Engineering, Department of Mathematics, P.O. Box 73, 18000 Niš, Yugoslavia

Abstract. The classical Markov (1889) and Bernstein (1912) inequalities and corresponding extremal problems were generalized for various domains, various norms and for various subclasses for polynomials, both algebraic and trigonometric. Beside some classical results in uniform norm, we give a short account of L^r inequalities of Markov type for algebraic polynomials, with a special attention to the case $r = 2$. We also study extremal problems of Markov's type

$$C_{n,m} = \sup_{P \in \mathcal{P}_n} \frac{\|\mathcal{D}_m P\|}{\|A^{m/2} P\|},$$

where $\mathcal{P}_n$ is the class of all algebraic polynomials of degree at most n, $d\lambda(t) = w(t)dt$ is a nonnegative measure corresponding to the classical orthogonal polynomials, $A \in \mathcal{P}_2$, $\|P\| = \left(\int_{\mathbb{R}} |P(t)|^2 d\lambda(t)\right)^{1/2}$, and $\mathcal{D}_m$ is a differential operator defined by

$$\mathcal{D}_m P = \frac{d^m}{dt^m} [A^m P] \qquad (P \in \mathcal{P}_n, \ m \geq 1).$$

1. Introduction and Notation

Let $\mathcal{P}_n$ be the set of all algebraic polynomials of degree at most n. We take

$$\|f\|_\infty := \max_{-1 \leq t \leq 1} |f(t)| \tag{1.1}$$

and

$$\|f\|_r := \left(\int_{\mathbb{R}} |f(t)|^r d\lambda(t)\right)^{1/r}, \quad r \geq 1, \tag{1.2}$$

1991 *Mathematics Subject Classification.* Primary 26C05, 26D05, 26D10, 33C45, 41A17, 41A44.
Key words and phrases. Extremal problems for polynomials; Inequalities; Eigenvalues; Best constants; Norm; Inner product; Weight function; Orthogonal polynomials; Classical weights; Recurrence relation; Differential operators.
This work was supported in part by the Serbian Scientific Foundation, grant number 04M03.

T.M. Rassias and H.M. Srivastava (eds.), Analytic and Geometric Inequalities and Applications, 245–264.

where $d\lambda(t)$ is a given nonnegative measure on the real line $\mathbb{R}$, with compact support or otherwise, for which all moments $\mu_k = \int_{\mathbb{R}} t^k \, d\lambda(t)$, $k = 0, 1, \ldots$, exist and are finite and $\mu_0 > 0$. In a special case $r = 2$, (1.2) reduces to

$$\|f\|_2 = \left(\int_{\mathbb{R}} |f(t)|^2 \, d\lambda(t) \right)^{1/2}. \tag{1.3}$$

In that case we have an inner product defined by

$$(f, g) = \int_{\mathbb{R}} f(t)\overline{g(t)} \, d\lambda(t)$$

such that $\|f\|_2 = \sqrt{(f, f)}$. Then also, there exists a unique set of (monic) orthogonal polynomials $\pi_k(\cdot) = \pi_k(\,\cdot\,; d\lambda)$, $k \geq 0$, with respect to $(\cdot, \cdot)$, such that

$$\pi_k(t) = t^k + \text{lower degree terms}, \qquad (\pi_k, \pi_m) = \|\pi_k\|_2^2 \, \delta_{km},$$

where δ_{km} is Kronecker's delta.

A standard case of orthogonal polynomials is when the measure $d\lambda$ can be express as $d\lambda(t) = w(t) \, dt$, where the weight function $t \mapsto w(t)$ is a non-negative and measurable in Lebesgue's sense for which all moments exist and $\mu_0 = \int_{\mathbb{R}} w(t) \, dt > 0$. A very important class of such orthogonal polynomials on an interval of orthogonality $(a, b) \in \mathbb{R}$ is constituted by so-called the *classical orthogonal polynomials*. They are distinguished by several particular properties (cf. [31]).

Without loss of generality, we can restrict our consideration only to the following three intervals of orthogonality: $(-1, 1)$, $(0, +\infty)$, $(-\infty, +\infty)$, with the inner product

$$(f, g) = \int_a^b w(t) f(t)\overline{g(t)} \, dt. \tag{1.4}$$

The orthogonal polynomials $\{Q_n\}$ on (a, b) with respect to the inner product (1.4) are called the *classical orthogonal polynomials* if their weight functions $t \mapsto w(t)$ satisfy the differential equation

$$\frac{d}{dt}(A(t)w(t)) = B(t)w(t),$$

where

$$A(t) = \begin{cases} 1 - t^2, & \text{if } (a, b) = (-1, 1), \\ t, & \text{if } (a, b) = (0, +\infty), \\ 1, & \text{if } (a, b) = (-\infty, +\infty), \end{cases}$$

and $B(t)$ is a polynomial of the first degree. For such classical weights we will write $w \in CW$.

Based on this definition, the classical orthogonal polynomials $\{Q_n\}$ on (a, b) can be specificated as the *Jacobi polynomials* $P_n^{(\alpha, \beta)}(t)$ $(\alpha, \beta > -1)$ on $(-1, 1)$, the *generalized Laguerre polynomials* $L_n^s(t)$ $(s > -1)$ on $(0, +\infty)$, and finally as the *Hermite polynomials* $H_n(t)$ on $(-\infty, +\infty)$. Their weight functions and the corresponding polynomials $A(t)$ and $B(t)$ are given in Table 1.1.

TABLE 1.1: The classification of the classical orthogonal polynomials

(a, b)	$w(t)$	$A(t)$	$B(t)$	λ_n
$(-1, 1)$	$(1-t)^\alpha(1+t)^\beta$	$1 - t^2$	$\beta - \alpha - (\alpha + \beta + 2)t$	$n(n + \alpha + \beta + 1)$
$(0, +\infty)$	$t^s e^{-t}$	t	$s + 1 - t$	n
$(-\infty, +\infty)$	e^{-t^2}	1	$-2t$	$2n$

The classical orthogonal polynomial $t \mapsto Q_n(t)$ is a particular solution of the second order linear differential equation of hyphergeometric type

$$L[y] = A(t)y'' + B(t)y' + \lambda_n y = 0, \tag{1.5}$$

where λ_n is given also in Table 1.1.

2. Classical Extremal Problems in Uniform Norm

The first result on the extremal problems of Markov type was connected with some investigations of the well-known Russian chemist Mendeleev [18]. Namely, the question was: *If $P(t)$ is an arbitrary quadratic polynomial defined on an interval $[a, b]$, with*

$$\max_{t \in [a,b]} P(t) - \min_{t \in [a,b]} P(t) = L,$$

how large can $P'(t)$ be on $[a, b]$?

Changing the horizontal scale and shifting the coordinate axis until $|P(t)| \leq 1$, the problem can be reduced to a simpler one: *If $P(t)$ is an arbitrary quadratic polynomial and $|P(t)| \leq 1$ on $[-1, 1]$, how large can $|P'(t)|$ be on $[-1, 1]$?* Mendeleev found that $|P'(t)| \leq 4$ on $[-1, 1]$. This result is the best possible because for $P(t) = 1 - 2t^2$ we have $P(t) \leq 1$ and $P'(\pm 1) = 4$.

The corresponding problem for polynomials of degree n was considered by A. A. Markov [16]. Taking the uniform norm (1.1) he solved the extremal problem

$$A_n = \sup_{P \in \mathcal{P}_n} \frac{\|P'\|_\infty}{\|P\|_\infty},$$

finding the best constant $A_n = n^2$ and the extremal polynomial $P^*(t) = cT_n(t)$, where T_n is the Chebyshev polynomial of the first kind of degree n and c is an arbitrary constant. The best constant can be expressed also as $A_n = T'_n(1)$. Thus, the classical Markov's inequality can be expressed in the form

$$\|P'\|_\infty \leq n^2 \|P\|_\infty \qquad (P \in \mathcal{P}_n).$$

In 1892, younger brother V. A. Markov [17] found the best possible inequality for k-th derivative,

$$\|P^{(k)}\|_\infty \le T_n^{(k)}(1)\|P\|_\infty \qquad (P \in \mathcal{P}_n),$$

where the extremal polynomial is also T_n. The best constant can be expressed in the form

$$T_n^{(k)}(1) = \|T_n^{(k)}\|_\infty = \frac{1}{(2k-1)!!} \prod_{i=0}^{k-1} (n^2 - i^2).$$

A version of this remarkable paper in German was published in 1916.

In 1912 Bernstein [4] considered another type of these inequalities taking $\|f\| = \max_{|z| \le 1} |f(z)|$. He proved the inequality

$$\|P'\| \le n\|P\| \qquad (P \in \mathcal{P}_n),$$

with equality case when $P(z) = cz^n$ (c is an arbitrary constant).

There are several different forms of this Bernstein's inequality. If we take $\mathcal{T}_n$ to be set of all trigonometric polynomials of degree at most n and

$$\|P\| = \max_{|z|=1} |P(z)| = \max_{-\pi < \theta \le \pi} |P(e^{i\theta})|,$$

then a trigonometric version can be stated in the following form: *Let $T \in \mathcal{T}_n$ and $|T(\theta)| \le M$, then $|T'(\theta)| \le nM$. The equality holds for $T(\theta) = \gamma \sin n(\theta - \theta_0)$, where $|\gamma| = 1$.*

The standard form of Bernstein's inequality can be done as:

Theorem 2.1. *Let $P \in \mathcal{P}_n$ and $|P(t)| \le 1$ $(-1 \le t \le 1)$, then*

$$|P'(t)| \le \frac{n}{\sqrt{1-t^2}}, \qquad -1 < t < 1.$$

The equality is attained at the points $t = t_\nu = \cos \frac{(2\nu-1)\pi}{2n}$, $\nu = 1, \ldots, n$, if and only if $P(t) = \gamma T_n(t)$, where $|\gamma| = 1$.

Combining the inequalities of Markov and Bernstein we can state the following result:

Theorem 2.2. *If $P \in \mathcal{P}_n$ then*

$$|P'(t)| \le \min\left\{n^2, \frac{n}{\sqrt{1-t^2}}\right\}\|P\|_\infty, \qquad -1 \le t \le 1.$$

A general question could be stated: *How large can $|P^{(k)}(t)|$ be, for a given t, when $|P(t)| \le 1$ on $[-1,1]$?* Let this maximum be $M_{n,k}(t)$, i.e.,

$$|P^{(k)}(t)| \le M_{n,k}(t) \qquad (1 \le k < n). \tag{2.1}$$

For $k = 1$ we put $M_{n,1}(t) = M_n(t)$. It is easy to see that the function M_n is even, i.e., $M_n(-t) = M_n(t)$.

The problem of finding $M_n(t)$ was stated by A. A. Markov himself, and solved for $n = 2$ and $n = 3$. He determined that

$$M_2(t) = \begin{cases} \dfrac{1}{1-t}, & 0 \le t \le \dfrac{1}{2}, \\[2mm] 4t, & \dfrac{1}{2} \le t \le 1, \end{cases}$$

and

$$M_3(t) = \begin{cases} 3(1 - 4t^2), & t \in [t_0, t_1], \\[2mm] \dfrac{7\sqrt{7} + 10}{9(1 + t)}, & t \in [t_1, t_2], \\[2mm] \dfrac{16t^3}{(9t^2 - 1)(1 - t^2)}, & t \in [t_2, t_3], \\[2mm] \dfrac{7\sqrt{7} - 10}{9(1 - t)}, & t \in [t_3, t_4], \\[2mm] 3(4t^2 - 1), & t \in [t_4, t_5], \end{cases}$$

where

$$t_0 = 0, \quad t_1 = \frac{1}{6}(\sqrt{7} - 2) \cong 0.1076, \quad t_2 = \frac{1}{9}(2\sqrt{7} - 1) \cong 0.4768,$$

$$t_3 = \frac{1}{9}(2\sqrt{7} + 1) \cong 0.6991, \quad t_4 = \frac{1}{6}(\sqrt{7} + 2) \cong 0.7743, \quad t_5 = 1.$$

The determination of $M_n(t)$, $n \ge 4$, is very complicated and it can be given by a technique of Voronovskaja (see [40]). Using the same method, Gusev [14] found the corresponding function $M_{n,k}(t)$ in the inequality (2.1).

Instead of the condition $|P(t)| \le 1$ on $[-1, 1]$, Bernstein [5] used a more general condition

$$|P(t)| \le \sqrt{H(t)} \qquad (-1 \le t \le 1),$$

where H is an arbitrary positive polynomial on $[-1, 1]$ of degree $s \, (\le 2n)$. We mention an interesting result of V. Videnskiĭ [39]:

Theorem 2.3. *Let* $P \in \mathcal{P}_n$ *and*

$$|P(t)| \le |\alpha t + i\sqrt{1 - t^2}| \qquad (\alpha \ge 0, \, -1 \le t \le 1).$$

Then, for $k = 1, \dots, n$ *and* $-1 \le t \le 1$, *we have that*

$$|P^{(k)}(t)| \le Q_n^{(k)}(1; \alpha),$$

where

$$Q_n(t; \alpha) = \frac{1}{2}(\alpha + 1)T_n(t) + \frac{1}{2}(\alpha - 1)T_{n-2}(t).$$

The equality is attained only for $P(t) = \gamma Q_n(t)$ *at the endpoints* $t = \pm 1$, *where* $|\gamma| = 1$.

Several inequalities of this type were given by Videnskiĭ, Duffin and Schaeffer, Turán, Rahman, Pirrre and Rahman, Rahman and Schmeisser (see Chapter 6 in [24]).

3. Extremal Problems in L^r-norm

The first results on extremal problems in L^2-norm given by (1.3),

$$\|P'\|_2 \leq A_n \|P\|_2 \qquad (P \in \mathcal{P}_n), \tag{3.1}$$

were given by E. Schmidt [27] and Turán [32]:

Theorem 3.1. (a) *Let* $(a, b) = (-\infty, +\infty)$ *and*

$$\|f\|_2^2 = \int_{-\infty}^{\infty} e^{-t^2} f(t)^2 \, dt.$$

Then the best constant in the inequality (3.1) *is* $A_n = \sqrt{2n}$. *An extremal polynomial is Hermite's polynomial* H_n.

(b) *Let* $(a, b) = (0, +\infty)$ *and*

$$\|f\|_2^2 = \int_0^{\infty} e^{-t} f(t)^2 \, dt.$$

Then

$$A_n = \left(2 \sin \frac{\pi}{4n + 2} \right)^{-1}.$$

The extremal polynomial is

$$P(t) = \sum_{\nu=1}^{n} \sin \frac{\nu \pi}{2n + 1} L_\nu(t),$$

where L_ν *is Laguerre polynomial.*

Theorem 3.1 (b), in this form, was formulated by Turán [32].

An important generalization of A. A. Markov's inequality for algebraic polynomials in an integral norm was given by Hille, Szegő, and Tamarkin [15]. Taking

$$\|f\|_r = \left(\int_{-1}^1 |f(t)|^r \, dt \right)^{1/r},$$

they proved the following theorem:

Theorem 3.2. *Let $r \geq 1$ and let $P \in \mathcal{P}_n$. Then*

$$\|P'\|_r \leq An^2 \|P\|_r, \tag{3.2}$$

where the constant $A = A(n, r)$ is given by

$$A(n, r) = 2(r - 1)^{1/r - 1}\left(r + \frac{1}{n}\right)\left(1 + \frac{r}{nr - r + 1}\right)^{n - 1 + 1/r}, \tag{3.3}$$

for $r > 1$, and

$$A(n, 1) = 2\left(1 + \frac{1}{n}\right)^{n+1}.$$

The factor n^2 in (3.2) cannot be replaced by any function tending to infinity more slowly. Namely, for each n, there exist polynomials $P(t)$ of degree n such that the left side of (3.2) is $\leq Bn^2$, where B is a constant of the same nature as $A = A(n, r)$.

The constant $A(n, r)$ in Theorem 3.2 is not the best possible. We can see that $A(n, r) \leq 6 \exp(1 + 1/e)$, for every n and $r \geq 1$. Also,

$$A(n, r) \rightarrow \begin{cases} 2(1 + 1/(n - 1))^{n-1} < 2e & (n \text{ fixed}, r \rightarrow +\infty), \\ 2e & (r = 1, n \rightarrow +\infty), \\ 2er(r - 1)^{(1/r) - 1} & (r > 1 \text{ fixed}, n \rightarrow +\infty). \end{cases}$$

Some improvements of the constant A have been obtained by Goetgheluck [8]. He found that

$$A = \bar{A}(n, 1) = \sqrt{\frac{8}{\pi}}\left(1 + \frac{3}{4n}\right)^2.$$

It is easy to see that for each $n \geq 1$,

$$\sqrt{8/\pi}\left(1 + \frac{3}{4n}\right)^2 < 2e < 2\left(1 + \frac{1}{n}\right)^{n+1}.$$

For $r > 1$ he found the following very complicated expression

$$A = \bar{A}(n, r) = \left(\frac{(2r + 1)^{2 + 1/r}}{r(r + 1)}\right)^{(r-1)/(r+1)}\left(2r\frac{r + 1}{r - 1}\right)^{1/r}\left(\frac{r - 1}{2}\right)^{2/r(r+1)} \times$$

$$\times \left(1 - \frac{3}{5n}\right)^{1 - 1/r}\left(1 + \frac{1}{nr}\right)^{n + 1/r}.$$

Remark 3.1. In [8, Theorem 2] there is a misprint in the last factor in $\bar{A}(n, r)$.

Numerical calculations show that this constant is less than the corresponding constant in (3.3). Typical graphics of $\bar{A}(n, r)$ and $A(n, r)$ are displyed in Figure 3.1. Also, one can see that $\bar{A}(n, r) \rightarrow 4(1 - 3/5n)$ as $r \rightarrow +\infty$, n being fixed.

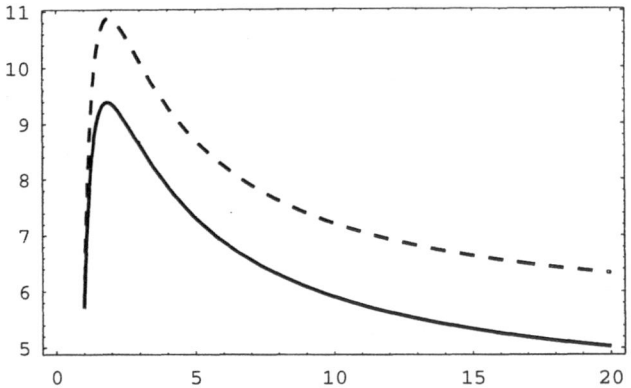

FIG. 3.1: Graphics $r \mapsto \bar{A}(n,r)$ (solid line) and $r \mapsto A(n,r)$ (broken line) for $n = 10$

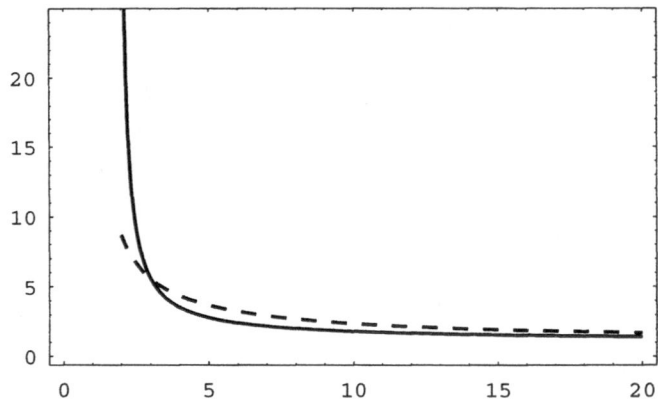

FIG. 3.2: Graphics $r \mapsto B(r)$ (solid line) and $r \mapsto C(r)$ (broken line)

Recently Baran [2] has given a new proof of the inequality (3.2) for $r > 2$, providing a better constant.

Theorem 3.3. *If $r > 2$ and $P \in \mathcal{P}_n$. Then (3.2) holds with*

$$A = B(r) = \left[2\nu(2/r)(r+3)^2\right]^{1/r},$$

where $\nu(t) = \pi t / \sin(\pi t)$, $t \in (0,1)$.

We note that $B(r) \to 1$ when $r \to +\infty$ and $B(r) \to +\infty$ when $r \to 2$. Since the function $\nu(t)$ is increasing on $(0, 1/2]$, it is easy to see that for $r \geq 4$, we have

$$B(r) = \pi^{1/r}(r+3)^{2/r}.$$

An improvement of the constant $B(r)$ for r near to 2 was also obtained by Baran [2]:

Theorem 3.4. *If $r \geq 2$ and $P \in \mathcal{P}_n$. Then (3.2) holds with*

$$A = C(r) = 2^{1/r} U_r^{1-2/r} \left(2U_r^2 + V_q U_r\right)^{1/r},$$

where

$$U_r = 2^{-1/r}(r+3)^{2/r}, \quad V_q = \left(8\nu(1/q)\right)^{1/q}, \quad \nu(t) = \pi t / \sin(\pi t), \quad 1/r + 1/q = 1.$$

This constant $C(r)$ can be expressed in the form

$$C(r) = (r+3)^{2/r} \left(2 + \frac{2^{1/r} V_q}{(r+3)^{2/r}}\right)^{1/r}.$$

Graphics of $B(r)$ and $C(r)$ are showed in Figure 3.2.

Applying the inequality $2^{1/r} V_q < 4\pi r(r+3)^{2/r}$ ($r \geq 2$, $q = r/(r-1)$), the constant $C(r)$ can be approximated by (cf. Baran [2, Corollary 2.10])

$$C(r) < \tilde{C}(r) = (r+3)^{2/r}(2+4\pi r)^{1/r}.$$

Graphics of $C(r)$ and $\tilde{C}(r)$ are presented in Figure 3.3 as well as the graphic of the Goetgheluck's estimate $\bar{A}(n,r)$ for $n = 100$.

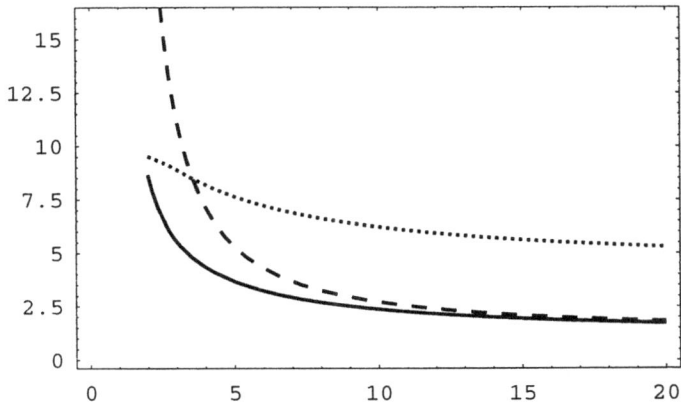

FIG. 3.3: Graphics $r \mapsto C(r)$ (solid line), $r \mapsto \tilde{C}(r)$ (broken line), and $r \mapsto \bar{A}(n,r)$ for $n = 100$ (dotted line)

The special case $r = 2$ has been investigated several times. Hille, Szegő, and Tamarkin [15] proved that $A(n,2) \to 1/\pi$ when $n \to +\infty$. Schmidt [27] also investigated an asymptotics case of $A(n,2)$ in (3.2). For $n \geq 5$, he obtained that

$$A(n,2) = \frac{(2+3/n)^2}{4\pi} \left(1 - \frac{\pi^2 - 3}{3(2n+3)^2} + \frac{16R}{(2n+3)^4}\right)^{-1},$$

where $-6 < R < 13$. Also, Bellman [3] proved that $A(n,2) \leq 1/\sqrt{2}$.

In 1987 Dörfler [6] considered the analogous problem for derivatives of higher order and gave a method for computing the best possible constant in the inequality of Markov type

$$\|P^{(k)}\| \leq C_{n,k}\|P\|, \tag{3.4}$$

where the norm $\|f\| = \|f\|_2$ is given by (1.4) and $d\lambda(t) = w(t)\,dt$. Here $w : (a,b) \to \mathbb{R}_+$ ($-\infty \leq a < b \leq +\infty$) is an arbitrary weight function for which all moments are finite.

Theorem 3.5. *Let $P \in \mathcal{P}_n$. Then the best possible constant $C_{n,k}$ in (3.4) is equal to the largest singular value of the matrix $A_n^{(k)}$, where*

$$A_n^{(k)} = \begin{bmatrix} e_{0,0}^{(k)} & \cdots & e_{n,0}^{(k)} \\ \vdots & & \\ e_{0,n-k}^{(k)} & & e_{n,n-k}^{(k)} \end{bmatrix}, \qquad e_{\nu,j}^{(k)} = \int_a^b \pi_\nu^{(k)}(t)\pi_j(t)w(t)\,dt,$$

and $\{\pi_\nu\}$ is a system of polynomials orthonormal with respect to the weight function w. Moreover,

$$\max_{0 \leq \nu \leq n} \|\pi_\nu^{(k)}\| \leq C_{n,k} \leq \left(\sum_{\nu=0}^n \|\pi_\nu^{(k)}\|^2\right)^{1/2}$$

holds.

Using Dörfler's method, Goetgheluck [8] calculated $C_{n,1} = A_n n^2$ for $n \leq 65$ and showed that A_n is a decreasing function in n and $1/\pi < A_n < 1/3$ for $n > 64$.

An alternative method for computing the best constant

$$C_{n,k} = C_{n,k}(d\lambda) = \sup_{P \in \mathcal{P}_n} \frac{\|P^{(k)}\|}{\|P\|} \qquad (1 \leq k \leq n) \tag{3.5}$$

was also given in 1987 by Milovanović [20] (see also [24]):

Theorem 3.6. *The best constant $C_{n,k}$ defined in (3.5) is equal to the spectral norm of one triangular matrix $Q_{n,k}^T$, where*

$$Q_{n,k} = \left[q_{ij}^{(k)}\right]_{k \leq i,j \leq n} \qquad (q_{i,j}^{(k)} = 0 \Leftarrow i > j),$$

i.e.,

$$C_{n,k} = \sigma(Q_{n,k}^T) = \left(\lambda_{\max}(Q_{n,k}Q_{n,k}^T)\right)^{1/2}. \tag{3.6}$$

The elements $q_{ij}^{(k)}$ are given by the following inner product

$$q_{ij}^{(k)} = (\pi_j^{(k)}, \pi_{i-k}) \qquad (k \le i, j \le n).$$

Alternatively, (3.6) can be expressed in the form

$$C_{n,k} = (\lambda_{\min}(M_{n,k}))^{-1/2}, \tag{3.7}$$

where $M_{n,k} = \left(Q_{n,k}Q_{n,k}^T\right)^{-1}$.

We mention now a case with a special even weight function which was also considered in [20]. Let $d\lambda(t) = w(t)dt$ on $(-a, a)$, $0 < a < \infty$, where $w(-t) = w(t)$. Then we have

$$\pi_i'(t) = \frac{1}{r_i} \sum_{j=1}^{[(i+1)/2]} q_{i,j}\pi_{i-2j+1}(t), \qquad r_i \ne 0.$$

We use a class of such weight functions for which $q_{i,j} = q_{i+2,j+1}$. For example, this property holds for Gegenbauer weight. In this case, for $P \in \mathcal{P}_n$, we have

$$P'(t) = \sum_{i=1}^{n} c_i\pi_i'(t) = \sum_{i=1}^{n} q_{i,1}\left(\sum_{j\ge0} c_{i+2j}r_{i+2j}^{-1}\right)\pi_{i-1}(t)$$

and

$$\|P'\|^2 = \sum_{i=1}^{n} y_i^2,$$

where

$$y_i = q_{i,1}\sum_{j\ge0} c_{i+2j}r_{i+2j}^{-1}, \qquad i = 1,\dots,n. \tag{3.8}$$

Putting $q_{i,1} = p_i$ and $y_{n+1} = y_{n+2} = 0$, from (3.8) follows

$$c_i = r_i\left(\frac{y_i}{p_i} - \frac{y_{i+2}}{p_{i+2}}\right), \qquad i = 1,\dots,n.$$

Then

$$\|P\|^2 = \sum_{i=1}^{n} c_i^2 = \frac{r_1^2}{p_1^2}y_1^2 + \frac{r_2^2}{p_2^2}y_2^2 + \sum_{i=3}^{n} \frac{r_i^2 + r_{i-2}^2}{p_i^2}y_i^2 - 2\sum_{i=1}^{n-2} \frac{r_i^2}{p_ip_{i+2}}y_iy_{i+2}.$$

The corresponding matrix $M_{n,1}$ (see (3.7) in Theorem 3.6) is given by

$$M_{n,1} = \begin{bmatrix} \alpha_1 & 0 & \beta_1 & & & & & & & O \\ 0 & \alpha_2 & 0 & \beta_2 & & & & & \\ \beta_1 & 0 & \alpha_3 & 0 & \beta_3 & & & & \\ & \beta_2 & 0 & \alpha_4 & 0 & \beta_4 & & & \\ & & \ddots & \ddots & \ddots & \ddots & \ddots & & \\ & & & \beta_{n-4} & 0 & \alpha_{n-2} & 0 & \beta_{n-2} \\ & & & & \beta_{n-3} & 0 & \alpha_{n-1} & 0 \\ O & & & & & \beta_{n-2} & 0 & \alpha_n \end{bmatrix},$$

where

$$\alpha_i = \frac{r_i^2 + r_{i-2}^2}{p_i^2}, \quad \beta_i = -\frac{r_i^2}{p_i p_{i+2}} \quad (r_{-1} = r_0 = 0).$$

We define now two sequences of polynomials $\{R_i\}$ and $\{S_i\}$ by the following three-term recurrence relations:

$$tR_{i-1}(t) = \beta_{2i-1} R_i(t) + \alpha_{2i-1} R_{i-1}(t) + \beta_{2i-3} R_{i-2}(t),$$
$$R_{-1}(t) = 0, \quad R_0(t) = R_0 = \text{const},$$

where $i = 1, \dots, [(n+1)/2]$ and

$$tS_{i-1}(t) = \beta_{2i} S_i(t) + \alpha_{2i} S_{i-1}(t) + \beta_{2i-2} S_{i-2}(t),$$
$$S_{-1}(t) = 0, \quad S_0(t) = S_0 = \text{const},$$

where $i = 1, \dots, [n/2]$.

Theorem 3.7. *The eigenvalues of the matrix $M_{n,1}$ are the zeros of polynomials*

(a) S_{m-1} *and* R_m, *when* $n = 2m - 1$,

(b) S_m *and* R_m, *when* $n = 2m$,

so that

$$C_{2m-1,1} = \left(\min\left(s_1^{(m-1)}, r_1^{(m)}\right)\right)^{-1/2} \quad \text{and} \quad C_{2m,1} = \left(\min\left(s_1^{(m)}, r_1^{(m)}\right)\right)^{-1/2},$$

where $s_1^{(k)}$ and $r_1^{(k)}$ are the minimal zeros of the polynomials S_k and R_k respectively.
The conditions $q_{i,j} = q_{i+2,j+1}$ are satisfied for Gegenbauer measure

$$d\lambda(t) = (1 - t^2)^{\lambda - 1/2} dt, \quad -1 < t < 1.$$

In fact, we have

$$\frac{d}{dt} \hat{C}_i^\lambda(t) = \frac{2}{h_i^{1/2}} \sum_{j=1}^{[(i+1)/2]} (i + \lambda - 2j + 1) h_{i-2j+1}^{1/2} \hat{C}_{i-2j+1}^\lambda(t),$$

where $\hat{C}_\nu^\lambda$ is the normalized Gegenbauer polynomial of degree ν, with

$$h_i = \|C_i^\lambda\|^2 = \sqrt{\pi}\frac{(2\lambda)_i\Gamma\left(\lambda + \frac{1}{2}\right)}{(i+\lambda)i!\Gamma(\lambda)}, \qquad (p)_i = p(p+1)\cdots(p+i-1).$$

Thus,

$$r_i = \frac{1}{2}\sqrt{h_i}, \qquad p_i = q_{i,1} = (i+\lambda-1)\sqrt{h_{i-1}}.$$

For $n = 1$ and $n = 2$, we have

$$C_{1,1} = \sqrt{2(\lambda+1)} \qquad \text{and} \qquad C_{2,1} = \sqrt{\frac{8(\lambda+1)(\lambda+2)}{2\lambda+1}},$$

respectively.

In a special case, when $\lambda = 1/2$ (Legendre case), we obtain

$$\alpha_1 = \frac{1}{3}, \quad \alpha_2 = \frac{1}{15}, \quad \alpha_i = \frac{2}{(2i+1)(2i-3)}, \quad i = 3,\dots,n;$$

$$\beta_i = -\frac{1}{(2i+1)\sqrt{(2i-1)(2i+3)}}, \quad i = 1,\dots,n-2.$$

Similarly, in the Chebyshev case ($\lambda = 0$), we have

$$\alpha_1 = \frac{1}{2}, \quad \alpha_2 = \frac{1}{16}, \quad \alpha_i = \frac{1}{4}\left(\frac{1}{i^2} + \frac{1}{(i-2)^2}\right), \quad i = 3,\dots,n;$$

$$\beta_1 = -\frac{\sqrt{2}}{4}, \quad \beta_i = -\frac{1}{4i^2}, \quad i = 2,\dots,n-2.$$

Numerical calculations of $C_{n,1}$ for some selected values of λ were given in [20].

4. Some Weighted Polynomial Inequalities in L^2-Norm

Guessab and Milovanović [11] have considered a weighted L^2-analogues of the Bernstein's inequality (see Theorem 2.1), which can be stated in the following form:

$$\|\sqrt{1-t^2}P'(t)\|_\infty \leq n\|P\|_\infty. \tag{4.1}$$

Let w be the weight of the classical orthogonal polynomials ($w \in CW$) and $A(t)$ be given as in Table 1.1. Using the norm $\|f\|_2^2 = (f,f)$, we consider the following problem connected with the Bernstein's inequality (4.1): *Determine the best constant $C_{n,m}(w)$ ($1 \leq m \leq n$) such that the inequality*

$$\|A^{m/2}P^{(m)}\|_2 \leq C_{n,m}(w)\|P\|_2 \tag{4.2}$$

holds for all $P \in \mathcal{P}_n$.

At first, we note if $w \in CW$, then the corresponding classical orthogonal polynomial $t \mapsto Q_n(t)$ is a particular solution of the differential equation of the second
order (1.5), i.e.,

$$\frac{d}{dt}\left(A(t)w(t)\frac{dy}{dt}\right) + \lambda_n w(t)y = 0,$$

where $\lambda_n = -n\left(\frac{1}{2}(n-1)A''(0) + B'(0)\right)$. The k-th derivative of Q_n is also the
classical orthogonal polynomial, with respect to the weight $t \mapsto w_k(t) = A(t)^k w(t)$,
and satisfies the following differential equation

$$\frac{d}{dt}\left(A(t)w_k(t)\frac{dy}{dt}\right) + \lambda_{n,k} w_k(t)y = 0,$$

where $\lambda_{n,k} = -(n-k)\left(\frac{1}{2}(n+k-1)A''(0) + B'(0)\right)$. We note that $\lambda_{n,0} = \lambda_n$.

A. Guessab and G. V. Milovanović [11] proved:

Theorem 4.1. *For all $P \in \mathcal{P}_n$ the inequality (4.2) holds, with the best constant*
$C_{n,m}(w) = \sqrt{\lambda_{n,0}\lambda_{n,1}\cdots\lambda_{n,m-1}}$. *The equality is attained in (4.2) if and only if*
P *is a constant multiple of the classical polynomial Q_n orthogonal with respect to*
the weight function $w \in CW$.

In some special cases we have:

(1) Let $w(t) = (1-t)^\alpha(1+t)^\beta$ $(\alpha, \beta > -1)$ on $(-1, 1)$ (Jacobi case). Then

$$\|(1-t^2)^{m/2}P^{(m)}\|_2 \leq \sqrt{\frac{n!\Gamma(n+\alpha+\beta+m+1)}{(n-m)!\Gamma(n+\alpha+\beta+1)}}\,\|P\|_2,$$

with equality if and only if $P(t) = cP_n^{(\alpha,\beta)}(t)$.

(2) Let $w(t) = t^s e^{-t}$ $(s > -1)$ on $(0, +\infty)$ (generalized Laguerre case). Then

$$\|t^{m/2}P^{(m)}\|_2 \leq \sqrt{n!/(n-m)!}\,\|P\|_2,$$

with equality if and only if $P(t) = cL_n^s(t)$.

(3) The Hermite case with the weight $w(t) = e^{-t^2}$ on $(-\infty, +\infty)$ is the simplest.
Then the best constant is $C_{n,m}(w) = 2^{m/2}\sqrt{n!/(n-m)!}$.

In connection with the previous results is also the following characterization of the
classiacal orthogonal polynomials given by Agarwal and Milovanović [1].

Theorem 4.2. *For all $P \in \mathcal{P}_n$ the inequality*

$$(2\lambda_n + B'(0))\|\sqrt{A}P'\|_2^2 \leq \lambda_n^2\|P\|_2^2 + \|AP''\|_2^2 \tag{4.3}$$

holds, with equality if only if $P(t) = cQ_n(t)$, where Q_n is the classical orthogonal
polynomial of degree n orthogonal to all polynomials of degree $\leq n-1$ with respect

to the weight function $w(t)$ on (a, b), and c is an arbitrary real constant. λ_n, $A(t)$ and $B(t)$ are given in Table 1.1.

The Hermite case was considered by Varma [37]. Then, the inequality (4.3) reduces to

$$\|P'\|_2^2 \leq \frac{1}{2(2n-1)}\|P''\|_2^2 + \frac{2n^2}{2n-1}\|P\|_2^2.$$

In the generalized Laguerre case, the inequality (4.3) becomes

$$\|\sqrt{t}P'\|_2^2 \leq \frac{n^2}{2n-1}\|P\|_2^2 + \frac{1}{2n-1}\|tP''\|_2^2,$$

where $w(t) = t^s e^{-t}$ on $(0, +\infty)$.

In the Jacobi case the inequality (4.3) reduces to the inequality

$$((2n-1)(\alpha+\beta) + 2(n^2 + n - 1))\|\sqrt{1-t^2}P'\|_2^2$$
$$\leq n^2(n+\alpha+\beta+1)^2\|P\|_2^2 + \|(1-t^2)P''\|_2^2.$$

In the simplest case, when $\alpha = \beta = 0$ (Legendre case), we obtain

$$\|\sqrt{1-t^2}P'\|_2^2 \leq \frac{n^2(n+1)^2}{2(n^2+n-1)}\|P\|_2^2 + \frac{1}{2(n^2+n-1)}\|(1-t^2)P''\|_2^2.$$

In the Chebyshev case ($\alpha = \beta = -1/2$), we get

$$\|\sqrt{1-t^2}P'\|_2^2 \leq \frac{n^4}{2n^2-1}\|P\|_2^2 + \frac{1}{2n^2-1}\|(1-t^2)P''\|_2^2,$$

where $\|f\|_2^2 = \int_{-1}^{1}(1-t^2)^{-1/2}f(t)^2\,dt$.

The corresponding result for trigonometric polynomials was obtained by Varma [38].

Recently Guessab [9] obtained sharp Markov-Bernstein inequalities in L^2 norms that are weighted with classical weights.

Theorem 4.3. *Let $P \in \mathcal{P}_n$ and $w \in CW$. Then*

$$\|w^{-1/2}(V(t)P)'\|_2^2 + \|\sqrt{A(t)C(t)}\,P\|_2^2 \leq \beta_n\|P\|_2^2, \tag{4.4}$$

where $A(t)$ and λ_n are given in Table 1.1, and $V(t) = \sqrt{A(t)w(t)}$,

$$C(t) = \begin{cases} \dfrac{1}{4}\left(\dfrac{\alpha^2-1}{(1-t)^2} + \dfrac{\beta^2-1}{(1+t)^2}\right), & \text{Jacobi case,} \\[3mm] \dfrac{1}{4}\left(\dfrac{s^2-1}{t^2} + 1\right), & \text{generalized Laguerre case,} \\[3mm] t^2, & \text{Hermite case,} \end{cases}$$

260

and

$$\beta_n = \lambda_n + \begin{cases} \dfrac{1}{2}(\alpha+1)(\beta+1), & \textit{Jacobi case,} \\[2mm] \dfrac{1}{2}(s+1), & \textit{generalized Laguerre case,} \\[2mm] 1, & \textit{Hermite case.} \end{cases}$$

The equality is attained in (4.4) if and only if P is a constant multiple of the classical polynomial Q_n orthogonal with respect to the weight function $t \mapsto w(t)$.

This elegant result is established by using the second-order Sturm-Liouville type differential equations satisfied by the classical orthogonal polynomials.

Using the method from [11], Guessab [10] has investigated the extremal problem

$$\max_{P \in \mathcal{P}_n^1} \left\| \left(\sqrt{A}/w_m\right)\left(w_m P^{(m)}\right)' \right\|_{w_m},$$

where $w \in CW$, $w_m = A^m w$, $\mathcal{P}_n^1 = \{P \in \mathcal{P}_n \mid \|P\|_{w_m} \le 1\}$, and

$$\|f\|_{w_m} = \left(\int_a^b w_m(t)|f(t)|^2 w_m(t)\,dt \right)^{1/2}.$$

Theorem 4.4. *Let $P \in \mathcal{P}_n^1$ and $w \in CW$. Then*

$$\left\| \left(\sqrt{A}/w_m\right)\left(w_m P^{(m)}\right)' \right\|_{w_m} \le \sqrt{\lambda_{n,0}\lambda_{n,1}\cdots\lambda_{n,m-1}\beta_{n,m}}, \tag{4.5}$$

where $\lambda_{n,\nu}$ is as in Theorem 4.1,

$$\beta_{n,m} = \lambda_{n,m} + B'(0) + (k-1)A''(0),$$

and $A(t)$ and $B(t)$ are given in Table 1.1.

The equality is attained in (4.5) if and only if P is a constant multiple of the classical polynomial Q_n orthogonal with respect to the weight function $t \mapsto w(t)$.

At the end we mention a result of Guessab and Milovanović [12]. They considered the extremal problems of Markov's type

$$C_{n,m}(d\lambda) = \sup_{P \in \mathcal{P}_n} \frac{\|\mathcal{D}_m P\|_2}{\|A^{m/2}P\|_2} \qquad (m \ge 1) \tag{4.6}$$

for the differential operator $\mathcal{D}_m$ defined by

$$\mathcal{D}_m P = \frac{d^m}{dt^m}\left[A^m P\right] \qquad (P \in \mathcal{P}_n), \tag{4.7}$$

where

$$\|P\|_2 = \left(\int_{\mathbb{R}} |P(t)|^2\,d\lambda(t) \right)^{1/2},$$

and found the best constant $C_{n,m}(d\lambda)$ in three following cases:

1° The *Legendre measure* $d\lambda(t) = dt$ on $[-1,1]$;

2° The *Laguerre measure* $d\lambda(t) = e^{-t}\,dt$ on $[0,+\infty)$.

3° The *Hermite measure* $d\lambda(t) = e^{-t^2}\,dt$ on $(-\infty,+\infty)$.

Some extremal problems for differential operators were investigated by Stein [29] and Džafarov [7].

Let $P \in \mathcal{P}_n$, $d\lambda(t) = w(t)\,dt$ on (a,b), and $\mathcal{D}_m$ be given by (4.7). An application of integration by parts gives

$$\|\mathcal{D}_m P\|_2^2 = \int_a^b (\mathcal{D}_m P)^2 w\,dt = (-1)^m \int_a^b A^m P[w\mathcal{D}_m P]^{(m)}\,dt.$$

Since

$$(-1)^m \int_a^b A^m P[w\mathcal{D}_m P]^{(m)}\,dt = \int_a^b (-1)^m \left(\sqrt{w}A^{m/2}P\right)\left(\frac{A^{m/2}[w\mathcal{D}_m P]^{(m)}}{\sqrt{w}}\right)dt,$$

using Cauchy-Schwarz-Buniakowsky inequality we obtain

$$\|\mathcal{D}_m P\|_2^2 \le \|A^{m/2}P\|_2 \left(\int_a^b \frac{A^m}{w}\left([w\mathcal{D}_m P]^{(m)}\right)^2 dt\right)^{1/2},$$

with equality if and only if

$$\mathcal{F}_m P = \frac{(-1)^m}{w}[w\mathcal{D}_m P]^{(m)} = \gamma P \qquad (P \in \mathcal{P}_n),$$

where γ is an arbitrary constant.

Taking a norm with respect to the measure $d\lambda_m(t) = A^m\,d\lambda(t) = A^m w\,dt$,

$$\|f\|_* = \left(\int_a^b |f(t)|^2\,d\lambda_m(t)\right)^{1/2},$$

we have

$$\frac{\|\mathcal{D}_m P\|_2}{\|A^{m/2}P\|_2} \le \left(\frac{\|\mathcal{F}_m P\|_*}{\|P\|_*}\right)^{1/2}, \tag{4.8}$$

with equality if and only if $\mathcal{F}_m P = \gamma P$, i.e.,

$$(-1)^m \frac{d^m}{dt^m}\left[w\frac{d^m}{dt^m}(A^m P)\right] = \gamma w P \qquad (P \in \mathcal{P}_n). \tag{4.9}$$

We are interested only in polynomial solutions of this equation. If they exist, then from the eigenvalue problem (4.9) and the inequality (4.8), we can determine the best constant in the extremal problem (4.6). Namely,

$$C_{n,m}(d\lambda) = \sqrt{\max_{0 \le \nu \le n} |\lambda_{\nu,m}|},$$

where $\lambda_{\nu,m}$ are eigenvalues of the operator $\mathcal{F}_m$. Then, the extremal polynomial is the eigenfunction corresponding to the maximal eigenvalue. Guessab and Milo-vanović [12] solved the following cases:

1° In the Legendre case $(d\lambda(t) = dt \text{ on } (-1,1))$,

$$C_{n,m}(d\lambda) = \sqrt{\frac{(n+2m)!}{n!}},$$

with the extremal polynomial $P^*(t) = \gamma C_n^{m+1/2}(t)$, where C_n^μ is the Gegenbauer polynomial of degree n.

2° In the Laguerre case $(d\lambda(t) = e^{-t} dt \text{ on } (0, +\infty))$,

$$C_{n,m}(d\lambda) = \sqrt{\frac{(n+m)!}{n!}},$$

with the extremal polynomial $P^*(t) = \gamma L_n^m(t)$, where L_n^m is the generalized La-guerre polynomial of degree n.

3° In the Hermite case $(d\lambda(t) = e^{-t^2} dt \text{ on the real line } \mathbb{R})$,

$$C_{n,m}(d\lambda) = 2^{m/2} \sqrt{n!/(n-m)!},$$

with the extremal polynomial $P^*(t) = \gamma H_n(t)$, where H_n is the Hermite poly-nomial of degree n. This result can be found in Ph. D. Thesis of Shampine [28] (see also, Dörfler [6] and Milovanović [20]). The case $m = 1$ was investigated by Schmidt [27] and Turán [32].

Remark 4.1. In the Jacobi case with the weight $t \mapsto (1-t)^\alpha (1+t)^\beta$ $(\alpha, \beta > -1)$ the equation (4.8) has no polynomial solution for $|\alpha| + |\beta| > 0$. Similarly, in the generalized Laguerre case with the weight function $t \mapsto t^s e^{-t}$ $(s > -1)$ the equation (4.8) has no polynomial solution for $s \neq 0$.

Remark 4.2. For extremal problems of Markov-Bernstein and Turán type on restricted polynomial classes in L^r norm see [13], [19], [21–26], [30], and [33–36].

References

1. R. P. Agarwal and G. V. Milovanović, *One characterization of the classical orthogonal poly-nomials*, Progress in Approximation Theory (P. Nevai and A. Pinkus, eds.), Academic Press, New York, 1991, pp. 1–4.

2. M. Baran, *New approach to Markov inequality in L^p norms*, Approximation Theory In Memory of A. K. Varma (N.K. Govil, R.N. Mohapatra, Z. Nashed, A. Sharma, and J. Szabados, eds.), Marcel Dekker, New York, 1998, pp. 75–95.

3. R. Bellman, *A note on an inequality of E. Schmidt*, Bull. Amer. Math. Soc. **50** (1944), 734–736.

4. S. N. Bernstein, *Sur l'ordre de la meilleure approximation des fonctions continues par des polynômes de degré donné*, Mémoires de l'Académie Royale de Belgique (2) **4** (1912), 1–103.

5. S. N. Bernstein, *Sur le théorème de W. Markov*, Trans. Lenin. Ind. Inst. (1938), 8–13.

6. P. Dörfler, *New inequalities of Markov type*, SIAM J. Math. Anal. **18** (1987), 490–494.

7. Ar. S. Džafarov, *Bernstein inequality for differential operators*, Analysis Math. **12** (1986), 251–268.

8. P. Goetgheluck, *On the Markov inequality in L^p-spaces*, J. Approx. Theory **62** (1990), 197–205.

9. A. Guessab, *Some weighted polynomial inequalities in L^2-norm*, J. Approx. Theory **79** (1994), 125–133.

10. A. Guessab, *Weighted L^2 Markoff type inequality for classical weights*, Acta Math. Hung. **66** (1995), 155–162.

11. A. Guessab and G. V. Milovanović, *Weighted L^2-analogues of Bernstein's inequality and classical orthogonal polynomials*, J. Math. Anal. Appl. **182** (1994), 244–249.

12. A. Guessab and G. V. Milovanović, *Extremal problems of Markov's type for some differential operators*, Rocky Mountain J. Math. **24** (1994), 1431–1438.

13. A. Guessab, G. V. Milovanović and O. Arino, *Extremal problems for nonnegative polynomials in L^p norm with generalized Laguerre weight*, Facta Univ. Ser. Math. Inform. **3** (1988), 1–8.

14. V. A. Gusev, *Functionals of derivatives of an algebraic polynomial and V. A. Markov's theorem*, Izv. Akad. Nauk SSSR Ser. Mat. **25** (1961), 367–384. (Russian)

15. E. Hille, G. Szegö and J. D. Tamarkin, *On some generalizations of a theorem of A. Markoff*, Duke Math. J. **3** (1937), 729–739.

16. A. A. Markov, *On a problem of D. I. Mendeleev*, Zap. Imp. Akad. Nauk, St. Petersburg **62** (1889), 1–24.

17. V. A. Markov, *On functions deviating least from zero in a given interval*, Izdat. Imp. Akad. Nauk, St. Petersburg, 1892 (Russian) [German transl. Math. Ann. **77** (1916), 218–258]..

18. D. Mendeleev, *Investigation of aqueous solutions based on specific gravity*, St. Petersburg, 1887..

19. G. V. Milovanović, *An extremal problem for polynomials with nonnegative coefficients*, Proc. Amer. Math. Soc. **94** (1985), 423–426.

20. G. V. Milovanović, *Various extremal problems of Markov's type for algebraic polynomials*, Facta Univ. Ser. Math. Inform. **2** (1987), 7–28.

21. G. V. Milovanović, *Extremal problems for restricted polynomial classes in L^r*, Approximation Theory In Memory of A. K. Varma (N.K. Govil, R.N. Mohapatra, Z. Nashed, A. Sharma, and J. Szabados, eds.), Marcel Dekker, New York, 1998, pp. 405–432.

22. G. V. Milovanović and R. Ž. Djordjević, *An extremal problem for polynomials with nonnegative coefficients. II*, Facta Univ. Ser. Math. Inform. **1** (1986), 7–11.

23. G. V. Milovanović and I. Ž. Milovanović, *An extremal problem for polynomials with nonnegative coefficients. III*, Constructive Theory of Functions '87 (Varna, 1987) (Bl. Sendov, P. Petrušev, K. Ivanov, R. Maleev, eds.), Bulgar. Acad. Sci., Sofia, 1988, pp. 315–321.

24. G. V. Milovanović, D. S. Mitrinović and Th. M. Rassias, *Topic in Polynomials: Extremal Problems, Inequalities, Zeros*, World Scientific, Singapore – New Jersey – London – Hong Kong, 1994.

25. G. V. Milovanović, D. S. Mitrinović and Th. M. Rassias, *On some Turán's extremal problems for algebraic polynomials*, Topics in Polynomials of One and Several Variables and Their Applications: A Mathematical Legacy of P. L. Chebyshev (1821–1894) (Th. M. Rassias, H. M. Srivastava, and A. Yanushauskas, eds.), World Scientific, Singapore – New Jersey – London – Hong Kong, 1993, pp. 403–433.

26. G. V. Milovanović and Th. M. Rassias, *New developments on Turán extremal problems for polynomials*, Approximation Theory In Memory of A. K. Varma (N.K. Govil, R.N. Mohapatra, Z. Nashed, A. Sharma, and J. Szabados, eds.), Marcel Dekker, New York, 1998, pp. 433–447.

27. E. Schmidt, *Über die nebst ihren Ableitungen orthogonalen Polynomensysteme und das zugehörige Extremum*, Math. Ann. **119** (1944), 165–204.

28. L. F. Shampine, *Asymptotic L_2 Inequalities of Markoff Type*, Ph. Thesis, California Institute of Technology, Pasadena, 1964.

264

29. E. M. Stein, *Interpolation in polynomial classes and Markoff's inequality*, Duke Math. J. **24** (1957), 467–476.

30. J. Szabados and A. K. Varma, *Inequalities for derivatives of polynomial having real zeros*, Approximation Theory III (E. W. Cheney, ed.), Academic Press, New York, 1980, pp. 881–887.

31. G. Szegő, *Orthogonal Polynomials*, Amer. Math. Soc. Colloq. Publ. vol. 23, 4th ed., Amer. Math. Soc., Providence, R. I., 1975.

32. T. Turán, *Remark on a theorem of Erhard Schmidt*, Mathematica **2** (25) (1960), 373–378.

33. A. K. Varma, *An analogue of some inequalities of P. Turán concerning algebraic polynomials having all zeros inside* [−1, 1], Proc. Amer. Math. Soc. **55** (1976), 305–309.

34. A. K. Varma, *An analogue of some inequalities of P. Erdős and P. Turán concerning algebraic polynomials satisfying certain conditions*, Fourier Analysis and Approximation Theory, Vol. II (Budapest 1976), Colloq. Math. Soc. János Bolyai 19, 1978, pp. 877–890.

35. A. K. Varma, *Some inequalities of algebraic polynomials having real zeros*, Proc. Amer. Math. Soc. **75** (1979), 243–250.

36. A. K. Varma, *Derivatives of polynomials with positive coefficients*, Proc. Amer. Math. Soc. **83** (1981), 107–112.

37. A. K. Varma, *A new characterization of Hermite polynomials*, Acta Math. Hung. **49** (1987), 169–172.

38. A. K. Varma, *Inequalities for trigonometric polynomials*, J. Approx. Theory **65** (1991), 273–278.

39. V. S. Videnskiĭ, *A generalization of an inequality of V. A. Markov*, Dokl. Akad. Nauk SSSR **120** (1958), 447–449.

40. E. V. Voronovskaja, *The Functional Method and Its Applications*, Trans. Math. Monographs 28, Amer. Math. Soc., Providence, 1970.

ON ALPHA-QUASI-CONVEX FUNCTIONS DEFINED BY CONVOLUTION

WITH INCOMPLETE BETA FUNCTIONS

KHALIDA INAYAT NOOR
Mathematics Department, College of Science,
King Saud University, P.O. Box 2455, Riyadh 11451, Saudi Arabia.

Abstract. We introduce a family $Q_\alpha(a, c)$, $\alpha \geq 0$ of functions $f : f(z) = z + \sum_{n=2}^{\infty} a_n z^n$, analytic in the unit disc E by using a well-known convolution operator $L(a, c)f = \phi(a, c) * f$ where $\phi(a, c)$ is an incomplete beta function. We investigate $Q_\alpha(a, c)$ and give some of its properties including integral representation, coefficient result, a covering theorem, and several inclusion results.

1. Introduction

The connections between the theory of univalent functions and the theory of special functions have received special attention after the surprising use of hypergeometric functions by de Branges in the proof of the 70-year old Bieberbach Conjecture, see [2]. Prior to this proof, the expansions and generating functions involving associated Lagurre Jacobi

1991 *Mathematics Subject Classification.* Primary 30C45.
   Key words and phrases. Incomplete beta function; convolution; convex; univalent; quasi-convex; integral operators; starlike.

T.M. Rassias and H.M. Srivastava (eds.), Analytic and Geometric Inequalities and Applications, 265–276.
© 1999 *Kluwer Academic Publishers.*

266

polynomials, Bessel functions and hypergeometric functions of one and several variables have been employed in many branches of Applied Sciences such as Physics, Engineering, Statistics, Probability and Operations Research.

Let A be the class of analytic functions f defined on the unit disc $E = \{z : | z | < 1\}$ and normalized by $f(0) = 0, f'(0) = 1$. By S, K, S^* and C denote the subclasses of A which are univalent, close-to-convex, starlike and convex in E respectively. The class A is closed under the Hadamard product or convolution

$$(f * g)(z) = \sum_{n=0}^{\infty} a_n b_n z^{n+1},$$

where $f(z) = \sum_{n=0}^{\infty} a_n z^{n+1}, \quad g(z) = \sum_{n=0}^{\infty} b_n z^{n+1}.$

In [3], a subclass C^* of univalent functions was introduced and studied. A function $f \in A$ is said to be in C^* if and only if there exists a function $g \in C$ such that, for $z \in E$, $Re\left\{ \dfrac{(zf'(z))'}{g'(z)} \right\} > 0$. The functions in C^* are called quasi-convex and $C \subset C^* \subset K \subset S$. It is also known that $f \in C^*$ if and only if $zf' \in K$.

A function $f \in A$ is in the class $Q_\alpha, \alpha \geq 0$ if and only if $F = [(1-\alpha)f + \alpha zf'] \in K$ for $z \in E$. We call Q_α the class of alpha-quasi-convex functions. Also a function $f \in A$ belongs to the class $K_\sigma^*(\gamma)$ of close-to-convex functions of order γ and type σ if and only if $Re\left\{ \dfrac{(zf'(z))'}{g(z)} \right\} > \gamma$, for some $g \in A$ with $Re\left\{ \dfrac{(zf'(z))'}{g(z)} \right\} > \sigma, z \in E$ and $0 \leq \gamma < 1$, $0 \leq \sigma < 1$. It is clear that $K_0^*(0) = K$.

Let $\phi(a, c, z), c \neq 0, -1, -2, \ldots$ be an incomplete beta function related to the Gauss hypergeometric functions by

$$\phi(a, c, z) = z\,{}_2F_1(1, a; c; z)$$

$$= \sum_{n=0}^{\infty} \frac{(a)_n}{(c)_n} z^{n+1}, \quad z \in E,$$

where

$$(a)_0 = 1, (a)_n = a(a+1)\ldots(a+n-1), \; n > 1$$
$$= \frac{\Gamma(a+n)}{\Gamma(a)}.$$

It has an analytic continuation to z-plane cut along the positive real line from 1 to ∞. We note that $\phi(a,1,z) = \dfrac{z}{(1-z)^{\alpha}}$ and $\phi(2,1,z)$ is the Koebe function. Carlson and Shaffer [1] defined a convolution operator on A involving an incomplete beta function as

$$L(a,c)f = \phi(a,c) * f, \qquad f \in A.$$

The Ruscheweyh derivatives of f are $L(n+1,1)f, n = 0,1,2,\ldots$. $L(a,a)$ is the identity and if $\alpha \neq 0, -1, -2, \ldots$, then $L(a,c)$ has a continuous inverse $L(c,a)$ and is a (1-1) mapping of A onto itself. $L(a,c)$ provides a convenient representation of differentiation and integration. If $g(z) = zf'(z)$ then $g = L(2,1)f$ and $f = L(1,2)g$.

We shall assume, unless otherwise stated, that $a \neq 0, -1, -2, \ldots$ and $c \neq 0, -1, -2, \ldots$.

We define the following.

Definition 1. Let $f \in A$. Then $f \in S^*(a,c)$ if and only if $L(a,c)f \in S^*$ in E. Similarly $f \in C(a,c)$ if and only if $L(a,c)f \in C$ in E.

Definition 2. Let $f \epsilon A$. Then $f \epsilon K(a,c)$ if and only if there exists a $g \epsilon S^*(a,c)$ such that, for $z \in E$. $Re \left\{ \dfrac{z(L(a,c)f(z))'}{L(a,c)g(z)} \right\} > 0$. That is $f \in K(a,c)$ if and only if $L(a,c)f \in C^*$.

Definition 3. Let $f \in A$. Then $f \in Q_\alpha(a, c)$, $\alpha \geq 0$ if and only if $F = \{(1 - \alpha)f + \alpha z f'\} \in K(a.c)$ for $z \in E$.

2. Preliminary Results.

Lemma 1. Let $f \in K(a, c)$ and ψ be a convex univalent function. Then $\psi * f \in K(a, c)$.

Lemma 2. Let a and c be real and $a > N(c)$ where

$$(2.1) \qquad N(c) = \begin{cases} \mid c \mid + \frac{1}{2}, & \text{if } \mid c \mid \geq \frac{1}{3} \\ \frac{3c^2}{2} + \frac{2}{3}, & \text{if } \mid c \mid \leq \frac{1}{3}, \end{cases}$$

Then, for $z \in E$ and $d \neq 0, -1, -2, \ldots$.

(i) $K(d, c) \subset K(d, a)$

(i) $K(a, d) \subset K(c, d)$.

The proofs of Lemmas 1 and 2 can be found in [5].

Lemma 3. [4]. Let p_0, p be analytic in E with $Re p_0(z) > 0$ for $z \in E$ and $p(0) = 1$. Let, for $z \in E$, $\alpha \geq 0$

$$Re\{[(1 - \alpha) + \alpha p_0(z)]p(z) + \alpha z p'(z)\} > 0.$$

Then $Re\, p(z) > 0$ in E.

Lemma 4. [6]. If $p(z)$ is analytic in E with $p(0) = 1$ and if λ is a complex number satisfying $Re\, \lambda \geq 0$ $(\lambda \neq 0)$, then

$$Re\{p(z) + \lambda z p'(z)\} > \alpha \quad (0 \leq \alpha < 1)$$

implies $Rep(z) > \alpha + (1 - \alpha)(2\gamma - 1)$, where γ is given by

$$\gamma = \gamma(Re\lambda) = \int_0^1 (1 + t^{Re\lambda})^{-1} dt$$

which is an increasing function of $Re\lambda$ and $\frac{1}{2} \le \gamma < 1$. The estimate is sharp in the sense that the bound can not be improved.

3. Main Results

Theorem 1. Let $f \epsilon Q_\alpha(a, c)$. Then there exists a function $g \in C(a, c)$ such that

$$Re\left\{(1 - \alpha)\frac{(L(a,c)f(z))'}{(L(a,c)g(z))'} + \alpha\frac{[z(L(a,c)f(z))']'}{(L(a,c)g(z))'}\right\} > 0, \quad z \in E,$$

and conversely.

The proof follows directly from the Definition 3.

Remark 1. By taking $L(a, c)g(z) = z$ in Theorem 1, we have

$$Re\{(La, c)f(z))' + \alpha[z(L(a, c)f(z))'']\} > 0, \quad z \in E.$$

We shall denote the class of such functions f as $R_\alpha(a, c)$.

Theorem 2. (Integral Representation). $f \in Q_\alpha(a, c)$, $\alpha > 0$ if and only if there exists a function $F \in K(a, c)$ such that, for $z \in E$

(3.1)
$$f(z) = \frac{1}{\alpha}z^{1-\frac{1}{\alpha}}\int_0^z t^{\frac{1}{\alpha}-2}F(t)dt$$

Proof. From (3.1), it follows that

$$\alpha(\frac{1}{\alpha} - 1)z^{\frac{1}{\alpha}-2}f(z) + \alpha z^{\frac{1}{\alpha}-1}f'(z) = z^{\frac{1}{\alpha}-2}F(z)$$

so that

$$(1 - \alpha)f(z) + \alpha z f'(z) = F(z)$$

and the result follows from the Definition 3.

Theorem 3. (Coefficient Result). Let $f \in Q_\alpha(a, c)$ and $f(z) = \sum_{n=0}^{\infty} a_{n+1} z^{n+1}$
($a_1 = 1$). Then

$$|a_n| \leq \frac{(n+1)}{\alpha n + 1} \left| \frac{(C)_n}{(a)_n} \right|, \qquad n \geq 1.$$

These bounds are sharp and the equality occurs for f_0 given by

$$(1 - \alpha)f_0(z) + \alpha z f_0'(z) = L(c, a)k(z),$$

where $k(z)$ is the Koebe function.

Proof. $f \in Q_\alpha(a, c)$ implies $(1 - \alpha)f + \alpha z f' = F \in K(a, c)$. Let $F(z) = \sum_{n=0}^{\infty} b_{n+1} z^{n+1}$
($b_1 = 1$). Then, from Definition 1.2 and the well-known coefficient result for the class K,
we have

(3.2)
$$|b_{n+1}| \leq \left| \frac{(n+1)(c)_n}{(a)_n} \right|, \quad n \geq 1$$

We now use (3.2) and Definition 1.3 to obtain the required result.

Theorem 4.

$$Q_\alpha(a, c) \subset K(a, c).$$

Proof. For $\alpha = 0$, $Q_0(a, c) = K(a, c)$ and the result is obvious. We shall assume $\alpha > 0$
and $f \in Q_\alpha(a, c)$. From (3.1), we have

$$
\begin{aligned}
f(z) &= \frac{1}{\alpha} z^{1 - \frac{1}{\alpha}} \int_0^z t^{\frac{1}{\alpha} - 2} F(t) dt, \quad F \in K(a, c) \\
&= \Psi_\alpha(z) * F(z),
\end{aligned}
$$

where $\Psi_\alpha(z) = \sum\limits_{n=1}^{\infty} \dfrac{1}{\alpha(n+1)+1} z^n$ is convex in E. We now use Lemma 1 to conclude $f \in K(a,c)$ and this proves the theorem.

Remark 2. If a and c satisfy the condition (2.1) of Lemma 2, then it shown in [5] that $K(a,c) \subset K \subset S$, and hence $f \in Q_\alpha(a,c)$ is univalent in E.

Theorem 5. (Covering Result). Let $f \in Q_\alpha(a,c)$ and let a and c satisfy (2.1). If B is the boundary of the image of E under f, then every point of B is at distance at least $\dfrac{a(\alpha+1)}{2[a(\alpha+1)+c]}$ from the origin.

Proof. Let w_0 be any non zero complex numbers such that $f(z) \neq w_0$ for $z \in E$. Then

$$\frac{w_0 f(z)}{w_0 - f(z)} = z + (a_2 + \frac{1}{w_0})z^2 + \cdots.$$

is univalent by Remark 2. Hence $|\frac{1}{w_0} + a_2| \leq 2$, and using Theorem 3.3 for $n = 1$, we obtain

$$|w_0| \geq \frac{a(\alpha+1)}{2[a(\alpha+1)+c]}.$$

We now prove that $Q_\alpha(a,c)$ is closed under convolution with convex univalent functions and give some of its applications as follows.

Theorem 6. Let $f \in Q_\alpha(a,c)$ and $\psi \in C$. Then $f * \psi \in Q_\alpha(a,c)$.

Proof. The Theorem is proved if we show that $\{(1-\alpha)(f * \psi) + \alpha z(f * \psi)'\} \in K(a,c)$.

Now

$$
\begin{aligned}
(1-\alpha)(f * \psi) + \alpha z(f * \psi)' &= (1-\alpha)(f * \psi) + \alpha z \frac{z f' * \psi}{z} \\
&= \psi * \{(1-\alpha)f + \alpha z f'\}
\end{aligned}
$$

$$= \psi * F, \quad F \in K(a,c)$$

$$= F_1, \quad F_1 \in K(a,c) \quad \text{by Lemma 1.}$$

Applications of Theorem 6.

(1) Let $f \in Q_\alpha(a,c)$. Then $Q_\alpha(a,c)$ is invariant under the following integral operators.

(i) $f_1(z) = \int_0^z \frac{f(t)}{t} dt$

(ii) $f_2(z) = \frac{2}{z} \int_0^z f(t) dt$

(iii) $f_3(z) = \int_0^z \frac{f(t)-f(xt)}{t-xt} dt, \quad |x| \le 1, \quad x \ne 1$

(iv) $f_4(z) = \frac{1+b}{z^b} \int_0^z t^{b-1} f(t) dt, \quad Re\, b > -1$

We note that (ii) is a special case of (iv) with $b = 1$.

In fact we can write

$$f_i(z) = (f * \psi_i)(z)$$

where $\psi_i \in C$, $i = 1,2,3,4$ and then apply Theorem 6 to obtain the required result.

(2) Let μ_1 and μ_2 be linear operators defined on the class A as follows,

$$\mu_1(f) = zf', \qquad \mu_2(f) = \frac{(zf)'}{2}.$$

Both of these operators can be written as $\mu_i(f) = h_i * f \quad i = 1,2$, where

$$h_1(z) = \sum_{n=1}^{\infty} nz^n = \frac{z}{(1-z)^2}, \quad h_2(z) = \sum_{n=1}^{\infty} \frac{n+1}{2} z^n = \frac{z-z^2}{(1-z)^2},$$

and we note that the radius of convexity $r_c(h_1) = 2 - \sqrt{3}$, $r_c(h_2) = 1/2$. Thus it follows from Theorem 3.6 that if $f \in Q(a,c)$ then $\mu_1(f) \in Q_\alpha(a,c)$ for $|z| < 2 - \sqrt{3}$ and $\mu_2(f) \in Q_\alpha(a,c)$ for $|z| < \frac{1}{2}$.

(3) Let $f \in Q_\alpha(a,c)$ and $\mu_b(f) = \psi_b * f$ where $\psi_b(z) = \sum_{n=1}^\infty \frac{n+b}{b+1} z^n$, $(b > -1)$. The function ψ_b is convex for $|z| < r_0$ where $r_0 = \frac{b+1}{2+\sqrt{3+b^2}}$. Then $\mu_b(f) \in Q_\alpha(a,c)$ for $|z| < r_0$.

We have shown that $Q_\alpha(a,c) \subset K(a,c)$. In the following we discuss the case in the opposite direction.

Theorem 7. Let $f \in K(a,c)$. Then, for $0 < \alpha \le 1$, $f \in Q_\alpha(a,c)$ for $|z| < r_\alpha$, where

(3.3) $$r_\alpha = 1/[2\alpha + \sqrt{4\alpha^2 - 2\alpha + 1}].$$

Proof. Let $f \in K(a,c)$. Define F as follows.

$$f(z) = \frac{1}{\alpha} z^{1-\frac{1}{\alpha}} \int_0^z t^{\frac{1}{\alpha}-2} F(t)dt, \quad 0 < \alpha \le 1.$$

Then we can write

$$F(z) = \psi_\alpha * f, \quad \psi_\alpha(z) = \sum_{n=1}^\infty [\alpha(n-1)+1]z^n.$$

ψ_α is convex for $|z| < r_\alpha$ where r_α is given by (3.3) and consequently it follows that $f \in Q_\alpha(a,c)$ for $|z| < r_\alpha$.

We now discuss some inclusion results.

Theorem 8. Let $0 \le \beta < \alpha$. Then $Q_\alpha(a,c) \subset Q_\beta(a,c)$.

Proof. For $\beta = 0$, the proof is immediate. Therefore we assume $\beta > 0$ and $f \in Q_\alpha(a,c)$.

Let

$$J_\alpha(a,c) = (1-\alpha)\frac{(La,c)f)'}{(L(a,c)g)'} + \alpha\frac{(z(L(a,c))')'}{(L(a,c)g)'}.$$

Then $ReJ_\alpha(a,c) > 0$, $z \in E$.

Also, with $\dfrac{(L(a,c)f)'}{(L(a,c)g)'} = p$, p analytic in E with $p(0) = 1$, we can write

$$J_\alpha(a,c) = [(1-\alpha) + \alpha p_0]p + \alpha z p', \quad \text{where} \quad Rep_0 = Re\frac{(z(L(a,c)g)')'}{(L(a,c)g)'} > 0, \quad z \in E,$$

Using Lemma 3, we see that $Rep > 0$ in E.

Now, for $0 < \beta < \alpha$,

$$J_\beta(a,c) = \frac{\beta}{\alpha}J_\alpha(a,c) + \left(1 - \frac{\beta}{\alpha}\right)\frac{(L(a,c)f)'}{(L(a,c)g)'},$$

and since the class P of functions with positive real part is a convex set, we conclude that $ReJ_\beta(a,c) > 0$ in E and hence $f \in Q_\beta(a,c)$.

It is proved in [5] that if $F \in K(a,c)$ and $f(z) = \frac{\beta+1}{z^\beta}\int_0^z t^{\beta-1}F(t)dt$ $(\beta > -1)$, then $L(a,c)f \in K_\sigma^*(\gamma)$ where

(3.4) $$\sigma = \frac{1}{4}[-(2\beta+1) + \sqrt{4\beta^2 + 4\beta + 9}],$$

and $\gamma : 0 \leq \gamma < 1$ is given as $\gamma \leq \gamma_A$,

(3.5) $$\gamma_A = \frac{J}{2M+J}, \quad J = (1-\sigma)p_1 + (\sigma+\beta),$$
$$M = J^2 + (1-\sigma)^2p_2^2 \quad (Rep_0 = Re(p_1 + ip_2) > 0)$$

Using this result with $\beta\frac{1}{\alpha} - 1$, $0 < \alpha \leq 1$, we have the following.

Theorem 9. For $0 < \alpha \leq 1$, let $f \in Q_\alpha(a,c)$. Then $L(a,c)f \in K_\sigma^*(\gamma)$, where σ and γ are given by (3.4) and (3.5) respectively.

Theorem 10. Let a and c satisfy (2.1). Then, for $z \in E$ and $d \neq 0, -1, -2, \ldots$,

$$Q_\alpha(d, c) \subset Q_\alpha(d, a).$$

Proof. Let $f \in Q_\alpha(d, c)$. Then $F = (1 - \alpha)f + \alpha z f' \in K(d, c)$, and the result follows immediately by using Lemma 2.

Remark 3. The inclusion result $Q_\alpha(a, d) \subset Q_\alpha(c, d)$ can also be proved similarly by imposing appropriate restrictions on a, c and d.

Theorem 11. For $\alpha \geq 1$, $Q_\alpha(a, c) \subset C^*(a, c)$.

Proof. Let $f \in Q_\alpha(a, c)$. Then there exists $g \in C(a, c)$ such that

$$
\begin{aligned}
\frac{(z(L(a, c)f)'}{(L(a, c)g)'} &= \frac{1}{\alpha}\{J_\alpha(a, c)\} + (1 - \frac{1}{\alpha})\frac{(L(a, c)f)'}{(L(a, c)g)'} \\
&= \frac{1}{\alpha}p_1 + (1 - \frac{1}{\alpha})p_2, \quad Re p_i > 0, \quad i = 1, 2, \quad z \in E.
\end{aligned}
$$

Since $p_1, p_2 \in P$ and P is a convex set, we obtain the required result that $f \in C^*(a, c)$.

Next we prove a result for the class $R_\alpha(a, c)$.

Theorem 12. Let $f \in R_\alpha(a, c)$, $\alpha > 0$. Then $Re\{L(a, c)f\}' > (2\lambda - 1)$, where $\lambda = \int_0^1 (1 + t^\alpha)^{-1}dt$. This bound is sharp.

Proof. Let $(L(a, c)f)' = p$. p is analytic in E and $p(0) = 1$. Therefore

$$(L(a, c)f)' + \alpha z(L(a, c)f)'' = p + \alpha z p'.$$

Since $f \in R_\alpha(a, c)$, $Re(p + \alpha z p') > 0$ in E. Using Lemma 4, we obtain the required result.

Corollary 1. Let $\alpha = 1$ and $f \in R_1(a, c)$. Then it follows from Theorem 2 that $Re(L(a, c)f)' > -1 + 2\log 2 = 0.39$ for $z \in E$. The constant $(-1 + 2\log 2)$ cannot be replaced by any larger one as can be seen from the function $F_0 \in R_1(a, c)$ and defined by

$$z(L(a, c)F_0)' = -z - 2\log(1 - z).$$

References

1. B.C. Carlson and D.B. Shaffer, Starlike and prestarlike hypergeometric functions, SIAM J. Math. Anal.**15**(1984), 737-745,

2. L,de Branges, A proof of the Bieberbach Conjecture, Acta Mathematica **154** (1985), 137-152.

3. K.I. Noor and D.K. Thomas, On quasi-convex univalent functions, Inter. J. Math. and Math. Sci. **3**(1980), 255-266.

4. K.I. Noor, On alpha quasi-convex functions, PanAmer. Math. J. **2**(1)(1992), 67-78.

5. K.I. Noor, On functions defined by convolution with incomplete beta functions, PanAmer. Math. J. **4**(4)(1994), 93-101.

6. S. Ponnusamy, Differential subordination and Bazilevič functions (Preprint).

FOURTH ORDER OBSTACLE BOUNDARY VALUE PROBLEMS

MUHAMMAD ASLAM NOOR
Department of Mathematics, Statistics, and Computer Science, Dalhousie University, Halifax, Nova Scotia, Canada B3H 3J5.

EISA A. AL-SAID
Mathematics Department, College of Science, King Saud University, P.O. Box 2455, Riyadh 11451, Saudi Arabia.

Abstract. It is known that contact, obstacle and unilateral problems arising in different branches of pure and applied sciences can be studied in the framework of variational inequalities. In this paper, we show that a class of variational inequalities related to contact problems in elastostatics can be characterized by a sequence of variational equations, which are solved using modified finite difference methods. Some examples are given to illustrate the efficiency of the new numerical methods.

1. Introduction

In recent years, variational inequality theory has emerged as an interesting and important branch of applicable mathematics. The ideas and techniques of variational inequalities are being applied in a variety of divers fields of mathematical and engineering sciences including fluid flow through porous media, elasticity, transportation, economics and optimal control, see, for example, [1-33] and the references therein. It is known that variational inequalities arise naturally in connection with the minimum of the convex

1991 *Mathematics Subject Classification.* Primary 49J40; Secondary 65L12.
Key words and phrases. Variational inequalitiy; Boundary value problems; Finite difference method.

T.M. Rassias and H.M. Srivastava (eds.), Analytic and Geometric Inequalities and Applications, 277–300.
© 1999 Kluwer Academic Publishers.

functions subject to certain constraints. The analogies with convex programming has been used to suggest many numerical methods for variational inequalities. For example, penalty methods have been used in the solution of variational inequalities. Since they reduce a constrained problem to a sequence of constrained problems to which standard numerical technique can be applied. It is well known that penalty methods are not efficient due to their inherent numerical instabilities, see [11] and the references therein. They have been modified by using some techniques that are based on augmented Lagrangian functions and include varying Lagrange multipliers value as well as the penalty parameter.

There is another approximation technique in which the original form of the energy functional associated with the unilateral problem does not change, but the original closed convex set of solutions is replaced by a sequence of sets with simpler structure such that the problem of minimizing one and the original functional in a certain space is solved at each step. We have noted that this technique is not practical for solving the obstacle problems.

In this paper, we discuss a simple, but a very powerful technique for solving both linear and nonlinear variational inequalities. This technique is very much similar to the penalty methods, but it does not have instability. This technique is known as the penalty function method. Such type of technique was used by Lewy and Stampacchia [8] to study the existence and regularity of the solution of the variational inequalities. The computational advantage of this method is its simple applicability for solving system of linear and nonlinear differential equations, which arise after using this penalty function technique. Once this is done, then any standard numerical method can be applied to find the numerical solution of the variational equations. Recently, this technique has been used as a basis for obtaining numerical solutions of some obstacle and contact problems successfully, see, for example, [1-7,16,21-23,25,26,33].

To convey an idea of this penalty function method, we describe briefly the approach of Kikuchi and Oden [17]. In general, penalty methods can be constructed for quite general minimization problems in which,

$I : H \to R$ is a weakly lower semicontinuous and coercive functional.

$P : H \to R$ is the penalty function, which satisfies :

(i) P is differentiable and weakly lower semicontinuous.

(ii) $P(v) \geq 0$ and $P(v) = 0$ if and only if $v \in K$; $v \in K \Rightarrow P(v) > 0$, where K is a closed convex set in H.

Then the penalty functional

$$I_\epsilon[v] = I[v] + \frac{1}{\epsilon} P(v),$$

has a minimizer u_ϵ for $\epsilon > 0$, and a subsequence of minimizer $\{u_\epsilon\}$ converges weakly to a solution u of the minimization problem

$$I[u] \leq I[v], \quad \text{for all } v \in K.$$

This type of penalty method can be generalized for other types of variational inequalities. Let P_K be the projection of H into the convex set K. Then the map $\beta : H \to H$ defined by

(1.1) $$\beta = v - P_K(v)$$

is monotone and continuous on H.

Using the fact that $P_K(v) = v$, if $v \in K$, one can show that

(1.2) $$\beta(v) = 0 \quad \text{if and only if} \quad v \in K.$$

Now, using (1.2) we get

(1.3) $< \beta(u) - \beta(v), u - v > \quad = \quad < u - v, u - v > - < P_K(u) - P_K(v), u - v >$

$$\geq \ \|u - v\|^2 - \|P_K(u) - P_K(v)\|\|u - v\|$$

$$\geq \ \|u - v\|^2 - \|u - v\|^2 \geq 0.$$

and

$$< \beta(u) - \beta(v), w > \ = \ < u - v, w > - < P_K(u) - P_K(v), w >$$

$$= \ < u - v - (P_K(u) - P_K(v)), w >$$

$$\leq \ \{\|u - v\| + \|P_K(u) - P_K(v)\|\}\|w\|$$

$$\leq \ 2\|u - v\|\|w\|.$$

Then, for every $\epsilon > 0$, there exists a unique solution $u_\epsilon \in H$ (if H is a finite dimensional space) to the problem

(1.4) $\qquad u_\epsilon \in H :< Tu_\epsilon, v > + \dfrac{1}{\epsilon} < \beta(u_\epsilon), v >= 0 \quad$ for all $v \in H$,

provided that T is assumed to be strictly monotone and coercive in the sense that

$$\lim_{\|v\| \to \infty} \left(\frac{< Tv - v_0, v >}{\|v\|} \right) = +\infty, \quad \text{for all} \quad v_0 \in K.$$

Furthermore, u_ϵ is uniformly bounded if $\epsilon > 0$. Then there exists a subsequence which converges to $w \in H$. By the definition of u_ϵ,

$$< \beta(u_\epsilon), v >\leq \epsilon \|A(u_\epsilon)\|\|v\| \quad \text{for all} \ v \in H$$

which implies that $\beta(u_\epsilon) \to 0$ as $\epsilon \to 0$.

Since β is continuous, $\beta(w) = 0$, that is, $w \in K$. Taking $v = v - u_\epsilon$ in (1.4), we have

$$< Tu_\epsilon, v - u_\epsilon > + \dfrac{1}{\epsilon} < \beta(u_\epsilon) - \beta(v), u_\epsilon - v >\geq 0.$$

Taking $\epsilon \to 0$, we have

$$< Tw, v - w >\geq 0,$$

provided that T is continuous.

We also know that the solution of the variational inequality (1.1) is unique, from which it follows that $w = u$, since for every convergent subsequence u_ϵ, the continuous are the same. Thus the sequence u_ϵ converges to the solution $u \in K$ of the variational inequality (1.1).

From the above discussions, we have the following result.

Lemma 1. Let $T : K \to H$ be a strongly monotone Lipschitz continuous operator. If β is the penalty function, then the sequence $u_\epsilon \in H$ such that

$$< Tu_\epsilon, u > + \frac{1}{\epsilon} < \beta u_\epsilon, v >= 0, \quad \text{for all } v \in H,$$

converges to the unique solution $u \in K$ of the variational inequality

$$< Tu, v - u >\geq 0, \quad \text{for all } v \in K \text{ as } \epsilon \to 0.$$

Using the penalty function technique of Lewy and Stampacchia [14], the variational inequality

$$< Tu, v - u >\geq 0, \quad \text{for all } v \in K$$

can be characterized by a sequence of variational equations as

$$< Tu, v > + < \nu(u - \psi)(u - \psi), v >= 0,$$

for all $v \in H$, where $\nu(t)$ is the discontinuous function defined by

$$\nu(t) = \begin{cases} 1, & \text{for } t \geq 0, \\ 0, & \text{for } t < 0, \end{cases}$$

is known as the penalty function and $\psi \leq 0$ on the boundary is the obstacle. For other type of penalty techniques, see Kikuchi and Oden [17].

In this paper, we consider the example of an elastic beam lying over an elastic obstacle. The formulation and the approximation of the elastic beam is very simple, however, it should emphasized that the type of numerical problems which occur for more complicated systems are the same. Our approach to these problems is to consider them in a general manner and specialize them later on. In section 2, we consider fourth order obstacle boundary value problem and study it via the variational inequality formulation. The penalty function technique is used to characterize the variational inequalities by a system of variational equations. In section 3, we investigate and develop some finite difference techniques for solving system of variational equations. Section 4 is devoted to the applications of the results given in section 3.

2. Formulation and Numerical Methods

We consider a system of the fourth order boundary value problem of the type

$$
(2.1) \qquad u^{(4)} = \begin{cases} f(x), & a \le x \le c, \\ g(x)u(x) + f(x) + r, & c \le x \le d, \\ f(x), & d \le x \le b, \end{cases}
$$

with the boundary conditions

$$
(2.2) \qquad u(a) = u(b) = \alpha_1, \quad u''(a) = u''(b) = \alpha_2
$$

and the continuity conditions

$$
(2.3) \qquad u(c) = u(d) = \alpha_3, \quad u''(c) = u''(d) = \alpha_4,
$$

where f and g are continuous on [a,b] and [c,d], respectively. The parameters r and α_i, $i = 1, 2, 3, 4$ are constants. Such type of systems arise in connection with the obstacle, unilateral and contact problems, see Section 4, for more details.

We shall investigate and develop finite difference methods for solving the coupled equations given by (2.1) over the whole interval [a,b]. Without loss of generality, we may take $c = \frac{1}{4}(3a + b)$ and $d = \frac{1}{4}(a + 3b)$. For this purpose we discritize the interval [a,b] using the equally spaced grid points

(2.4)
$$x_i = a + ih, \quad i = 0, 1, 2, \ldots, 4n$$
$$x_0 = a, \quad x_{4n} = b \quad \text{and} \quad h = \frac{b-a}{4n},$$

where n is a positive integer.

When $g(x) \equiv \lambda$ is a constant, then problem (2.1) reduces to

(2.5)
$$u^{(4)} = \begin{cases} f(x), & a \leq x \leq c, \\ \lambda u(x) + f(x) + r, & c \leq x \leq d, \\ f(x), & d \leq x \leq b, \end{cases}$$

with the boundary and continuity conditions (2.2) and (2.3). This problem was solved by Al-Said and Noor [3,4] using finite difference schemes based on the second order difference formula

(2.6)
$$h^4 u_i^{(4)} = u_{i-2} - 4u_{i-1} + 6u_i - 4u_{i+1} + u_{i+2} - \frac{h^6}{6} u_i^{(6)},$$

and the eighth order (Pade') difference formula

(2.7)
$$\begin{cases} -2u_{i-1} + 5u_i - 4u_{i+1} + u_{i+2} = -h^2 u_{i-1}'' + \frac{h^4}{360}[28u_{i-1}^{(4)} + 245u_i^{(4)} + 56u_{i+1}^{(4)} + u_{i+2}^{(4)}] \\ \qquad\qquad - \frac{h^8}{11!}[7568u_{i-1}^{(8)} + 136,070u_i^{(8)} + 16016u_{i+1}^{(8)} - 594u_{i+2}^{(8)}] + T_{i-1}, \\[4pt] \text{for } i = 1, n+1 \text{ and } 3n+1, \\[6pt] u_{i-2} - 4u_{i-1} + 6u_i - 4u_{i+1} + u_{i+2} \\ \qquad = \frac{h^4}{6}[u_{i-1}^{(4)} + 4u_i^{(4)} + u_{i+1}^{(4)}] - \frac{h^8}{10080}[-u_{i-1}^{(8)} + 16u_i^{(8)} - u_{i+1}^{(8)}] + T_i, \\[4pt] \text{for } 2 \leq i \leq n-2, \ n+2 \leq 3n-2 \text{ and } 3n+2 \leq i \leq 4n-2, \\[6pt] u_{i-2} - 4u_{i-1} + 5u_i - 2u_{i+1} = -h^2 u_{i+1}'' + \frac{h^4}{360}[u_{i-2}^{(4)} + 56u_{i-1}^{(4)} + 245u_i^{(4)} + 28u_{i+1}^{(4)}] \\ \qquad\qquad - \frac{h^8}{11!}[-594u_{i-2}^{(8)} + 16016u_{i-1}^{(8)} + 136070u_i^{(8)} + 7568u_{i+1}^{(8)}] + T_i, \\[4pt] \text{for } i = n-1, \ 3n-1 \text{ and } 4n-1, \end{cases}$$

Although the numerical results obtained by the scheme based on (2.7) are better than those computed by the method based on (2.6), both methods give second order accurate approximations for the solution of (2.5).

Khalifa and Noor [16] have used collocation method with quintic spline as the bases function to solve the special form of (2.1), namely,

$$(2.8) \qquad u^{(4)} = \begin{cases} 0, & a \le x \le c, \\ -4u(x) + 1, & c \le x \le d, \\ 0, & d \le x \le b. \end{cases}$$

It has been shown in [16] that the collocation-quintic spline method is a second order convergent technique. We mention here in passing that the finite difference technique used in [3,4] produce better results than those obtained by the collocation-quintic spline method discussed in [16].

In the present paper, we use the five point approximations of the fourth derivative

$$(2.9) \qquad \frac{h^4}{6}[u_{i-1}^{(4)} + 4u_i^{(4)} + u_{i+1}^{(4)}] = \delta^4 u_i - \frac{h^8}{720} u_i^{(8)},$$

and

$$(2.10) \qquad \frac{h^4}{720}[-u_{i-2}^{(4)} + 124u_{i-1}^{(4)} + 474u_i^{(4)} + 124u_{i+1}^{(4)} - u_{i+2}^{(4)}] = \delta^4 u_i - \frac{h^{10}}{3024} u_i^{(10)},$$

where

$$(2.11) \qquad \delta^4 u_i = u_{i-2} - 4u_{i-1} + 6u_i - 4u_{i+1} + u_{i+2},$$

and Taylor series expansions along with the method of undetermined coefficients, boundary and continuity conditions to develop the following finite difference schemes

$$(2.12) \quad \begin{cases} -2u_{i-1} + 5u_i - 4u_{i+1} + u_{i+2} = -h^2 u_{i-1}'' + \frac{h^4}{360}[28u_{i-1}^{(4)} + 245u_i^{(4)} \\ \qquad\qquad\qquad +56u_{i+1}^{(4)} + u_{i+2}^{(4)}] + t_{i-1}, \\ \text{for } i = 1, n+1 \text{ and } 3n+1, \\[6pt] u_{i-2} - 4u_{i-1} + 6u_i - 4u_{i+1} + u_{i+2} = \frac{h^4}{6}[u_{i-1}^{(4)} + 4u_i^{(4)} + u_{i+1}^{(4)}] + t_i, \\ \text{for } 2 \le i \le n-2, \ n+2 \le 3n-2 \text{ and } 3n+2 \le i \le 4n-2, \\[6pt] u_{i-2} - 4u_{i-1} + 5u_i - 2u_{i+1} = -h^2 u_{i+1}'' + \frac{h^4}{360}[u_{i-2}^{(4)} + 56u_{i-1}^{(4)} \\ \qquad\qquad\qquad +245u_i^{(4)} + 28u_{i+1}^{(4)}] + t_i \\ \text{for } i = n-1, \ 3n-1 \text{ and } 4n-1, \end{cases}$$

and

$$\begin{cases} -2u_{i-1} + 5u_i - 4u_{i+1} + u_{i+2} = -h^2 u''_{i-1} + \frac{h^4}{60480}[4233u^{(4)}_{i-1} + 43274u^{(4)}_i \\ \qquad\qquad +5662u^{(4)}_{i+1} + 3432u^{(4)}_{i+2} - 1391u^{(4)}_{i+3} + 230u^{(4)}_{i+4}] + t_{i-1}, \\ \text{for } i = 1, n+1 \text{ and } 3n+1, \\ u_{i-2} - 4u_{i-1} + 6u_i - 4u_{i+1} + u_{i+2} = \frac{h^4}{720}[-u^{(4)}_{i-2} + 124u^{(4)}_{i-1} + 474u^{(4)}_i \\ \qquad\qquad +124u^{(4)}_{i+1} - u^{(4)}_{i+2}] + t_i, \\ \text{for } 2 \le i \le n-2, \ n+2 \le 3n-2 \text{ and } 3n+2 \le i \le 4n-2, \\ u_{i-2} - 4u_{i-1} + 5u_i - 2u_{i+1} = -h^2 u''_{i+1} + \frac{h^4}{60480}[230u^{(4)}_{i-4} - 1391u^{(4)}_{i-3} \\ \qquad\qquad +3432u^{(4)}_{i-2} + 5662u^{(4)}_{i-1} + 43274u^{(4)}_i + 4233u^{(4)}_{i+1}] + t_i \\ \text{for } i = n-1, \ 3n-1 \text{ and } 4n-1, \end{cases}$$

(2.13)

where

(2.14) $\quad u^{(4)}_i = \begin{cases} f_i, & 0 \le i \le n-1 \text{ and } 3n+1 \le i \le 4n-1 \\ g_i u_i + f_i + r, & n \le i \le 3n, \end{cases}$

$u_i = u(x_i)$, $f_i = f(x_i)$ and $g_i = g(x_i)$. The local truncation errors t_i are given by

(2.15) $\quad t_i = \begin{cases} -\frac{241}{60480}h^8 u^{(8)}_{i-1}, & i = 1, n+1 \text{ and } 3n+1 \\ -\frac{1}{720}h^8 u^{(8)}_i, & 2 \le i \le n-2, n+2 \le i \le 3n-2 \\ & \text{and } 3n+2 \le i \le 4n-2 \\ -\frac{241}{60480}h^8 u^{(8)}_{i+1}, & i = n-1, 3n-1 \text{ and } 4n-1. \end{cases}$

and

(2.16) $\quad t_i = \begin{cases} -\frac{501}{151200}h^{10} u^{(10)}_{i-1}, & i = 1, n+1 \text{ and } 3n+1 \\ -\frac{1}{3024}h^{10} u^{(10)}_i, & 2 \le i \le n-2, n+2 \le i \le 3n-2 \\ & \text{and } 3n+2 \le i \le 4n-2 \\ -\frac{501}{151200}h^{10} u^{(10)}_{i+1}, & i = n-1, 3n-1 \text{ and } 4n-1. \end{cases}$

for methods (2.12) and (2.13) respectively.

The equations (2.15) and (2.16) suggest that the schemes (2.12) and (2.13) are fourth and sixth order accurate over each subinterval, respectively. However, due to the extra continuity conditions at $x = c$ and $x = d$, the numerical results given in Section 4 indicate that the accuracy of both methods is not better than second order accurate over the whole interval $[a, b]$. The numerical experiments also indicate that the accuracy of the second method is slightly better than that of former one. From the above discussions we can notice that all finite difference methods give second order accurate results regardless of the size of the truncation errors involved. This is most probably is due to the fact that the fourth derivative is not continuous across the interfaces, and all the above methods use grid values of the fourth derivatives at these points.

Motivated by the above discussions, we now develop some new finite difference schemes which approximate the solution of (2.1) at midknots. Using Taylor polynomial along with the method of undetermined coefficients and the boundary and continuity conditions we introduce the finite difference methods

$$(2.17) \quad \begin{cases} -6u_{i-1} + 10u_{i-\frac{1}{2}} - 5u_{i+\frac{1}{2}} + u_{i+\frac{3}{2}} = -\frac{5}{4}h^2 u''_{i-1} + \frac{h^4}{384}[154u^{(4)}_{i-\frac{1}{2}} + 75u^{(4)}_{i+\frac{1}{2}} + u^{(4)}_{i+\frac{3}{4}}], \\[4pt]
\text{for } i = 1, n+1 \text{ and } 3n+1 \\[8pt]
2u_{i-2} - 5u_{i-\frac{3}{2}} + 6u_{i-\frac{1}{2}} - 4u_{i+\frac{1}{2}} + u_{i+\frac{3}{2}} \\[4pt]
\qquad = -\frac{h^2}{4}u''_{i-2} + \frac{h^4}{384}[75u^{(4)}_{i-\frac{3}{2}} + 230u^{(4)}_{i-\frac{1}{2}} + 76u^{(4)}_{i+\frac{1}{2}} + u^{(4)}_{i+\frac{3}{2}}], \\[4pt]
\text{for } i = 2, n+2 \text{ and } 3n+2 \\[8pt]
u_{i-\frac{5}{2}} - 4u_{i-\frac{3}{2}} + 6u_{i-\frac{1}{2}} - 4u_{i+\frac{1}{2}} + u_{i+\frac{3}{2}} \\[4pt]
\qquad = \frac{h^4}{384}[u^{(4)}_{i-\frac{5}{2}} + 76u^{(4)}_{i-\frac{3}{2}} + 230u^{(4)}_{i-\frac{1}{2}} + 76u^{(4)}_{i+\frac{1}{2}} + u^{(4)}_{i+\frac{3}{2}}], \\[4pt]
\text{for } 3 \leq i \leq n-2, \ n+3 \leq 3n-2 \text{ and } 3n+3 \leq i \leq 4n-2, \\[8pt]
u_{i-\frac{5}{2}} - 4u_{i-\frac{3}{2}} + 6u_{i-\frac{1}{2}} - 5u_{i+\frac{1}{2}} + 2u_{i+1} \\[4pt]
\qquad = -\frac{h^2}{4}u''_{i+1} + \frac{h^4}{384}[u^{(4)}_{i-\frac{5}{2}} + 76u^{(4)}_{i-\frac{3}{2}} + 230u^{(4)}_{i-\frac{1}{2}} + 75u^{(4)}_{i+\frac{1}{2}}], \\[4pt]
\text{for } i = n-1, \ 3-1 \text{and } 4n-1, \\[8pt]
u_{i-\frac{5}{2}} - 5u_{i-\frac{3}{2}} + 10u_{i-\frac{1}{2}} - 6u_i = -\frac{5}{4}h^2 u''_i + \frac{h^4}{384}[u^{(4)}_{i-\frac{5}{2}} + 75u^{(4)}_{i-\frac{3}{2}} + 154u^{(4)}_{i-\frac{1}{2}}], \\[4pt]
\text{for } i = n, 3n \text{ and } 4n,
\end{cases}$$

and

$$(2.18) \begin{cases} -6u_{i-1} + 10u_{i-\frac{1}{2}} - 5u_{i+\frac{1}{2}} + u_{i+\frac{3}{2}} = -\frac{5h^2}{4}u''_{i-1} + \frac{h^4}{384}[154u^{(4)}_{i-\frac{1}{2}} + 75u^{(4)}_{i+\frac{1}{2}} + u^{(4)}_{i+\frac{3}{2}}], \\[4pt] \text{for } i = 1, n+1 \text{ and } 3n+1, \\[6pt] 2u_{i-2} - 5u_{i-\frac{3}{2}} + 6u_{i-\frac{1}{2}} - 4u_{i+\frac{1}{2}} + u_{i+\frac{3}{2}} \\[4pt] \qquad\qquad = -\frac{h^2}{4}u''_{i-2} + \frac{h^4}{384}[75u^{(4)}_{i-\frac{3}{2}} + 230u^{(4)}_{i-\frac{1}{2}} + 76u^{(4)}_{i+\frac{1}{2}} + u^{(4)}_{i+\frac{3}{2}}], \\[4pt] \text{for } i = 2, n+2 \text{ and } 3n+2, \\[6pt] u_{i-\frac{5}{2}} - 4u_{i-\frac{3}{2}} + 6u_{i-\frac{1}{2}} - 4u_{i+\frac{1}{2}} + u_{i+\frac{3}{2}} = \frac{h^4}{6}[u^{(4)}_{i-\frac{3}{2}} + 4u^{(4)}_{i-\frac{1}{2}} + u^{(4)}_{i+\frac{1}{2}}], \\[4pt] \text{for } 3 \leq i \leq n-2, \ n+3 \leq i \leq 3n-2 \text{ and } 3n+3 \leq i \leq 4n-2, \\[6pt] u_{i-\frac{5}{2}} - 4u_{i-\frac{3}{2}} + 6u_{i-\frac{1}{2}} - 5u_{i+\frac{1}{2}} + 2u_{i+1} \\[4pt] \qquad\qquad = -\frac{h^2}{4}u''_{i+1} + \frac{h^4}{384}[u^{(4)}_{i-\frac{5}{2}} + 76u^{(4)}_{i-\frac{3}{2}} + 230u^{(4)}_{i-\frac{1}{2}} + 75u^{(4)}_{i+\frac{1}{2}}], \\[4pt] \text{for } i = n-1, 3n-1 \text{ and } 4n-1, \\[6pt] u_{i-\frac{5}{2}} - 5u_{i-\frac{3}{2}} + 10u_{i-\frac{1}{2}} - 6u_i = -\frac{5}{4}h^2 u''_i + \frac{h^4}{384}[u^{(4)}_{i-\frac{5}{2}} + 75u^{(4)}_{i-\frac{3}{2}} + 154u^{(4)}_{i-\frac{1}{2}}], \\[4pt] \text{for } i = n, 3n \text{ and } 4n, \end{cases}$$

where

$$u^{(4)}_{i+\frac{1}{2}} = \begin{cases} f_{i+\frac{1}{2}}, & \text{for } 0 \leq i \leq n-1 \text{ and } 3n \leq i \leq 4n-1 \\[6pt] -4u_{i+\frac{1}{2}} + f_{i+\frac{1}{2}} + 1, & \text{for } n \leq i \leq 3n-1, \end{cases}$$

$f_{i+\frac{1}{2}} = f(x_{i+\frac{1}{2}}), i = 0, 1, 2, \ldots, 4n-1$. The relations (2.17) and (2.18) forms two systems of $4n$ linear equations in the unknowns $u_{i-\frac{1}{2}}, i = 1, 2, \ldots, 4n$.

The local truncation errors related to (2.17) and (2.18) are given by

$$(2.19) \quad t_i = \begin{cases} \frac{-41}{2304}h^6 u^{(6)}_{i-1} + O(h^7), & i = 1, n+1 \text{ and } 3n+1 \\[6pt] \frac{-473}{11520}h^6 u^{(6)}_{i-2} + O(h^7), & i = 2, n+2 \text{ and } 3n+2 \\[6pt] \frac{-1}{24}h^6 u^{(6)}_i + O(h^7), & \begin{array}{l} 3 \leq i \leq n-2, n+3 \leq i \leq 3n-2 \\ \text{and } 3n+3 \leq i \leq 4n-2 \end{array} \\[6pt] \frac{-473}{11520}h^6 u^{(6)}_{i+1} + O(h^7), & i = n-1, 3n-1 \text{ and } 4n-1 \\[6pt] \frac{-41}{2304}h^6 u^{(6)}_i + O(h^7), & i = n, 3n \text{ and } 4n. \end{cases}$$

and

$$(2.20) \quad t_i = \begin{cases} \frac{-41}{2304} h^6 u_{i-1}^{(6)} + O(h^7), & \text{for } i = 1, n+1 \text{ and } 3n+1 \\[2ex] \frac{-473}{11520} h^6 u_{i-2}^{(6)} + O(h^7), & \text{for } i = 2, n+2 \text{ and } 3n+2 \\[2ex] \frac{-1}{720} h^8 u_i^{(8)} + O(h^9), & \text{for } 3 \le i \le n-2, \ n+3 \le i \le 3n-2 \\[1ex] & \text{and } 3n+3 \le i \le 4n-2 \\[2ex] \frac{-473}{11520} h^6 u_{i+1}^{(6)} + O(h^7), & \text{for } i = n-1, 3n-1 \text{ and } 4n-1 \\[2ex] \frac{-41}{2304} h^6 u_i^{(6)} + O(h^7), & \text{for } i = n, 3n \text{ and } 4n. \end{cases}$$

respectively. It is clear from (2.19) and (2.20) that method (2.17) is a second order convergent over the whole interval [-1,1], and that method (2.18) is a fourth order accurate method at the interior points of the interval and of order two at and nearby the boundaries. However, our numerical experiments indicate that (2.18) is of order four over the whole interval [-1,1].

3. Convergence Analysis

In this Section, we discuss the convergence criteria of the finite difference scheme (2.17) developed in Section 2. The discussions of the convergent analysis for the other methods can be studied in a similar way.

Let $e_{i+\frac{1}{2}} = u_{i+\frac{1}{2}} - w_{i+\frac{1}{2}}, i = 0, 1, 2, \ldots, 4n-1$ be the discretization error, where $w_{i+\frac{1}{2}}$ is the numerical approximation to $u_{i+\frac{1}{2}}$ obtained from (2.17) by neglecting the truncation error (2.19). Let $\mathbf{U} = (u_{i+\frac{1}{2}})$, $\mathbf{W} = (w_{i+\frac{1}{2}})$, $\mathbf{C} = (c_i)$, $\mathbf{T} = (t_i)$ and $\mathbf{E} = (e_{i+\frac{1}{2}})$ be $4n$-dimensional vectors. We also define $\|\mathbf{E}\| = \max_i |e_{i+\frac{1}{2}}|$, where $\|\cdot\|$ represent the ∞-norm of matrix vector. Using these notations, we can rewrite equations (2.17) and (2.19) as follows:

$$(3.1) \qquad \mathbf{AU} = \mathbf{C} + \mathbf{T}$$

$$(3.2) \qquad \mathbf{AW} = \mathbf{C}$$

(3.3)
$$\mathbf{AE} = \mathbf{T}$$

where $\mathbf{T}$ is the local truncation error defined in (2.19), $\mathbf{C}$ is given by

$$c_i = \begin{cases} 6u_{i-1} - \frac{5}{4}h^2 u''_{i-1} + \frac{h^4}{384}[154f_{i-\frac{1}{2}} + 75f_{i+\frac{1}{2}} + f_{i+\frac{3}{2}}], & i = 1, \quad i = 3n+1 \\[2mm] 6u_{i-1} - \frac{5h^2}{4}u''_{i-1} + \frac{h^4}{384}[154f_{i-\frac{1}{2}} + 75f_{i+\frac{1}{2}} + f_{i+\frac{3}{2}} + 230r], & i = n+1 \\[2mm] -2u_{i-2} - \frac{h^2}{4}u''_{i-2} + \frac{h^4}{384}[75f_{i-\frac{3}{2}} + 230f_{i-\frac{1}{2}} + 76f_{i+\frac{1}{2}} + f_{i+\frac{3}{2}}], & i = 2, \quad i = 3n+2 \\[2mm] -2u_{i-2} - \frac{h^2}{4}u''_{i-2} + \frac{h^4}{384}[75f_{i-\frac{3}{2}} + 230f_{i-\frac{1}{2}} + 76f_{i+\frac{1}{2}} + f_{i+\frac{3}{2}} + 382r], & i = n+2 \\[2mm] \frac{h^4}{384}[f_{i-\frac{5}{2}} + 76f_{i-\frac{3}{2}} + 230f_{i-\frac{1}{2}} + 76f_{i+\frac{1}{2}} + f_{i+\frac{3}{2}}], & \begin{array}{l} 3 \le i \le n-2 \\ \text{and } \frac{3N}{4} + 3 \le i \le N-2, \end{array} \\[4mm] \frac{h^4}{384}[f_{i-\frac{5}{2}} + 76f_{i-\frac{3}{2}} + 230f_{i-\frac{1}{2}} + 76f_{i+\frac{1}{2}} + f_{i+\frac{3}{2}}] + h^4 r, & n+3 \le i \le 3n-2, \\[2mm] -2u_{i+1} - \frac{h^2}{4}u''_{i+1} + \frac{h^4}{384}[f_{i-\frac{5}{2}} + 76f_{i-\frac{3}{2}} + 230f_{i-\frac{1}{2}} + 75f_{i+\frac{1}{2}} + 382r], & i = 3n-1, \\[2mm] 6u_{i-1} - \frac{5}{4}h^2 u''_i + \frac{h^4}{384}[154f_{i-\frac{5}{2}} + 75f_{i-\frac{3}{2}} + 154f_{i-\frac{1}{2}} + 230r], & i = 3n, \\[2mm] 6u_{i-1} - \frac{5}{4}h^2 u''_i + \frac{h^4}{384}[f_{i-\frac{5}{2}} + 75f_{i+\frac{3}{2}} + 154f_{i-\frac{1}{2}}], & i = \frac{N}{4}, \quad i = 4n, \end{cases}$$

and

(3.4)
$$\mathbf{A} = \mathbf{B} + \frac{h^4}{384}\mathbf{HD}.$$

Here $\mathbf{B}$ is an $N \times N$ matrix given by

$$\mathbf{B} = \begin{bmatrix} \mathbf{B_1} & 0 & 0 \\ 0 & \mathbf{B_2} & 0 \\ 0 & 0 & \mathbf{B_3} \end{bmatrix},$$

where $\mathbf{B_1}$, $\mathbf{B_2}$ and $\mathbf{B_3}$ are five-band matrices of order M each given by $\mathbf{B_k} = \mathbf{P}^2$, $k = 1, 2, 3$, where $\mathbf{P} = (p_{ij})$ is a tridiagonal matrix defined by

$$p_{ij} = \begin{cases} 3, & i = j = 1, M \\ 2, & i = j = 2, 3, \ldots, M-1 \\ -1, & |i-j| = 1 \\ 0, & \text{otherwise,} \end{cases}$$

with $M = n$, $2n$ and n for $\mathbf{B_1}$, $\mathbf{B_2}$ and $\mathbf{B_3}$, respectively. The matrix $\mathbf{H}$ is also a $4n \times 4n$ matrix which has the form

$$
\mathbf{H} =
\begin{bmatrix}
0 & 0 & & & & & & & & & 0 \\
& \ddots & & & & & & & & & \\
0 & & 0 & & & & & & & & 0 \\
& & 154 & 75 & 1 & & & & & & \\
& & 75 & 230 & 76 & 1 & & & & & \\
& & 1 & 76 & 230 & 76 & 1 & & & & \\
& & \ddots & \ddots & \ddots & \ddots & \ddots & & & & \\
& & & \ddots & \ddots & \ddots & \ddots & \ddots & & & \\
& & & & \ddots & \ddots & \ddots & \ddots & \ddots & & \\
& & & & & 1 & 76 & 230 & 76 & 1 & \\
& & & & & & 1 & 76 & 230 & 75 & \\
& & & & & & & 1 & 75 & 154 & \\
& & & & & & & & & 0 & \\
& & & & & & & & & & \ddots \\
& & & & & & & & & & \ddots \\
0 & & & & & & & 0 & & & 0
\end{bmatrix}
\begin{matrix} \\ \\ \\ n+1 \\ \\ \\ \\ \\ \\ \\ \\ 3n \\ \\ \\ \\ \end{matrix}
$$

and $\mathbf{D} = (d_{ij})$ is a diagonal matrix of order $4n$ satisfies

$$
d_{ij} =
\begin{cases}
g_{i-\frac{1}{2}} & i = j = n+1, \ldots, 3n \\
0 & \text{otherwise.}
\end{cases}
$$

In fact, the matrix $\mathbf{A}$ is given in (3.4) takes the form

$$
\mathbf{A} =
\begin{bmatrix}
\mathbf{A}_1 & 0 & 0 \\
0 & \mathbf{A}_2 & 0 \\
0 & 0 & \mathbf{A}_3
\end{bmatrix},
$$

where

(3.5)
$$
\begin{aligned}
\mathbf{A}_1 &= \mathbf{B}_1 \\
\mathbf{A}_2 &= \mathbf{B}_2 + \tfrac{h^4}{384}\mathbf{H}_2\mathbf{D}_2 \\
\mathbf{A}_3 &= \mathbf{B}_3,
\end{aligned}
$$

$$
\mathbf{H}_2 =
\begin{bmatrix}
154 & 75 & 1 & & & & & & \\
75 & 230 & 76 & 1 & & & & & \\
1 & 76 & 230 & 76 & 1 & & & & \\
& \ddots & \ddots & \ddots & \ddots & \ddots & & & \\
& & \ddots & \ddots & \ddots & \ddots & \ddots & & \\
& & & \ddots & \ddots & \ddots & \ddots & \ddots & \\
& & & & 1 & 76 & 230 & 76 & 1 \\
& & & & & 1 & 76 & 230 & 75 \\
& & & & & & 1 & 75 & 154
\end{bmatrix}
$$

and $\mathbf{D}_2 = (d_{ij})$ is a $2n \times 2n$ diagonal matrix with $d_{ii} = g_{i-\frac{1}{2}}$ for $i = n+1, n+2, \ldots, 3n$.

Following the technique of Usmani [32], we get

$$(3.6) \qquad \|\mathbf{B}_k^{-1}\| \leq \frac{5(c-a)^4 + 10(c-a)^2 h^2 + 9h^4}{384 h^4}, \qquad k = 1, 3$$

$$(3.7) \qquad \|\mathbf{B}_2^{-1}\| \leq \frac{5(d-c)^4 + 10(d-c)^2 h^2 + 9h^4}{384 h^4},$$

and $\mathbf{A}_2$ is nonsingular provided that

$$(3.8) \qquad \frac{\lambda}{384}[5(d-c)^4 + 10(d-c)h^2 + 9h^4] < 1.$$

Here we have used the fact that $c - a = b - d$ in (3.6).

For the convergence analysis of (2.17) we need the following well known result.

Theorem 1. Let $\mathbf{P}, \mathbf{Q}$ and $\mathbf{R}$ be three matrices of order l, m and n respectively. Then the matrix

$$\mathbf{S} = \begin{bmatrix} \mathbf{P} & 0 & 0 \\ 0 & \mathbf{Q} & 0 \\ 0 & 0 & \mathbf{R} \end{bmatrix}$$

is nonsingular if and only if $\mathbf{P}, \mathbf{Q}$ and $\mathbf{R}$ are nonsingular. The inverse of $\mathbf{S}$ is

$$\mathbf{S}^{-1} = \begin{bmatrix} \mathbf{P}^{-1} & 0 & 0 \\ 0 & \mathbf{Q}^{-1} & 0 \\ 0 & 0 & \mathbf{R}^{-1} \end{bmatrix}.$$

Using Theorem 1 and the properties of the five-band matrix, one can easily show that

$$(3.9) \qquad \|\mathbf{B}^{-1}\| = \|\mathbf{B}_2^{-1}\| > \|\mathbf{B}_1^{-1}\| = \|\mathbf{B}_3^{-1}\|.$$

Hence, from (3.7) and (3.9) we get

$$(3.10) \qquad \|\mathbf{B}^{-1}\| \leq \frac{5(d-c)^4 + 10(d-c)^2 h^2 + 9h^4}{384 h^4}.$$

From (3.3) we have

$$\mathbf{E} = \mathbf{A}^{-1}\mathbf{T} = (\mathbf{B} + \frac{h^4}{384}\mathbf{HD})^{-1}\mathbf{T} = (\mathbf{I} + \frac{h^4}{384}\mathbf{B}^{-1}\mathbf{HD})^{-1}\mathbf{B}^{-1}\mathbf{T}$$

and it follows that

(3.11)
$$\|\mathbf{E}\| \leq \frac{\|\mathbf{B}^{-1}\|\|\mathbf{T}\|}{1 - \frac{h^4}{384}\|\mathbf{B}^{-1}\|\|\mathbf{H}\|\|\mathbf{D}\|},$$

provided that $\frac{h^4}{384}\|\mathbf{B}^{-1}\|\|\mathbf{H}\|\|\mathbf{D}\| < 1$.

Now, since $\|\mathbf{T}\| = \frac{h^6}{24}M_6$, where $M_6 = \max|u^{(6)}(x)|$, $\|\mathbf{H}\| = 384$ and $\|\mathbf{D}\| = \max|g(x)|$, then (3.10) and (3.11) give

(3.12)
$$\|\mathbf{E}\| \leq \frac{[5(d-c)^4 + 10(d-c)^2h^2 + 9h^4]M_6h^2}{24\{384 - \lambda[5(d-c)^4 + 10(d-c)^2h^2 + 9h^4]\}} = O(h^2),$$

which shows that (2.17) is a second order convergent method.

4. Applications and Numerical Results

To illustrate the application of the numerical method developed in Section 2. We consider the contact boundary value problem of finding u such that

(4.1)
$$\begin{cases} u^{(4)} \geq f(x), & \text{on } \Omega = [-1, 1] \\ u \geq \psi(x), & \text{on } \Omega = [-1, 1] \\ \left[u^{(4)} - f(x)\right]\left[u - \psi(x)\right] = 0 & \text{on } \Omega = [-1, 1] \\ u(-1) = u(1) = 0, \quad u''(-1) = u''(1) = 0, \end{cases}$$

where f is a given force acting on the beam and $\psi(x)$ is the elastic obstacle. Equation (4.1) describes the equilibrium configuration of an elastic beam, pulled at the ends and lying over an elastic obstacle.

We study the problem (4.1) in the framework of variational inequality approach. To do so, we define the set K as

(4.2)
$$K = \{v : v \in H_0^2(\Omega) : v \geq \psi \text{ on } \Omega\},$$

which is a closed convex set in $H_0^2(\Omega)$, where $H_0^2(\Omega)$ is a Sobolev space, which is in fact a Hilbert space. For the definitions of the spaces $H_0^2(\Omega)$, see [8]. Now using the technique of Tonti [31], we can easily show that the energy functional associated with the obstacle problem (4.1) is

(4.3)
$$I[v] = \int_{-1}^{1} \left\{ \frac{d^4 v}{dx^4} - 2f(x) \right\} v(x) dx, \quad \text{for all } v \in H_0^2(\Omega)$$
$$= \int_{-1}^{1} \left(\frac{d^2 v}{dx^2} \right)^2 dx - 2 \int_{-1}^{1} f(x) v(x) dx$$
$$= a(v, v) - 2 < f, v >,$$

where

(4.4)
$$a(u, v) = \int_{-1}^{1} \left(\frac{d^2 u}{dx^2} \right) \left(\frac{d^2 v}{dx^2} \right) dx$$

and

(4.5)
$$< f, v >= \int_{-1}^{1} f(x) v(x) dx.$$

It can be easily proved that the form $a(u, v)$ defined by (4.4) is bilinear, symmetric and positive (in fact, coercive) and the functional f defined by (4.5) is a linear continuous functional. It is well known [8,17,18,27] that the minimum u of the functional $I[v]$ defined by (4.3) on the closed convex set K in $H_0^2(\Omega)$ can be characterized by the variational inequality

(4.6)
$$a(u, v - u) \geq < f, v - u >, \quad \text{for all } v \in K.$$

Thus we conclude that the obstacle problem (4.1) is equivalent to solving the variational inequality problem (4.6). This equivalence has been used to study the existence of a unique solution of (4.1), see [8,17,18,27]. Now using the idea of Lewy and Stampacchia [14], the problem (4.6) can be written as

(4.7)
$$u^{(4)} + \nu(u - \psi)(u - \psi) = f,$$

where ψ is the obstacle function and $\nu(t)$ is penalty function defined by

(4.8)
$$\nu(t) = \begin{cases} 4, & t \geq 0 \\ 0, & t < 0. \end{cases}$$

We assume that the obstacle function $\psi(x)$ is defined by

(4.9)
$$\psi(x) = \begin{cases} -\frac{1}{4} & \text{for } -1 \leq x \leq -\frac{1}{2}, \quad \frac{1}{2} \leq x \leq 1 \\ \frac{1}{4} & \text{for } -\frac{1}{2} \leq x \leq \frac{1}{2}. \end{cases}$$

From (4.7) - (4.9), we obtain the following system of equations

$$(4.10) \qquad u^{(4)} = \begin{cases} f, & \text{for } -1 \le x \le -\frac{1}{2} \text{ and } \frac{1}{2} \le x \le 1 \\ 1 - 4u + f, & \text{for } -\frac{1}{2} \le x \le \frac{1}{2} \end{cases}$$

with the boundary conditions

$$(4.11) \qquad \begin{aligned} u(-1) &= u(-\tfrac{1}{2}) = u(\tfrac{1}{2}) = u(1) = 0 \\ u''(-1) &= u''(-\tfrac{1}{2}) = u''(\tfrac{1}{2}) = u''(1) = 0 \end{aligned}$$

and the conditions of continuity of u and u'' at $x = -\frac{1}{2}$ and $\frac{1}{2}$.

Example 1. [16]   When $f = 0$, the problem (4.10) become

$$(4.12) \qquad u^{(4)} = \begin{cases} 0, & \text{for } -1 \le x \le -\frac{1}{2} \text{ and } \frac{1}{2} \le x \le 1 \\ 1 - 4u, & \text{for } -\frac{1}{2} \le x \le \frac{1}{2} \end{cases}$$

with the given boundary conditions (4.11). The analytic solution for this problem is

$$(4.13) \qquad u(x) = \begin{cases} 0, & \text{for } -1 \le x \le -\frac{1}{2} \text{ and } \frac{1}{2} \le x \le 1 \\ 0.25 - \frac{1}{2\beta_3}[\beta_1 \sin x \sinh x + \beta_2 \cos x \cosh x], \end{cases}$$

for $-\frac{1}{2} \le x \le \frac{1}{2}$, where $\beta_1 = \sin \frac{1}{2} \sinh \frac{1}{2}$, $\beta_2 = \cos \frac{1}{2} \cosh \frac{1}{2}$ and $\beta_3 = \cos 1 + \cosh 1$.

The problem (4.12) was solved using the finite difference techniques(2.12), (2.13), (2.17) and (2.18) with a variety of h values and the observed maximum errors (in absolute values) are listed in Table 1.

Table 1: Maximum errors (in absolute values) for example 1.

h	Method (2.12)	Method (2.13)	Method (2.17)	Method (2.18)
$\frac{1}{8}$	1.9×10^{-5}	1.7×10^{-5}	3.1×10^{-6}	6.1×10^{-7}
$\frac{1}{16}$	4.8×10^{-6}	4.3×10^{-6}	8.1×10^{-7}	4.9×10^{-8}
$\frac{1}{32}$	1.3×10^{-6}	1.1×10^{-6}	2.1×10^{-7}	5.1×10^{-9}

It can be noted from Table 1 that if the stepsize h is reduced by a factor $1/2$, then the observed maximum errors for methods (2.12), (2.13) and (2.17) are reduced by a factor $1/4$, which clearly indicate that (2.12), (2.13) and (2.17) are second order convergent procedures. Also, it is clear from the table that method (2.17) is the most accurate of the three. This is due to the fact that (2.17) approximates the solution at midknots and the approximate values of the solution at $x = 1/2$ and $-1/2$ are reached from left and right sides.

From the numerical results given in Table 1 we may also notice that halving the stepsize h would reduce the maximum error obtained by method (2.18) by a factor $1/16$, which indicate that method (2.18) gives fourth order accurate results over the whole interval.

The problem (4.12) with the boundary conditions

(4.14)
$$\begin{array}{ll} u(-1) & = u(-\tfrac{1}{2}) = u(\tfrac{1}{2}) = u(1) = 0 \\ u''(-1) & = -u''(-\tfrac{1}{2}) = u''(\tfrac{1}{2}) = -u''(1) = \epsilon \end{array}$$

where $\epsilon \to 0$, was solved by Al-Said and Noor [3,4] using a finite difference scheme based on (2.6) and (2.7), and by Khalifa and Noor [16] using collocation method with quintic B-spline as basis functions. Some of their numerical results for the case when $\epsilon = 10^{-6}$ are tabulated in Table 2. We mention here in passing that when we used our present methods to solve (4.12) with the conditions (4.14), we have better results than that given in Table 2. In fact we have the same results that are listed in Table 1.

Table 2: Observed maximum errors in absolute value.

h	Scheme (2.6) [3]	Scheme (2.7) [4]	Quintic B-spline [16]
$\frac{1}{8}$	1.4×10^{-4}	1.9×10^{-5}	3×10^{-4}
$\frac{1}{16}$	3.6×10^{-5}	4.8×10^{-6}	7×10^{-5}
$\frac{1}{32}$	8.9×10^{-6}	1.2×10^{-6}	1.4×10^{-5}

Example 2. [3] In this example we use schemes (2.12), (2.13), (2.17) and (2.18) to solve problem (4.11) with $f = 1$,

(4.15) $\qquad u^{(4)} = \begin{cases} 1, & \text{for } -1 \le x \le -\tfrac{1}{2} \text{ and } \tfrac{1}{2} \le x \le 1 \\ 2 - 4u, & \text{for } -\tfrac{1}{2} \le x \le \tfrac{1}{2} \end{cases}$

and the boundary conditions (4.11). The analytical solution for this problem is

(4.16) $\quad u(x) = \begin{cases} \frac{1}{24}x^4 + \frac{1}{8}x^3 + \frac{1}{8}x^2 + \frac{3}{64}x + \frac{1}{192}, & \text{for } -1 \le x \le -\tfrac{1}{2} \\[2mm] 0.5 - \frac{1}{\beta_3}[\beta_1 \sin x \sinh x + \beta_2 \cos x \cosh x], & \text{for } -\tfrac{1}{2} \le x \le \tfrac{1}{2} \\[2mm] \frac{1}{24}x^4 - \frac{1}{8}x^3 + \frac{1}{8}x^2 - \frac{3}{64}x + \frac{1}{192}, & \text{for } -\tfrac{1}{2} \le x \le 1 \, , \end{cases}$

where β_1, β_2 and β_3 are as defined in example 4.1. The the observed maximum errors (in absolute values) for different values of h are shown in Table 3, where it is clear that it agree with the results of example 1 given in Table 1. The problem (4.15) along with the boundary conditions (4.11) was also solved in [3,4] by a finite difference schemes based on (2.6) and (2.7), and the results are listed in Table 4. Again, it is clear form Tables 3 and 4 that the results produced by using our present schemes are better than those given in [3,4].

Table 3: Maximum errors (in absolute values) for example 4.2.

h	Scheme (2.12)	scheme (2.13)	Scheme (2.17)	Scheme (2.18)
$\frac{1}{12}$	8.4×10^{-6}	7.6×10^{-6}	3.9×10^{-6}	3.6×10^{-7}
$\frac{1}{24}$	2.2×10^{-6}	1.9×10^{-6}	9.8×10^{-7}	3.8×10^{-8}
$\frac{1}{48}$	5.4×10^{-7}	4.9×10^{-7}	2.6×10^{-7}	2.5×10^{-9}

Table 4: Maximum error (in absolute values) for example 4.2.

h	Scheme (2.6) [3]	Scheme (2.7) [4]
$\frac{1}{12}$	1.8×10^{-4}	8.4×10^{-6}
$\frac{1}{24}$	7.2×10^{-5}	2.2×10^{-6}
$\frac{1}{48}$	4.6×10^{-5}	5.4×10^{-7}

5. CONCLUSION In this paper, we have developed some finite difference schemes for solving a system of fourth order value problems. We have shown that a wide class of obstacle, unilateral and contact problems can be studied in the framework of variational inequalities. Using the penalty function method it is shown that the obstacle problems can be characterized by a system of differential equations, which is solved by using the new developed schemes. The results obtained are better than the previous one and represent a refinement and an improvement. A detailed analysis of such systems of fourth order boundary value problems in pure and applied sciences (both analytical and numerical) will constitute an interesting subject of the future study. Much work is needed to develop efficient numerical techniques for solving such special systems of boundary value problems related with obstacle, unilateral, contact, free and moving boundary problems.

References

1. E.A. Al-Said, Spline solutions for system of second order boundary value problems, Inter. J. Computer Math. **62** (1996), 143-154.

2. E.A. Al-Said, Smooth spline solutions for a system of second order boundary value problems, J. nat. Geometry (1999), to appear.

3. E.A. Al-Said, and M.A. Noor, Computational methods for fourth order obstacle boundary value problems, Commun. Appl. Nonlinear Anal. **2**, 3 (1995), 73-83.

4. E.A. Al-Said and M.A. Noor, Numerical solutions of a system of fourth order boundary value problems, Intern. J. Computer Math. **71** (1998).

5. E.A. Al-Said, M.A. Noor and A. Al-Shejari, Numerical solutions for system of boundary value problems, Korean J. Comput. Appl. Math. **5** (3) (1998).

6. E.A. Al-Said, M.A. Noor and A.K. Khalifa, Finite difference scheme for variational inequalities, J. Opt. Theory Applic. **89** (2) (1996), 453-459.

7. E.A. Al-Said, M.A. Noor and Th. Rassias, Numerical solutions of third-order obstacle problems, Inter. J. Computer Math. **69** (1998).

8. C. Baiocchi and A. Capelo, Variational and Quasi- Variational Inequalities, John Wiley and Sons, New York, 1984.

9. R.W. Cottle, F. Giannessi and J.L. Lions, Variational Inequalities and Complementarity Problems: Theory and Applications, J. Wiley and Sons, New York, 1980.

10. J. Crank, Free and Moving Boundary Problems, Clarendon Press, Oxford, U.K. 1984.

11. J.P. Dussault, Numerical stability and efficiency of penalty algorithms, SIAM J. Numer. Anal. **32** (1995), 296-317.

12. R. Glowinski, Numerical Methods for Nonlinear Problems, Springer-Verlag, Berlin, 1984.

13. R. Glowinski, J.L. Lions and R. Tremolieres, Numerical Analysis of Variational Inequalities, North-Holland, Amsterdam, 1981.

14. H. Lewy and G. Stampacchia, On the regularity of the solutions of the variational inequalities, Comm. Pure and Appl. Math. **22** (1969), 153-188.

15. J.L. Lions and G. Stampacchia, Variational Inequalities, Comm. Pure Appl. Math. **20** (1967), 493-519.

16. A.K. Khalifa and M.A. Noor, Quintic splines solutions of a class of contact problems, Math. Comput. Modelling, **13** (1990), 51-58.

17. N. Kikuchi and J.T. Oden, Contact Problems in Elasticity, SIAM Publishing Co., Philadelphia, 1988.

18. M. Aslam Noor. Some recent advances in variational inequalities Part I, Basic Concepts, New Zealand J. Math. **26** (1997), 53-80.

19. M. Aslam Noor, Some recent advances in variational inequalities, Part II, Other concepts, New Zealand J. Math. **26** (1997), 229-255.

20. M.A. Noor, Auxiliary principle for generalized mixed variational inequalities, J. Math. Anal. Appl. **215** (1997), 75-85.

21. M.A. Noor and R.A. Ashrafi, On a numerical method for solving obstacle problems, Math. Computers in Simulation, **36** (1994), 49-55.

22. M.A. Noor and Khalifa, Cubic spline collocation methods for unilateral problems, Int. J. Engng. Sci. **25** (1987), 1527-1530.

23. M.A. Noor and Khalifa, A numerical approach for odd-order obstacle problems, Intern. J. Computer Math. **54** (1994), 159-166.

24. M.A. Noor and W. Oettli, On general nonlinear complementarity problems and quasi-equilibria, Le Matematiche, **49** (1994), 313-331.

25. M.A. Noor and S.I. Tirmizi, Finite difference techniques for solving obstacle problems, Appl. Math. Lett. **1** (1988), 267-271.

26. M.A. Noor and S.I. Tirmizi, Numerical methods for a class of contact problems, Int. J. Engng. Sci. **29** (4) (1991), 513-521.

27. M.A. Noor, K.I. Noor and Th.M. Rassias, Some aspects of variational inequalities, J. Comput. Appl. Math. **47** (1993), 285-312.

28. M.A. Noor, K.I. Noor and Th.M. Rassias, Invitation to variational inequalities, in: Analysis, Geometry and Groups: A Riemann Legacy Volume, (ed. H.M. Srivastava and Th.M. Rassias), Hadronic Press, U.S.A., (1993), 373-448.

29. P.D. Panagiotopoulos, Inequality Problems in Mechanics and Applications, Birkhauser, Boston, 1985.

30. G. Stampacchia, Formes bilineaires coercitives sur les ensembles convexes, C.R. Acad. Sci. Paris, **258** (1964),4413-4416.

31. E. Tonti, Variational formulation for every nonlinear problem, Int. J. Engng. Sci. **22** (1984),1343-1371.

32. R.A. Usmani, The use of quartic spline in numerical solution of fourth-order boundary value problem, J. Comput. and Applied Math. **44** (1992), 187-199.

300

33. P. Villaggio, The ritz method in solving unilateral problems in elasticity, Meccanica, september 1981, 123-127.

ON SOME GENERALIZED OPIAL TYPE INEQUALITIES

B.G. PACHPATTE
57 Shri Niketan Colony
Near Abhinay Talkies
Aurangabad 431 001
Maharashtra, India

Abstract. In this paper we establish some new Opial type inequalities involving functions of one and many independent variables. Our results in the special cases yield some of the recent generalizations of the Opial's inequality and also provide new estimates on inequalities of this type.

1. Introduction

In a classic paper [8] Z. Opial proved the following remarkable inequality.

If $f \in C^1[0,h]$ satisfies $f(0) = f(h) = 0$ and $f'(x) > 0$ for $x \in [0,h]$, then

$$\int_0^h |f(x)\, f'(x)|\, dx \le \frac{h}{4} \int_0^h |f'(x)|^2\, dx, \tag{1}$$

where the constant factor $\frac{h}{4}$ is best possible.

Since its discovery in the 1960's, Opial's inequality has proven to be one of the most useful inequalities in analysis. Its applications in proving the existence and uniqueness of initial and boundary value problems for differential equations have been particularly striking, see [1, 7, 19, 20]. The inequality (1) has received considerable attention and a large number of papers dealing with new proofs, extensions, generalizations, variants and discrete analogues of Opial's inequality have appeared in the literature. For an extensive survey on these inequalities, see [1, 5, 6] which contain a large number of references on this subject. The main purpose of this paper is to establish some new Opial type inequalities involving functions of one and many independent variables which in the special cases yield the inequalities of the form given by Godunova and Levin in [3] and also the inequalities recently established by Pachpatte in [9–13], Pečarić in [14] and Pečarić and Brnetić in [15, 16]. The analysis used in the proofs is elementary and our results provide new estimates on these types of inequalities.

T.M. Rassias and H.M. Srivastava (eds.), Analytic and Geometric Inequalities and Applications, 301–322.
© *1999 Kluwer Academic Publishers.*

2. Main Results

In this section we establish some new Opial type inequalities involving functions and their derivatives. In fact, the inequalities that we propose here are motivated by the interesting generalizations of the Opial's inequality given by Pachpatte in [9–13], Pečarić in [14] and Pečarić and Brnetić in [15, 16]. In what follows, R denotes the set of real numbers, $R_+ = [0, \infty)$ and $I = [a, b]$, $a, b \in R_+$. The derivative of a function f defined on R_+ or I will be denoted by $f'(t)$ and the first partial derivatives of a function $F(x_1, \cdots, x_i, \cdots, x_n)$ with respect to the ith component will be denoted by $F_i'(x_1, \cdots, x_i, \cdots, x_n)$ for $i = 1, \cdots, n$.

Our first result deals with an interesting generalization of the inequality established by Pachpatte in [9, Theorem 1].

Theorem 1. *Let u_i, v_i, $i = 1, \cdots, n$ be real-valued absolutely continuous functions defined on I with $u_i(a) = v_i(a) = 0$, $i = 1, \cdots, n$. Let F, G be real-valued nonnegative continuous and nondecreasing functions on R_+^n with $F(0, \cdots, 0) = 0$, $G(0, \cdots, 0) = 0$ such that all their first order partial derivatives F_i', G_i', $i = 1, \cdots, n$ are nonnegative continuous and nondecreasing on R_+^n. Then the following integral inequality holds:*

$$\int_a^b \left[F\left(|u_1(t)|, \cdots, |u_n(t)|\right) \sum_{i=1}^n G_i'\left(|v_1(t)|, \cdots, |v_n(t)|\right) |v_i'(t)| \right.$$

$$\left. + G\left(|v_1(t)|, \cdots, |v_n(t)|\right) \sum_{i=1}^n F_i'\left(|u_1(t)|, \cdots, |u_n(t)|\right) |u_i'(t)| \right] dt$$

$$\tag{2}$$

$$\leq F\left(\int_a^b |u_1'(t)|\, dt, \cdots, \int_a^b |u_n'(t)|\, dt \right)$$

$$\cdot G\left(\int_a^b |v_1'(t)|\, dt, \cdots, \int_a^b |v_n'(t)|\, dt \right).$$

Proof. From the hypotheses on u_i, v_i, $i = 1, \cdots, n$ we have

$$|u_i(t)| = \left| \int_a^t u_i'(s)\, ds \right| \leq \int_a^t |u_i'(s)|\, ds, \tag{3}$$

$$|v_i(t)| = \left| \int_a^t v_i'(s)\, ds \right| \leq \int_a^t |v_i'(s)|\, ds, \tag{4}$$

for $t \in I$. From (3), (4) and using the hypotheses on F, F_i', G, G_i', $i = 1, \cdots, n$ we observe that

$$\int_a^b \left[F(|u_1(t)|, \cdots, |u_n(t)|) \sum_{i=1}^n G_i'(|v_1(t)|, \cdots, |v_n(t)|) |v_i'(t)| \right.$$

$$\left. + G(|v_1(t)|, \cdots, |v_n(t)|) \sum_{i=1}^n F_i'(|u_1(t)|, \cdots, |u_n(t)|) |u_i'(t)| \right] dt$$

$$\leq \int_a^b \left[F\left(\int_a^t |u_1'(s)| \, ds, \cdots, \int_a^t |u_n'(s)| \, ds \right) \right.$$

$$\cdot \sum_{i=1}^n G_i'\left(\int_a^t |v_1'(s)| \, ds, \cdots, \int_a^t |v_n'(s)| \, ds \right) |v_i'(t)|$$

$$+ G\left(\int_a^t |v_1'(s)| \, ds, \cdots, \int_a^t |v_n'(s)| \, ds \right)$$

$$\left. \cdot \sum_{i=1}^n F_i'\left(\int_a^t |u_1'(s)|, \cdots, \int_a^t |u_n'(s)| \, ds \right) |u_i'(t)| \right] dt$$

$$= \int_a^b \frac{d}{dt} \left[F\left(\int_a^t |u_1'(s)| \, ds, \cdots, \int_a^t |u_n'(s)| \, ds \right) \right.$$

$$\left. \cdot G\left(\int_a^t |v_1'(s)| \, ds, \cdots, \int_a^t |v_n'(s)| \, ds \right) \right] dt$$

$$= F\left(\int_a^b |u_1'(t)| \, dt, \cdots, \int_a^b |u_n'(t)| \, dt \right)$$

$$\cdot G\left(\int_a^b |v_1'(t)| \, dt, \cdots, \int_a^b |v_n'(t)| \, dt \right).$$

This completes the proof of inequality (2).

A slight variant of Theorem 1 which in turn is a further generalization of the inequality established by Pachpatte in [9, Theorem 2] is given in the following theorem.

Theorem 2. *Let* u_i, v_i, F, F_i', G, G_i', $i = 1, \cdots, n$ *be as in Theorem 1. Let* p_i, q_i, $i = 1, \cdots, n$ *be real-valued positive functions defined on* I *and*

$$\int_a^b p_i(t) \, dt = 1, \qquad \int_a^b q_i(t) \, dt = 1 \qquad (i = 1, \cdots, n).$$

Let h_i, w_i, $i = 1, \cdots, n$ *be real-valued positive convex and increasing functions on*

(0, ∞). Then the following integral inequality holds:

$$\int_a^b \left[F\left(|u_1(t)|, \cdots, |u_n(t)|\right) \sum_{i=1}^n G_i'\left(|v_1(t)|, \cdots, |v_n(t)|\right) |v_i'(t)| \right.$$

$$\left. + G\left(|v_1(t)|, \cdots, |v_n(t)|\right) \sum_{i=1}^n F_i'\left(|u_1(t)|, \cdots, |u_n(t)|\right) |u_i'(t)| \right] dt \qquad (5)$$

$$\leq F\left(h_1^{-1}\left(\int_a^b p_1(t) h_1\left(\frac{|u_1'(t)|}{p_1(t)}\right) dt\right), \cdots, h_n^{-1}\left(\int_a^b p_n(t) h_n\left(\frac{|u_n'(t)|}{p_n(t)}\right) dt\right)\right)$$

$$\cdot G\left(w_1^{-1}\left(\int_a^b q_1(t) w_1\left(\frac{|v_1'(t)|}{q_1(t)}\right) dt\right), \cdots, w_n^{-1}\left(\int_a^b q_n(t) w_n\left(\frac{|v_n'(t)|}{q_n(t)}\right) dt\right)\right).$$

Proof. From the assumptions we write

$$\int_a^b |u_i'(t)| \, dt = \frac{\displaystyle\int_a^b p_i(t) \frac{|u_i'(t)|}{p_i(t)} \, dt}{\displaystyle\int_a^b p_i(t) \, dt}, \qquad (6)$$

$$\int_a^b |v_i'(t)| \, dt = \frac{\displaystyle\int_a^b q_i(t) \frac{|v_i'(t)|}{q_i(t)} \, dt}{\displaystyle\int_a^b q_i(t) \, dt}, \qquad (7)$$

for $i = 1, \cdots, n$. From (6), (7) and using the hypotheses on h_i, w_i, $i = 1, \cdots, n$ and Jensen's inequality [4, p. 133] we obtain

$$h_i\left(\int_a^b |u_i'(t)| \, dt\right) \leq \int_a^b p_i(t) \, h_i\left(\frac{|u_i'(t)|}{p_i(t)}\right) dt, \qquad (8)$$

$$w_i\left(\int_a^b |v_i'(t)| \, dt\right) \leq \int_a^b q_i(t) \, w_i\left(\frac{|v_i'(t)|}{q_i(t)}\right) dt, \qquad (9)$$

for $i = 1, \cdots, n$. From (8), (9) we observe that

$$\int_a^b |u_i'(t)| \, dt \leq h_i^{-1}\left(\int_a^b p_i(t) \, h_i\left(\frac{|u_i'(t)|}{p_i(t)}\right) dt\right), \qquad (10)$$

$$\int_a^b |v_i'(t)| \, dt \leq w_i^{-1}\left(\int_a^b q_i(t) \, w_i\left(\frac{|v_i'(t)|}{q_i(t)}\right) dt\right), \qquad (11)$$

for $i = 1, \cdots, n$. Since all the hypotheses of Theorem 1 are among those of Theorem 2, we see that the inequality (2) holds. Now using (10), (11) on the right hand side of (2), we get the required inequality in (5) and the proof is complete.

Remark 1.
  (i) In the special cses when $n = 1$, the inequalities established in (2) and (5) reduce to the inequalities established by Pachpatte in [9].
 (ii) If we take $u_i = v_i$, $G = F$ and hence $G'_i = F'_i$, $i = 1, \cdots, n$ and $n = 1$ in (2), then it reduces to the inequality given by Pachpatte in [9, p. 22].
(iii) If we take $G = 1$ and hence $G'_i = 0$, $i = 1, \cdots, n$ and

$$F(x_1, \cdots, x_n) = \prod_{i=1}^{n} f_i(x_i), \quad h_i = h, \quad i = 1, \cdots, n,$$

then (2) reduces to the most general result established by Pachpatte in [12], where the functions f_i must satisfy some suitable conditions.
 (iv) If we take $G = 1$ and hence $G'_i = 0$, $i = 1, \cdots, n$ in (2) and (5), then we get the inequalities recently given by Pečarić [14], see also [15].
  (v) In the special case when $G = 1$ and hence $G'_i = 0$, $i = 1, \cdots, n$ and $n = 1$, the inequality (2) reduces to the inequality

$$\int_a^b F'_1\left(|u_1(t)|\right) |u'_1(t)|\, dt \le F\left(\int_a^b |u'_1(t)|\, dt\right). \tag{12}$$

The inequality (12) under the hypotheses that F is convex and increasing function with $F(0) = 0$ is first given by Godunova and Levin in [3]. If we take $F(u) = u^2$ in (12), then we get the following version of the Opial's inequality (see [5, Theorem 2', p. 154])

$$\int_a^b |u_1(t)|\, |u'_1(t)|\, dt \le \frac{1}{2}(b-a)\int_a^b |u'_1(t)|^2\, dt. \tag{13}$$

For various other special versions of the inequalities (2) and (5), see [9, 12].

Our next two theorems deal with the generalized versions of the inequalities established by Pachpatte in [13].

Theorem 3. *Let u_i, v_i, F, F'_i, G, G'_i, $i = 1, \cdots, n$ be as in Theorem 1. Let ϕ_i, ψ_i, $i = 1, \cdots, n$ be real-valued positive convex and increasing functions on $(0, \infty)$. Let $r_i(x) \ge 0$, $r'_i(x) > 0$, $r_i(a) = 0$, $e_i(x) \ge 0$, $e'_i(x) > 0$, $e_i(a) = 0$, $i = 1, \cdots, n$. Then the following integral inequality holds:*

$$\int_a^b \left[F\left(r_1(t)\phi_1\left(\frac{|u_1(t)|}{r_1(t)}\right), \cdots, r_n(t)\phi_n\left(\frac{|u_n(t)|}{r_n(t)}\right)\right) \right.$$

$$\left. \cdot \sum_{i=1}^{n} G'_i\left(e_1(t)\psi\left(\frac{|v_1(t)|}{e_1(t)}\right), \cdots, e_n(t)\psi_n\left(\frac{|v_n(t)|}{e_n(t)}\right)\right) e'_i(t)\psi_i\left(\frac{|v'_i(t)|}{e'_i(t)}\right)\right.$$

$$+ G \left(e_1(t) \psi \left(\frac{|v_1(t)|}{e_1(t)} \right), \cdots, e_n(t) \psi_n \left(\frac{|v_n(t)|}{e_n(t)} \right) \right) \tag{14}$$

$$\cdot \sum_{i=1}^{n} F_i' \left(r_1(t) \phi_1 \left(\frac{|u_1(t)|}{r_1(t)} \right), \cdots, r_n(t) \phi_n \left(\frac{|u_n(t)|}{r_n(t)} \right) \right) r_i'(t) \phi_i \left(\frac{|u_i'(t)|}{r_i'(t)} \right) \Bigg] dt$$

$$\leq F \left(\int_a^b r_1'(t) \phi_1 \left(\frac{|u_1'(t)|}{r_1'(t)} \right), \cdots, \int_a^b r_n'(t) \phi_n \left(\frac{|u_n'(t)|}{r_n'(t)} \right) dt \right)$$

$$\cdot G \left(\int_a^b e_1'(t) \psi_1 \left(\frac{|v_1'(t)|}{e_1'(t)} \right) dt, \cdots, \int_a^b e_n'(t) \psi_n \left(\frac{|v_n'(t)|}{e_n'(t)} \right) dt \right).$$

Proof. From the hypotheses on u_i, v_i, r_i, e_i, $i = 1, \cdots, n$ we have

$$|u_i(t)| = \left| \int_a^t u_i'(s) \, ds \right| \leq \int_a^t |u_i'(s)| \, ds, \tag{15}$$

$$|v_i(t)| = \left| \int_a^t v_i'(s) \, ds \right| \leq \int_a^t |v_i'(s)| \, ds, \tag{16}$$

$$r_i(t) = \int_a^t r_i'(s) \, ds, \tag{17}$$

$$e_i(t) = \int_a^t e_i'(s) \, ds, \tag{18}$$

for $t \in I$. From (15)–(18) and using the hypotheses on ϕ_i, ψ_i, $i = 1, \cdots, n$ and Jensen's inequality [4, p. 133] we obtain

$$\phi_i \left(\frac{|u_i(t)|}{r_i(t)} \right) \leq \phi_i \left(\frac{\int_a^t \frac{r_i'(s)|u_i'(s)|}{r_i'(s)} \, ds}{\int_a^t r_i'(s) \, ds} \right) \tag{19}$$

$$\leq \frac{1}{r_i(t)} \int_a^t r_i'(s) \, \phi_i \left(\frac{|u_i'(s)|}{r_i'(s)} \right) \, ds,$$

$$\psi_i \left(\frac{|v_i(t)|}{e_i(t)} \right) \leq \psi_i \left(\frac{\int_a^t \frac{e_i'(s) |v_i'(s)|}{e_i'(s)} \, ds}{\int_a^t r_i'(s) \, ds} \right) \tag{20}$$

$$\leq \frac{1}{e_i(s)} \int_a^t e_i'(s) \, \psi_i \left(\frac{|v_i'(s)|}{e_i'(s)} \right) \, ds,$$

for $t \in I$. From (19), (20) and using the hypotheses on F, F_i', G, G_i', $i = 1, \cdots, n$, we observe that

$$\int_a^b \left[F\left(r_1(t)\,\phi_1\left(\frac{|u_1(t)|}{r_1(t)}\right), \cdots, r_n(t)\,\phi_n\left(\frac{|u_n(t)|}{r_n(t)}\right) \right) \right.$$

$$\cdot \sum_{i=1}^n G_i'\left(e_1(t)\psi_1\left(\frac{|v_1(t)|}{e_1(t)}\right), \cdots, e_n(t)\psi_n\left(\frac{|v_n(t)|}{e_n(t)}\right) \right) e_i'(t)\,\psi_i\left(\frac{|v_i'(t)|}{e_i'(t)}\right)$$

$$+ G\left(e_1(t)\psi\left(\frac{|v_1(t)|}{e_1(t)}\right), \cdots, e_n(t)\psi_n\left(\frac{|v_n(t)|}{e_n(t)}\right) \right)$$

$$\left. \cdot \sum_{i=1}^n F_i'\left(r_1(t)\phi_1\left(\frac{|u_1(t)|}{r_1(t)}\right), \cdots, r_n(t)\phi_n\left(\frac{|u_n(t)|}{r_n(t)}\right) \right) r_i'(t)\phi_i\left(\frac{|u_i'(t)|}{r_i'(t)}\right) \right] dt$$

$$\leq \int_a^b \left[F\left(\int_a^t r_1'(s)\phi_1\left(\frac{|u_1'(s)|}{r_1'(s)}\right) ds, \cdots, \int_a^t r_n'(s)\phi_n\left(\frac{|u_n'(s)|}{r_n'(s)}\right) ds \right) \right.$$

$$\cdot \sum_{i=1}^n G_i'\left(\int_a^t e_1'(s)\psi_1\left(\frac{|v_1'(s)|}{e_1'(s)}\right) ds, \cdots, \int_a^t e_n'(s)\psi_n\left(\frac{|v_n'(s)|}{e_n'(s)}\right) ds \right)$$

$$\cdot e_i'(t)\psi_i\left(\frac{|v_i'(t)|}{e_i'(t)}\right)$$

$$+ G\left(\int_a^t e_1'(s)\psi_1\left(\frac{|v_1'(s)|}{e_1'(s)}\right) ds, \cdots, \int_a^t e_n'(s)\psi_n\left(\frac{|v_n'(s)|}{e_n'(s)}\right) ds \right)$$

$$\cdot \sum_{i=1}^n F_i'\left(\int_a^t r_1'(s)\phi_1\left(\frac{|u_1'(s)|}{r_1'(s)}\right) ds, \cdots, \int_a^t r_n'(s)\phi_n\left(\frac{|u_n'(s)|}{r_n'(s)}\right) ds \right)$$

$$\left. \cdot r_i'(t)\phi_i\left(\frac{|u_i'(t)|}{r_i'(t)}\right) \right] dt$$

$$= \int_a^b \frac{d}{dt}\left[F\left(\int_a^t r_1'(s)\phi_1\left(\frac{|u_1'(s)|}{r_1'(s)}\right) ds, \cdots, \int_a^t r_n'(s)\phi_n\left(\frac{|u_n'(s)|}{r_n'(s)}\right) ds \right) \right.$$

$$\left. \cdot G\left(\int_a^t e_1'(s)\psi_1\left(\frac{|v_1'(s)|}{e_1'(s)}\right) ds, \cdots, \int_a^t e_n'(s)\psi_n\left(\frac{|v_n'(s)|}{e_n'(s)}\right) ds \right) \right] dt$$

$$= F\left(\int_a^b r_1'(t)\phi_1\left(\frac{|u_1'(t)|}{r_1'(t)}\right) dt, \cdots, \int_a^b r_n'(t)\phi_n\left(\frac{|u_n'(t)|}{r_n'(t)}\right) dt \right)$$

$$\cdot G\left(\int_a^b e_1'(t)\psi_1\left(\frac{|v_1'(t)|}{e_1'(t)}\right) dt, \cdots, \int_a^b e_n'(t)\psi_n\left(\frac{|v_n'(t)|}{e_n'(t)}\right) dt \right).$$

This completes the proof of the inequality (14).

Theorem 4. *Let u_i , v_i , F, F_i' , G, G_i' , ϕ_i , ψ_i , r_i , e_i , $i = 1, \cdots, n$, be as in Theorem 3. Let p_i , q_i , h_i , w_i , $i = 1, \cdots, n$ be as in Theorem 2. Then the following integral inequality holds:*

$$
\int_a^b \left[F\left(r_1(t)\,\phi_1\left(\frac{|u_1(t)|}{r_1(t)}\right), \cdots, r_n(t)\,\phi_n\left(\frac{|u_n(t)|}{r_n(t)}\right)\right) \right.
$$

$$
\cdot \sum_{i=1}^n G_i'\left(e_1(t)\,\psi_1\left(\frac{|v_1(t)|}{e_1(t)}\right), \cdots, e_n(t)\psi_n\left(\frac{|v_n(t)|}{e_n(t)}\right)\right) e_i'(t)\psi_i\left(\frac{|v_i'(t)|}{e_i'(t)}\right)
$$

$$
+ G\left(e_1(t)\,\psi_1\left(\frac{|v_1(t)|}{e_1(t)}\right), \cdots, e_n(t)\psi_n\left(\frac{|v_n(t)|}{e_n(t)}\right)\right)
$$

$$
\left. \cdot \sum_{i=1}^n F_i'\left(r_1(t)\,\phi_1\left(\frac{|u_1(t)|}{r_1(t)}\right), \cdots, r_n(t)\phi_n\left(\frac{|u_n(t)|}{r_n(t)}\right)\right) r_i'(t)\phi_i\left(\frac{|u_i'(t)|}{r_i'(t)}\right) \right] dt
$$

$$
\leq F\left(h_1^{-1}\left(\int_a^b p_1(t)h_1\left(\frac{r_1'(t)\phi_1\left(\frac{|u_1'(t)|}{r_1'(t)}\right)}{p_1(t)} \right) dt \right), \cdots, \right. \tag{21}
$$

$$
\left. h_n^{-1}\left(\int_a^b p_n(t)\, h_n\left(\frac{r_n'(t)\,\phi_n\left(\frac{|u_n'(t)|}{r_n'(t)}\right)}{p_n(t)} \right) dt \right) \right)
$$

$$
\cdot G\left(w_1^{-1}\left(\int_a^b q_1(t)\,w_1\left(\frac{e_1'(t)\,\psi_1\left(\frac{|v_1'(t)|}{e_1'(t)}\right)}{q_1(t)} \right) dt \right), \cdots, \right.
$$

$$
\left. w_n^{-1}\left(\int_a^b q_n(t)w_n\left(\frac{e_n'(t)\psi_n\left(\frac{|v_n'(t)|}{e_n'(t)}\right)}{q_n(t)} \right) dt \right) \right) .
$$

Proof. From the assumptions we write

$$
\int_a^b r_i'(t)\phi_i\left(\frac{|u_i'(t)|}{r_i'(t)}\right) dt = \frac{\displaystyle\int_a^b p_i(t)\, \frac{r_i'(t)\phi_i\left(\frac{|u_i'(t)|}{r_i'(t)}\right)}{p_i(t)}\, dt}{\displaystyle\int_a^b p_i(t)\,dt} \tag{22}
$$

$$\int_a^b e_i'(t)\psi_i\left(\frac{|v_i'(t)|}{e_i'(t)}\right)dt = \frac{\displaystyle\int_a^b q_i(t)\frac{e_i'(t)\psi_i\left(\frac{|v_i'(t)|}{e_i'(t)}\right)}{q_i(t)}dt}{\displaystyle\int_a^b q_i(t)dt}, \tag{23}$$

for $i = 1, \cdots, n$. From (22), (23) and using the hypotheses on h_i, w_i, $i = 1, \cdots, n$ and Jensen's inequality [4, p. 133] we obtain

$$h_i\left(\int_a^b r_i'(t)\phi_i\left(\frac{|u_i'(t)|}{r_i'(t)}\right)dt\right) \le \int_a^b p_i(t)h_i\left(\frac{r_i'(t)\phi_i\left(\frac{|u_i'(t)|}{r_i'(t)}\right)}{p_i(t)}\right)dt \tag{24}$$

$$w_i\left(\int_a^b e_i'(t)\psi_i\left(\frac{|v_i'(t)|}{e_i'(t)}\right)dt\right) \le \int_a^b q_i(t)w_i\left(\frac{e_i'(t)\psi_i\left(\frac{|v_i'(t)|}{e_i'(t)}\right)}{q_i(t)}\right)dt, \tag{25}$$

for $i = 1, \cdots, n$. From (24), (25) we observe that

$$\int_a^b r_i'(t)\phi_i\left(\frac{|u_i'(t)|}{r_i'(t)}\right) \le h_i^{-1}\left(\int_a^b p_i(t)\,h_i\left(\frac{r_i'(t)\phi_i\left(\frac{|u_i'(t)|}{r_i'(t)}\right)}{p_i(t)}\right)dt\right), \tag{26}$$

$$\int_a^b e_i'(t)\psi_i\left(\frac{|v_i'(t)|}{e_i'(t)}\right)dt \le w_i^{-1}\left(\int_a^b q_i(t)\,w_i\left(\frac{e_i'(t)\psi_i\left(\frac{|v_i'(t)|}{e_i'(t)}\right)}{q_i(t)}\right)dt\right), \tag{27}$$

for $i = 1, \cdots, n$. Since all the hypotheses of Theorem 3 are among those of Theorem 4, we see that the inequality (14) holds. Now using (26), (27) on the right-hand side of (14) we get the desired inequality in (21) and the proof is complete.

Remark 2.
(i) In the special cases when $n = 1$, the inequalities established in Theorems 3 and 4 reduce to the inequalities recently established by Pachpatte in [13].
(ii) If we take $G = 1$ and hence $G_i' = 0$, $i = 1, \cdots, n$ in (14) and (21), then we get the inequalities recently established by Pečarić and Brnetić in [15]. For a detailed discussion on the further special versions of the inequalities given in (14) and (21), see [13, 17].

3. Multidimensional Generalizations

First, we will introduce notation which we will use throughout this section. Let

$$B = \prod_{i=1}^m [a_i, b_i] \qquad a_i, b_i \in R_+.$$

310

Let $t = (t_1, \cdots, t_m)$ be a general point in B, $B_t = \prod_{i=1}^{m} [a_i, t_i]$ and dt the volume form $dt_1 \cdots dt_m$. Let $Dh(u)$ stand for $dh(u)/du$, $D_k h(t_1, \cdots, t_m)$ stand for $\left(\frac{\partial}{\partial t_k}\right) h(t_1, \cdots, t_m)$, $1 \le k \le m$, and $D^k h(t_1, \cdots, t_m)$ stand for

$$\left(\frac{\partial^k}{\partial t_1 \cdots \partial t_k}\right) h(t_1, \cdots, t_m), \qquad 1 \le k \le m.$$

By using the above notation we have $D_h = D^1 h$.

Our next two theorems deal with the multidimensional generalizations of the inequalities given in Theorems 1 and 2.

Theorem 5. Let $m \ge 2$ and $f_i, D^1 f_i, \cdots, D^m f_i, g_i, D^1 g_i, \cdots, D^m g_i, i = 1, \cdots, n$ be real-valued continuous functions on B with

$$f_i(t)|_{t_j=a_j} = 0, \quad i = 1, \cdots, n, \quad j = 1, \cdots, m, \tag{28}$$

$$g_i(t)|_{t_j=a_j} = 0, \quad i = 1, \cdots, n, \quad j = 1, \cdots, m. \tag{29}$$

Let H, L be real-valued, nonnegative, continuous, nondecreasing and differentiable functions on R_+^n with $H(0, \cdots, 0) = 0$, $L(0, \cdots, 0) = 0$ such that $D_i H, D_i L$, $i = 1, \cdots, n$ are nonnegative, continuous and nondecreasing on R_+^n. Then the following integral inequality holds:

$$\int_B \left[H(|f_1(t)|, \cdots, |f_n(t)|) \sum_{i=1}^{n} D_i L(|g_1(t)|, \cdots, |g_n(t)|) |D^m g_i(t)| \right.$$

$$\left. + L(|g_1(t)|, \cdots, |g_n(t)|) \sum_{i=1}^{n} D_i H(|f_1(t)|, \cdots, |f_n(t)|) |D^m f_i(t)| \right] dt \tag{30}$$

$$\le H\left(\int_B |D^m f_1(t)| \, dt, \cdots, \int_B |D^m f_n(t)| \, dt \right)$$

$$\cdot L\left(\int_B |D^m g_1(t)| \, dt, \cdots, \int_B |D^m g_n(t)| \, dt \right).$$

Proof. First we define

$$z_i(t) = \int_{B_t} |D^m f_i(u)| \, du, \tag{31}$$

$$c_i(t) = \int_{B_t} |D^m g_i(u)| \, du, \tag{32}$$

for any $t = (t_1, \cdots, t_m) \in B$ and $i = 1, \cdots, n$. From the hypotheses it is easy to observe that

$$|f_i(t)| = \left| \int_{B_t} D^m f_i(u) du \right| \le \int_{B_t} |D^m f_i(u)| \, du = z_i(t), \tag{33}$$

$$|g_i(t)| = \left| \int_{B_t} D^m g_i(u) du \right| \le \int_{B_t} |D^m g_i(u)| \, du = c_i(t), \tag{34}$$

for $i = 1, \cdots, n$. Let $B' = \prod_{i=2}^m [a_i, b_i]$. Now, using (33), (34), the facts that the functions z_i, c_i, $i = 1, \cdots, n$ are nondecreasing in each variable and the hypotheses of theorem, we have

$$\int_B \left[H\left(|f_1(t)|, \cdots, |f_n(t)|\right) \sum_{i=1}^n D_i L\left(|g_1(t)|, \cdots, |g_n(t)|\right) |D^m g_i(t)| \right.$$

$$\left. + L\left(|g_1(t)|, \cdots, |g_n(t)|\right) \sum_{i=1}^n D_i H\left(|f_1(t)|, \cdots, |f_n(t)|\right) |D^m f_i(t)| \right] dt$$

$$\le \int_B \left[H\left(z_1(t), \cdots, z_n(t)\right) \sum_{i=1}^n D_i L\left(c_1(t), \cdots, c_n(t)\right) |D^m g_i(t)| \right.$$

$$\left. + L\left(c_1(t), \cdots, c_n(t)\right) \sum_{i=1}^n D_i H\left(z_1(t), \cdots, z_n(t)\right) |D^m f_i(t)| \right] dt$$

$$\le \int_{a_1}^{b_1} \left[H\left(z_1(t_1, b_2, \cdots, b_m), \cdots, z_n(t_1, b_2, \cdots, b_m)\right) \right.$$

$$\cdot \sum_{i=1}^n D_i L\left(c_1(t_1, b_2, \cdots, b_m), \cdots, c_n(t_1, b_2, \cdots, b_m)\right) \int_{B'} |D^m g_i(t)| \, dt_2 \cdots dt_m$$

$$+ L\left(c_1(t_1, b_2, \cdots, b_m), \cdots, c_n(t_1, b_2, \cdots, b_m)\right)$$

$$\cdot \sum_{i=1}^n D_i H\left(z_1(t_1, b_2, \cdots, b_m), \cdots, z_n(t_1, b_2, \cdots, b_m)\right)$$

$$\left. \int_{B'} |D^m f_i(t)| \, dt_2 \cdots dt_m \right] dt_1$$

$$\le \int_{a_1}^{b_1} \left[H\left(z_1(t_1, b_2, \cdots, b_m), \cdots, z_n(t_1, b_2, \cdots, b_m)\right) \right.$$

$$\cdot \sum_{i=1}^n D_i L\left(c_1(t_1, b_2, \cdots, b_m), \cdots, c_n(t_1, b_2, \cdots, b_m)\right) D^1 c_i(t_1, b_2, \cdots, b_m)$$

$$+ L\left(c_1(t_1, b_2, \cdots, b_m), \cdots, c_n(t_1, b_2, \cdots, b_m)\right)$$

$$\cdot \sum_{i=1}^n D_i H\left(z_1(t_1, b_2, \cdots, b_m), \cdots, z_n(t_1, b_2, \cdots, b_m)\right) D^1 z_i(t_1, b_2, \cdots, b_m) \right] dt_1$$

$$= \int_{a_1}^{b_1} \frac{d}{dt_1} \Big[H \left(z_1(t_1, b_2, \cdots, b_m), \cdots, z_n(t_1, b_2, \cdots, b_m) \right)$$

$$\cdot L \left(c_1(t_1, b_2, \cdots, b_m), \cdots, c_n(t_1, b_2, \cdots, b_m) \right) \Big] dt_1$$

$$= H \left(z_1(b_1, \cdots, b_m), \cdots, z_n(b_1, \cdots, b_m) \right) L \left(c_1(b_1, \cdots, b_m), \cdots, c_n(b_1, \cdots, b_m) \right)$$

$$= H \left(\int_B |D^m f_1(t)| \, dt, \cdots, \int_B |D^m f_n(t)| \, dt \right)$$

$$\cdot L \left(\int_B |D^m g_1(t)| \, dt, \cdots, \int_B |D^m g_n(t)| \, dt \right)$$

which is the required inequality in (30) and the proof of the theorem is complete.

Theorem 6. *Let $m \geq 2$ and $f_i, D^1 f_i, \cdots, D^m f_i, g_i, D^1 g_i, \cdots, D^m g_i, H, D_i H, L, D_i L, i = 1, \cdots, n$ be as in Theorem 5. Let $p_i, q_i, i = 1, \cdots, n$ be positive continuous functions defined on B and*

$$\int_B p_i(t) dt = 1, \quad \int_B q_i(t) dt = 1 \quad (i = 1, \cdots, n).$$

Let $h_i, w_i, i = 1, \cdots, n$ be positive convex and increasing functions on $(0, \infty)$. Then the following integral inequality holds:

$$\int_B \Big[H \left(|f_1(t)|, \cdots, |f_n(t)| \right) \sum_{i=1}^{n} D_i L \left(|g_1(t)|, \cdots, |g_n(t)| \right) |D^m g_i(t)|$$

$$+ L \left(|g_1(t)|, \cdots, |g_n(t)| \right) \tag{35}$$

$$\cdot \sum_{i=1}^{n} D_i H \left(|f_1(t)|, \cdots, |f_n(t)| \right) |D^m f_i(t)| \Big] dt$$

$$\leq H \left(h_1^{-1} \left(\int_B p_1(t) h_1 \left(\frac{|D^m f_1(t)|}{p_1(t)} \right) dt \right), \cdots, \right.$$

$$h_n^{-1} \left(\int_B p_n(t) h_n \left(\frac{|D^m f_n(t)|}{p_n(t)} \right) dt \right) \Big)$$

$$\cdot L \left(w_1^{-1} \left(\int_B q_1(t) w_1 \left(\frac{|D^m g_1(t)|}{q_1(t)} \right) dt \right), \cdots, \right.$$

$$w_n^{-1} \left(\int_B q_n(t) w_n \left(\frac{|D^m g_n(t)|}{q_n(t)} \right) dt \right) \Big).$$

Proof. From the assumptions we write

$$\int_B |D^m f_i(t)|\, dt = \frac{\int_B \dfrac{p_i(t)\, |D^m f_i(t)|}{p_i(t)}\, dt}{\int_B p_i(t)\, dt}, \tag{36}$$

$$\int_B |D^m g_i(t)|\, dt = \frac{\int_B \dfrac{q_i(t)\, |D^m g_i(t)|}{q_i(t)}\, dt}{\int_B q_i(t)\, dt}, \tag{37}$$

for $i = 1, \cdots, n$. From (36), (37) and using the hypotheses on h_i, w_i, $i = 1, \cdots, n$ and Jensen's inequality [4, p. 133] we obtain:

$$h_i \left(\int_B |D^m f_i(t)|\, dt \right) \le \int_B p_i(t) h_i \left(\frac{|D^m f_i(t)|}{p_i(t)} \right) dt, \tag{38}$$

$$w_i \left(\int_B |D^m g_i(t)|\, dt \right) \le \int_b q_i(t) w_i \left(\frac{|D^m g_i(t)|}{q_i(t)} \right) dt, \tag{39}$$

for $i = 1, \cdots, n$. From (38), (39) we observe that

$$\int_B |D^m f_i(t)|\, dt \le h_i^{-1} \left(\int_B p_i(t) h_i \left(\frac{|D^m f_i(t)|}{p_i(t)} \right) dt \right), \tag{40}$$

$$\int_B |D^m g_i(t)|\, dt \le w_i^{-1} \left(\int_B q_i(t) w_i \left(\frac{|D^m g_i(t)|}{q_i(t)} \right) dt \right), \tag{41}$$

for $i = 1, \cdots, n$. Since all the hypotheses of Theorem 5 are among those of Theorem 6, we see that the inequality (30) holds. Now, using (40), (41) on the right hand side of (30) we get the desired inequality in (35) and the proof is complete.

Remark 3.

(i) If we take $L = 1$ and hence $D_i L = 0$, $i = 1, \cdots, n$ in (30), then it reduces to the following inequality:

$$\int_B \sum_{i=1}^{n} D_i H\left(|f_1(t)|, \cdots, |f_n(t)| \right) |D^m f_i(t)|\, dt$$
$$\le F \left(\int_B |D^m f_1(t)|\, dt, \cdots, \int_B |D^m f_n(t)|\, dt \right). \tag{42}$$

The inequality (42) was recently established by Pečarić and Brnetić in [16, Theorem 1].

(ii) By taking $g_i(t) = f_i(t)$, $L = H$ and hence $D_i L = D_i H$, $i = 1, \cdots, n$ in (30), we get the following inequality:

$$\int_B H\left(|f_1(t)|, \cdots, |f_n(t)|\right)$$

$$\cdot \sum_{i=1}^{n} D_i H\left(|f_1(t)|, \cdots, |f_n(t)|\right) |D^m f_i(t)| \, dt \qquad (43)$$

$$\leq \frac{1}{2} \left[H\left(\int_B |D^m f_1(t)| \, dt, \cdots, \int_B |D^m f_n(t)| \, dt\right)\right]^2.$$

The inequality obtained in (43) for functions involving two independent variables is established by Pachpatte in [11, p. 232]. The special versions of Theorem 6 can be discussed similarly, see also [10, 11].

In the following two theorems we establish the multidimensional versions of the inequalities established in Theorems 3 and 4.

Theorem 7. *Let $m \geq 2$ and f_i, $D^1 f_i, \cdots, D^m f_i$, $D^1 g_i, \cdots, D^m g_i$, H, $D_i H$, L, $D_i L$ be as in Theorem 5. Let r_i, e_i, $i = 1, \cdots, n$ be nonnegative continuous functions defined on B such that*

$$D^1 r_i, \cdots, D^m r_i, \quad D^1 e_i, \cdots, D^m e_i, \quad i = 1, \cdots, n,$$

are continuous positive functions defined on B and let the condition (28) be satisfied for functions r_i, e_i. Let ϕ_i, ψ_i, $i = 1, \cdots, n$ be real-valued positive, convex and increasing functions on $(0, \infty)$. Then the following integral inequality holds:

$$\int_B \left[H\left(r_1(t)\phi_1\left(\frac{|f_1(t)|}{r_1(t)}\right), \cdots, r_n(t)\phi_n\left(\frac{|f_n(t)|}{r_n(t)}\right)\right) \right.$$

$$\cdot \sum_{i=1}^{n} D_i L\left(e_1(t)\psi_1\left(\frac{|g_1(t)|}{e_1(t)}\right), \cdots, e_n(t)\psi_n\left(\frac{|g_n(t)|}{e_n(t)}\right)\right)$$

$$\cdot D^m e_i(t)\psi_i\left(\frac{|D^m g_i(t)|}{D^m e_i(t)}\right)$$

$$+ L\left(e_1(t)\psi_1\left(\frac{|g_1(t)|}{e_1(t)}\right), \cdots, e_n(t)\psi_n\left(\frac{|g_n(t)|}{e_n(t)}\right)\right)$$

$$\cdot \sum_{i=1}^{n} D_i H\left(r_1(t)\phi_1\left(\frac{|f_1(t)|}{r_1(t)}\right), \cdots, r_n(t)\phi_n\left(\frac{|f_n(t)|}{r_n(t)}\right)\right) \qquad (44)$$

$$\left. \cdot D^m r_i(t)\phi_i\left(\frac{|D^m f_i(t)|}{D^m r_i(t)}\right)\right] dt$$

$$\leq H \left(\int_B D^m r_1(t) \phi_1 \left(\frac{|D^m f_1(t)|}{D^m r_1(t)} \right) dt, \cdots, \int_B D^m r_n(t) \phi_n \left(\frac{|D^m f_n(t)|}{D^m r_n(t)} \right) dt \right)$$

$$\cdot L \left(\int_B D^m e_1(t) \psi_1 \left(\frac{|D^m g_1(t)|}{D^m e_1(t)} \right) dt, \cdots, \int_B D^m e_n(t) \psi_n \left(\frac{|D^m g_n(t)|}{D^m e_n(t)} \right) dt \right).$$

Proof. From the hypotheses it is easy to observe that

$$|f_i(t)| = \left| \int_{B_t} D^m f_i(u) du \right| \leq \int_{B_t} |D^m f_i(u)| \, du, \tag{45}$$

$$|g_i(t)| = \left| \int_{B_t} D^m g_i(u) du \right| \leq \int_{B_t} |D^m g_i(u)| \, du, \tag{46}$$

$$r_i(t) = \int_{B_t} D^m r_i(u) du, \tag{47}$$

$$e_i(t) = \int_{B_t} D^m e_i(u) du, \tag{48}$$

for $t \in B$ and $i = 1, \cdots, n$. From (45), (47) and using Jensen's inequality [4, p. 133] we observe that

$$\phi_i \left(\frac{|f_i(t)|}{r_i(t)} \right) \leq \phi_i \left(\frac{\displaystyle\int_{B_t} \frac{D^m r_i(u) \, |D^m f_i(u)|}{D^m r_i(u)} du}{\displaystyle\int_{B_t} D^m r_i(u) du} \right) \tag{49}$$

$$\leq \frac{1}{r_i(t)} \int_{B_t} D^m r_i(u) \phi_i \left(\frac{|D^m f_i(u)|}{D^m r_i(u)} \right) du,$$

for $i = 1, \cdots, n$. Similarly, from (46), (48) and using Jensen's inequality we have

$$\psi_i \left(\frac{|g_i(t)|}{e_i(t)} \right) \leq \frac{1}{e_i(t)} \int_{B_t} D^m e_i(u) \psi_i \left(\frac{|D^m g_i(u)|}{D^m e_i(u)} \right) du, \tag{50}$$

for $i = 1, \cdots, n$. Define

$$z_i(t) = \int_{B_t} D^m r_i(u) \phi_i \left(\frac{|D^m f_i(u)|}{D^m r_i(u)} \right) du, \tag{51}$$

$$c_i(t) = \int_{B_t} D^m e_i(u) \psi_i \left(\frac{|D^m g_i(u)|}{D^m e_i(u)} \right) du, \tag{52}$$

for $i = 1, \cdots, n$. The rest of the proof follows by the similar arguments as in the proof of Theorem 5 with suitable changes and hence we omit the further details.

316

Theorem 8. *Let $m \geq 2$ and $f_i,\ D^1 f_i, \cdots, D^m f_i,\ g_i,\ D^1 g_i, \cdots, D^m g_i,\ H,\ D_i H,$
$L,\ r_i,\ D^1 r_i, \cdots, D^m r_i,\ e_i,\ D^1 e_i, \cdots, D^m e_i,\ \phi_i,\ \psi_i,\ i = 1, \cdots, n$ be as in Theorem
7. Let $p_i,\ q_i,\ h_i,\ w_i,\ i = 1, \cdots, n$ be as in Theorem 6. Then the following integral
inequality holds:*

$$\int_B \left[H\left(r_1(t)\phi_1\left(\frac{|f_1(t)|}{r_1(t)}\right), \cdots, r_n(t)\phi_n\left(\frac{|f_n(t)|}{r_n(t)}\right) \right) \right.$$

$$\cdot \sum_{i=1}^n D_i L\left(e_1(t)\psi_i\left(\frac{|g_1(t)|}{e_1(t)}\right), \cdots, e_n(t)\psi_n\left(\frac{|g_n(t)|}{e_n(t)}\right) \right)$$

$$\cdot D^m e_i(t)\psi_i\left(\frac{|D^m g_i(t)|}{D^m e_i(t)}\right)$$

$$+ L\left(e_1(t)\psi_1\left(\frac{|g_1(t)|}{e_i(t)}\right), \cdots, e_n(t)\psi_n\left(\frac{|g_n(t)|}{e_n(t)}\right) \right)$$

$$\cdot \sum_{i=1}^n D_i H\left(r_1(t)\psi_1\left(\frac{|f_1(t)|}{r_1(t)}\right), \cdots, r_n(t)\phi_n\left(\frac{|f_n(t)|}{r_n(t)}\right) \right)$$

$$\left. \cdot D^m r_i(t)\phi_i\left(\frac{|D^m f_i(t)|}{D^m r_i(t)}\right) \right] dt \tag{53}$$

$$\leq H\left(h_1^{-1}\left(\int_B p_1(t) h_1\left(\frac{D^m r_1(t)\phi_1\left(\frac{|D^m f_1(t)|}{D^m r_1(t)}\right)}{p_1(t)} \right) dt \right), \cdots, \right.$$

$$\left. h_n^{-1}\left(\int_B p_n(t) h_n\left(\frac{D^m r_n(t)\phi_n\left(\frac{|D^m f_n(t)|}{D^m r_n(t)}\right)}{p_n(t)} \right) dt \right) \right)$$

$$\cdot L\left(w_1^{-1}\left(\int_B q_1(t) w_1\left(\frac{D^m e_1(t)\psi_1\left(\frac{|D^m g_1(t)|}{D^m e_1(t)}\right)}{q_1(t)} \right) dt \right), \cdots, \right.$$

$$\left. w_n^{-1}\left(\int_B q_n(t) w_n\left(\frac{D^m e_n(t)\psi_n\left(\frac{|D^m g_n(t)|}{D^m e_n(t)}\right)}{q_n(t)} \right) dt \right) \right).$$

The proof of this theorem follows by the similar arguments as in the proof of
Theorem 6 and closely looking at the proof of Theorem 4, with suitable modifications and hence we omit the details.

Remark 4. If we take $g_i = f_i$, $\psi_i = \phi_i$, $e_i = r_i$, $L = 1$ and hence $D_i L = 0$, $i = 1, \cdots, n$, in Theorem 8, then the inequality established in (53) reduces to the inequality recently given by Pečarić and Brnetić in [16, Theorem 3]. We also note that in the special cases the inequalities established in Theorems 7 and 8 yield the various new inequalities as discussed in Remarks 1 and 2.

4. Further Inequalities

In a recent paper [2] Bloom established some Opial type inequalities involving generalized Hardy operator. The results given in [2] are based on the simple observation made by Sinnamon in [18] and also discussed by Bloom in [2, p. 28]. In this section we present some interesting variants of the inequalities given in Section 2 involving Hardy operator, which in fact are motivated by the recent results given by Bloom in [2] and by Sinnamon in [18].

An operator T acting on R is called a Hardy operator, if T has the form

$$Tf(t) = \int_a^t f(s)ds, \tag{54}$$

for $t \in I$, where $f(t)$ is a real-valued continuous function defined on I. We say that the function $f(t)$ belongs to the class U if it can be represented in the form (54). We note that the results given below also hold, if the Hardy operator T has the form

$$Tf(t) = \int_t^b f(s)ds, \tag{55}$$

for $t \in I$, where $f(t)$ is a continuous function on I.

An interesting variant of Theorem 1 is given in the following theorem:

Theorem 9. *Let* f_i, $g_i \in U$, $i = 1, \cdots, n$ *and* F, F_i', G, G_i', $i = 1, \cdots, n$ *be as in Theorem 1. Then the following integral inequality holds:*

$$\int_a^b \left[F\left(|Tf_1(t)|, \cdots, |Tf_n(t)|\right) \sum_{i=1}^n G_i'\left(|Tg_1(t)|, \cdots, |Tg_n(t)|\right) |g_i(t)| \right.$$

$$\left. + G\left(|Tg_1(t)|, \cdots, |Tg_n(t)|\right) \sum_{i=1}^n F_i'\left(|Tf_1(t)|, \cdots, |Tf_n(t)|\right) |f_i(t)| \right] dt$$

$$\tag{56}$$

$$\leq F\left(\int_a^b |f_1(t)|\, dt, \cdots, \int_a^b |f_n(t)|\, dt\right)$$

$$\cdot G\left(\int_a^b |g_1(t)|\, dt, \cdots, \int_a^b |g_n(t)|\, dt\right).$$

Proof. From the hypotheses it is easy to observe that

$$|Tf_i(t)| = \left| \int_a^t f_i(s)ds \right| \le \int_a^t |f_i(s)|\,ds, \tag{57}$$

$$|Tg_i(t)| = \left| \int_a^t g_i(s)ds \right| \le \int_a^t |g_i(s)|\,ds, \tag{58}$$

for $t \in I$, $i = 1, \cdots, n$. From (57), (58) and using the assumptions, we observe that

$$\int_a^b \left[F\left(|Tf_1(t)|, \cdots, |Tf_n(t)|\right) \sum_{i=1}^n G_i'\left(|Tg_1(t)|, \cdots, |Tg_n(t)|\right) |g_i(t)| \right.$$

$$+ G\left(|Tg_1(t)|, \cdots, |Tg_n(t)|\right)$$

$$\left. \cdot \sum_{i=1}^n F_i'\left(|Tf_1(t)|, \cdots, |Tf_n(t)|\right) |f_i(t)| \right] dt$$

$$\le \int_a^b \left[F\left(\int_a^t |f_1(s)|\,ds, \cdots, \int_a^t |f_n(s)|\,ds \right) \right.$$

$$\cdot \sum_{i=1}^n G_i'\left(\int_a^t |g_1(s)|\,ds, \cdots, \int_a^t |g_n(s)|\,ds \right) |g_i(t)|$$

$$+ G\left(\int_a^t |g_1(s)|\,ds, \cdots, \int_a^t |g_n(s)|\,ds \right)$$

$$\left. \cdot \sum_{i=1}^n F_i'\left(\int_a^t |f_1(s)|\,ds, \cdots, \int_a^t |f_n(s)|\,ds \right) |f_i(t)| \right] dt$$

$$= \int_a^b \frac{d}{dt} \left[F\left(\int_a^t |f_1(s)|\,ds, \cdots, \int_a^t |f_n(s)|\,ds \right) \right.$$

$$\left. \cdot G\left(\int_a^t |g_1(s)|\,ds, \cdots, \int_a^t |g_n(s)|\,ds \right) \right] dt$$

$$= F\left(\int_a^b |f_1(t)|\,dt, \cdots, \int_a^b |f_n(t)|\,dt \right)$$

$$\cdot G\left(\int_a^b |g_1(t)|\,dt, \cdots, \int_a^b |g_n(t)|\,dt \right).$$

This is the required inequality in (56) and hence the proof is complete.

In the following theorems we present the variants of Theorems 2 to 4 by following the idea of Theorem 9.

Theorem 10. *Let f_i, g_i, F, F_i', G, G_i' be as in Theorem 9, and p_i, q_i, h_i, w_i, for $i = 1, \cdots, n$ be as in Theorem 2. Then the following integral inequality holds:*

$$\int_a^b \left[F\left(|Tf_1(t)|, \cdots, |Tf_n(t)|\right) \sum_{i=1}^n G_i'\left(|Tg_1(t)|, \cdots, |Tg_n(t)|\right) |g_i(t)| \right.$$

$$+ G\left(|Tg_1(t)|, \cdots, |Tg_n(t)|\right)$$

$$\left. \cdot \sum_{i=1}^n F_i'\left(|Tf_1(t)|, \cdots, |Tf_n(t)|\right) |f_i(t)| \right] dt$$

$$\leq F\left(h_1^{-1}\left(\int_a^b p_1(t) h_1\left(\frac{|f_1(t)|}{p_1(t)}\right) dt\right), \cdots, \right. \tag{59}$$

$$\left. h_n^{-1}\left(\int_a^b p_n(t) h_n\left(\frac{|f_n(t)|}{p_n(t)}\right) dt\right) \right)$$

$$\cdot G\left(w_1^{-1}\left(\int_a^b q_1(t) w_1\left(\frac{|g_1(t)|}{q_1(t)}\right) dt\right), \cdots, \right.$$

$$\left. w_n^{-1}\left(\int_a^b q_n(t) w_n\left(\frac{|g_n(t)|}{q_n(t)}\right) dt\right) \right).$$

Theorem 11. *Let f_i, g_i, F, F_i', G, G_i', $i = 1, \cdots, n$ be as in Theorem 9. Let ϕ_i, ψ_i, $i = 1, \cdots, n$ be as in Theorem 3. Let $r_i(t)$, $e_i(t)$ be positive continuous functions defined on I and let*

$$R_i(t) = \int_a^t r_i(s)ds, \quad E_i(t) = \int_a^t e_i(s)ds \qquad (t \in I;\ i = 1, \cdots, n).$$

Then the following integral inequality holds:

$$\int_a^b \left[F\left(R_1(t)\phi_1\left(\frac{|Tf_1(t)|}{R_1(t)}\right), \cdots, R_n(t)\phi_n\left(\frac{|Tf_n(t)|}{R_n(t)}\right) \right) \right.$$

$$\cdot \sum_{i=1}^n G_i'\left(E_1(t)\psi_1\left(\frac{|Tg_1(t)|}{E_1(t)}\right), \cdots, E_n(t)\psi_n\left(\frac{|Tg_n(t)|}{E_n(t)}\right) \right)$$

$$\cdot e_i(t)\psi_i\left(\frac{|g_i(t)|}{e_i(t)}\right)$$

$$+ G\left(E_1(t)\psi_1\left(\frac{|Tg_1(t)|}{E_1(t)}\right), \cdots, E_n(t)\psi_n\left(\frac{|Tg_n(t)|}{E_n(t)}\right)\right) \tag{60}$$

$$\cdot \sum_{i=1}^{n} F_i'\left(R_1(t)\phi_1\left(\frac{|Tf_1(t)|}{R_1(t)}\right), \cdots, R_n(t)\phi_n\left(\frac{|Tf_n(t)|}{R_n(t)}\right)\right)$$

$$\cdot r_i(t)\phi_i\left(\frac{|f_i(t)|}{r_i(t)}\right)\bigg] dt$$

$$\leq F\left(\int_a^b r_1(t)\phi_1\left(\frac{|f_1(t)|}{r_1(t)}\right) dt, \cdots, \int_a^b r_n(t)\phi_n\left(\frac{|f_n(t)|}{r_n(t)}\right) dt\right)$$

$$\cdot G\left(\int_a^b e_1(t)\psi_1\left(\frac{|g_1(t)|}{e_1(t)}\right) dt, \cdots, \int_a^b e_n(t)\psi_n\left(\frac{|g_n(t)|}{e_n(t)}\right) dt\right).$$

Theorem 12. *Let f_i, g_i, F, F_i', G, G_i', ϕ_i, ψ_i, r_i, R_i, e_i, E_i, $i = 1,\cdots,n$ be as in Theorem 11. Let p_i, q_i, h_i, w_i, $i = 1,\cdots,n$ be as in Theorem 2. Then the following integral inequality holds:*

$$\int_a^b\bigg[F\left(R_1(t)\phi_1\left(\frac{|Tf_1(t)|}{R_1(t)}\right), \cdots, R_n(t)\phi_n\left(\frac{|Tf_n(t)|}{R_n(t)}\right)\right)$$

$$\cdot \sum_{i=1}^{n} G_i'\left(E_1(t)\psi_1\left(\frac{|Tg_1(t)|}{E_1(t)}\right), \cdots, E_n(t)\psi_n\left(\frac{|Tg_n(t)|}{E_n(t)}\right)\right)$$

$$\cdot e_i(t)\psi_i\left(\frac{|g_i(t)|}{e_i(t)}\right)$$

$$+ G\left(E_1(t)\psi_1\left(\frac{|Tg_1(t)|}{E_1(t)}\right), \cdots, E_n(t)\psi_n\left(\frac{|Tg_n(t)|}{E_n(t)}\right)\right)$$

$$\cdot \sum_{i=1}^{n} F_i'\left(R_1(t)\phi_1\left(\frac{|Tf_1(t)|}{R_1(t)}\right), \cdots, R_n(t)\phi_n\left(\frac{|Tf_n(t)|}{R_n(t)}\right)\right)$$

$$\cdot r_i(t)\phi_i\left(\frac{|f_i(t)|}{r_i(t)}\right)\bigg] dt \tag{61}$$

$$\leq F\left(h_1^{-1}\left(\int_a^b p_1(t)h_1\left(\frac{r_1(t)\phi_1\left(\frac{|f_1(t)|}{r_1(t)}\right)}{p_1(t)}\right) dt\right), \cdots, \right.$$

$$\left. h_n^{-1}\left(\int_a^b p_n(t)h_n\left(\frac{r_n(t)\phi_n\left(\frac{|f_n(t)|}{r_n(t)}\right)}{p_n(t)}\right) dt\right)\right)$$

$$
\cdot G\left(w_1^{-1}\left(\int_a^b q_1(t)w_1\left(\frac{e_1(t)\psi_1\left(\frac{|g_1(t)|}{e_1(t)}\right)}{q_1(t)}\right)dt\right),\cdots,\right.
$$

$$
\left.w_n^{-1}\left(\int_a^b q_n(t)w_n\left(\frac{e_n(t)\psi_n\left(\frac{|g_n(t)|}{e_n(t)}\right)}{q_n(t)}\right)dt\right)\right).
$$

The proofs of Theorems 10 to 12 can be completed following the proof of Theorem 9 and closely looking at the proofs of Theorems 2 to 4 with suitable modifications. Here we omit the details.

In concluding this paper we note that, as discussed in Remarks 1 and 2, in the special cases the inequalities given in Theorems 10 to 12 yield various new and interesting inequalities of the Opial type which are different from those of recently given by Bloom in [2]. We also note that there is no essential difficulty in obtaining the multidimensional variants of the inequalities in Theorems 5 to 8 by following the idea used to establish our results in Theorems 9 to 12. The precise formulation of these results is very close to that of the results given in Theorems 9 to 12 with suitable modifications and hence we do not discuss them here.

References

[1] R.P. Agarwal and P.Y.H. Pang, *Opial Inequalities with Applications in Differential and Difference Equations*, Kluwer Academic Publishers, Dordrecht, 1995.

[2] S. Bloom, First and second order Opial inequalities, *Studia Math.* **126**(1997), 27–50.

[3] E.K. Godunova and V.I. Levin, On an inequality of Maroni (Russian), *Mat. Zametki* **2**(1967), 221–224.

[4] A. Kufner, O. John and S. Fučik, *Function Spaces*, Noordhoff International Publishing, Leiden, 1977.

[5] D.S. Mitrinovič, *Analytic Inequalities*, Springer-Verlag, Berlin, New York, 1970.

[6] D.S. Mitrinovič, J.E. Pečarić and A.M. Fink, *Inequalities Involving Functions and Their Integrals and Derivatives*, Kluwer Academic Publishers, Dordrecht, 1991.

[7] J. Myjak, Boundary value problems for nonlinear differential and difference equations of the second order, *Zeszyty Nauk Uniw. Jagiellon. Prace Mat.* **15**(1971), 113–123.

[8] Z. Opial, Sur une inégalité, *Ann. Polon. Math.* **8**(1960), 29–32.

[9] B.G. Pachpatte, On integral inequalities similar to Opial's inequality, *Demonstratio Math.* **22**(1989), 21–27.

[10] B.G. Pachpatte, On certain two dimensional integral inequalities, *Chinese J. Math.* **17** (1989), 273–279.

[11] B.G. Pachpatte, On a generalized Opial type inequality in two independent variables, *An. Ştiinţ. Univ. "Al. I. Cuza" Iaşi. I a Mat. (N.S.)* **35**(1989), 231–235.

[12] B.G. Pachpatte, Some inequalities similar to Opial's inequality, *Demonstratio Math.* **26** (1993), 643–647.

[13] B.G. Pachpatte, A note on generalized Opial type inequalities, *Tamkang J. Math.* **24**(1993), 229–235.

[14] J.E. Pečarić, An integral inequality, in *Analysis, Geometry and Groups: A Riemann Legacy Volume* (H.M. Srivastava and Th. M. Rassias, Editors), Part II, Hadronic Press, Palm Harbor, Florida, 1993, pp. 471–478.

[15] J.E. Pečarić and I. Brnetić, Note on generalization of Godunova-Levin-Opial inequality, *Demonstratio Math.* **30**(1997), 545–549.

[16] J.E. Pečarić and I. Brnetić, Note on the generalization of the Godunova-Levin-Opial inequality in several independent variables, *J. Math. Anal. Appl.* **215**(1997), 274–282.

[17] G.I. Rozanova, Integral inequalities with derivatives and with arbitrary convex functions (Russian), *Moskov. Gos. Ped. Inst. Vcen Zap.* **460**(1972), 58–65.

[18] G.J. Sinnamon, Weighted Hardy and Opial-type inequalities, *J. Math. Anal. Appl.* **160** (1991), 434–445.

[19] J. Traple, On a boundary value problem for systems of ordinary differential equations of second order, *Zeszyty Nauk. Uniw. Jagiellon. Prace Mat.* **15**(1971), 159–168.

[20] D. Willett, The existence-uniqueness theorem for an nth order linear ordinary differential equation, *Amer. Math. Monthly* **75**(1968), 174–178.

Editors' Note. The interested reader will find each of the following *additional* references to be useful and relevant in connection with the Opial-type inequalities which are considered in this paper.

[a] R.P. Agarwal and P.Y.H. Pang, Remarks on the generalization of Opial's inequality, *J. Math. Anal. Appl.* **190**(1995), 559–577.

[b] R.P. Agarwal, J. Pečarić and I. Brnetić, Improved integral inequalities in n independent variables, *Comput. Math. Appl.* **33**(8) (1997), 27–38.

[c] R.P. Agarwal and Q. Sheng, Sharp integral inequalities in n independent variables, *Nonlinear Anal.* **26**(2) (1996), 179–210.

[d] I. Brnetić and J. Pečarić, Some new Opial-type inequalities, *Math. Inequal. Appl.* **1**(1998), 385–390.

[e] W.S. Cheung, On integral inequalities of the Sobolev type, *Acquationes Math.* **49**(1995), 153–159.

In particular, Agarwal *et al.* [b] improved many known Opial-type inequalities in n independent variables, while (in a sequel to [b]) Brnetić and Pečarić [d] similarly improve and generalize several Opial-type inequalities (including, for example, those given earlier by Agarwal and Pang [a]).

Controlling the Velocity of Brownian Motion by its Terminal Value

GORAN PESKIR*

Institute of Mathematics, University of Aarhus, Ny Munkegade, 8000 Aarhus, Denmark;
goran@imf.au.dk; http://www.imf.au.dk/~goran;
(Department of Mathematics, University of Zagreb, Bijenička 30, 10000 Zagreb, Croatia)

Abstract. Let $V = (V_t)_{t\geq 0}$ be the Ornstein-Uhlenbeck velocity process solving

$$dV_t = -\beta V_t \, dt + \sigma \, dB_t$$

with $V_0 = 0$, where $\beta > 0$, $\sigma > 0$ and $B = (B_t)_{t\geq 0}$ is a standard Brownian motion. Then the following inequality is satisfied:

$$E\left(\max_{0\leq t\leq \tau} |V_t| \right) \leq s_*(\kappa; \beta, \sigma) + \frac{\kappa}{\beta} E\left(e^{(\beta/\sigma^2)V_\tau^2} - 1 \right)$$

for all bounded stopping times τ of V and all $\kappa > 0$, where $s_*(\kappa; \beta, \sigma) > 0$ is the unique zero point of the maximal solution of the equation

$$y' = \frac{\sigma^2}{2\kappa \displaystyle\int_y^x e^{(\beta/\sigma^2)z^2} \, dz}$$

staying strictly below the diagonal in $\mathbb{R}^2$. This inequality is sharp, and equality can be attained for each $\kappa > 0$. The following estimate is established:

$$s_*(\kappa; \beta, \sigma) \leq \frac{\sigma}{\sqrt{\beta}} \, \Psi^{-1}\left(\frac{\sqrt{\beta}\,\sigma}{2\kappa} \right)$$

where $\Psi(x) = \int_0^x e^{z^2} dz$. In particular, this yields the existence of a universal constant $C \geq \sqrt{2}$ such that

$$E\left(\max_{0\leq t\leq \tau} |V_t| \right) \leq C \frac{\sigma}{\sqrt{\beta}} \sqrt{\log E\left(e^{(\beta/\sigma^2)V_\tau^2} \right)}$$

for all stopping times τ of V for which $(e^{(\beta/\sigma^2)V_{\tau \wedge t}^2})_{t\geq 0}$ is uniformly integrable. This inequality shows that the question of controlling the velocity of Brownian motion by its terminal value can be answered positively. Better versions of this inequality are also derived which go beyond the best value for C.

1. The Ornstein-Uhlenbeck process

The Ornstein-Uhlenbeck *velocity process* $V = (V_t)_{t\geq 0}$ is a diffusion process which is a unique strong solution of the *Langevin* stochastic differential equation

(1.1)
$$dV_t = -\beta V_t \, dt + \sigma \, dB_t$$

where $\beta > 0$, $\sigma > 0$ and $B = (B_t)_{t\geq 0}$ is a standard Brownian motion (see [7] and [1]). This

---

*Centre for Mathematical Physics and Stochastics, supported by the Danish National Research Foundation.

AMS 1980 subject classifications. Primary 60J65, 60E15, 34A34. Secondary 60G40, 60J60, 60G15.

Key words and phrases: The Ornstein-Uhlenbeck velocity process, maximum process, terminal value, optimal stopping, maximal inequality, Brownian motion, the Ornstein-Uhlenbeck position process, Newton's law, Einstein-Smoluchowski theory, first-order nonlinear differential equation, stopping time, diffusion process, Gaussian process, the Langevin stochastic differential equation.

323

T.M. Rassias and H.M. Srivastava (eds.), Analytic and Geometric Inequalities and Applications, 323–333.
© 1999 *Kluwer Academic Publishers.*

process was initially introduced and later used by Ornstein and Uhlenbeck [13] as a model for the velocity of a Brownian particle in order to develop the theory of Brownian motion in accordance with Newtonian particle mechanics; the Ornstein-Uhlenbeck *position process*

$$(1.2) \qquad X_t = x + \int_0^t V_r \, dr$$

is then another model for Brownian motion; if m is the mass of the Brownian particle, then in accordance with *Newton's law* $F = ma$, we may formally rewrite (1.1) as follows

$$(1.3) \qquad m \frac{d^2 X_t}{dt^2} = -m\beta V_t + m\sigma \frac{dB_t}{dt}$$

where $-m\beta V_t$ represents a "frictional" force towards the origin (equilibrium state of velocity zero), and $m\sigma \, (dB_t/dt)$ represents a "fluctuating" white-noise force due to the media. The theory of Brownian motion $B_t \sim N(0, t)$, established earlier by Einstein and Smoluchowski, does not rely upon Newton's law, and does, moreover, make the concept of velocity meaningless (the trajectories of B are nowhere differentiable). For a Brownian motion under no influence of external force, it is known that both theories are in agreement with experiment, and to a large extent offer predictions which are numerically indistinguishable. For Brownian motion under influence of external force, however, the Einstein-Smoluchowski theory breaks down, while the Ornstein-Uhlenbeck theory remains successful (see [9] p.53-78).

1. The solution of (1.1) started at $v \in \mathbb{R}$ is explicitly given by

$$(1.4) \qquad V_t = e^{-\beta t} v + \sigma e^{-\beta t} \int_0^t e^{\beta r} \, dB_r$$

which is easily verified by Itô formula (see also [12] p.361). This shows that the Ornstein-Uhlenbeck velocity process $V = (V_t)_{t \geq 0}$ is a *Gaussian* process satisfying

$$(1.5) \qquad V_t \sim N\left(e^{-\beta t} v, \frac{\sigma^2}{2\beta}(1 - e^{-2\beta t})\right)$$

for all t. Thus, no matter what the initial value v is, the equilibrium state of the process at infinity is distributed according to a single Gaussian law; that is:

$$(1.6) \qquad V_t \to N\left(0, \frac{\sigma^2}{2\beta}\right)$$

as $t \to \infty$. Moreover, if $v \sim N(0, \frac{\sigma^2}{2\beta})$ in (1.4) is taken independently from B, then $V = (V_t)_{t \geq 0}$ is a *stationary* Gaussian process (sometimes called a *coloured noise*). In fact, these processes are the only stationary Gaussian Markov processes (see [12] p.81).

2. By (1.4) we find that the Ornstein-Uhlenbeck position process (1.2) is explicitly given by

$$(1.7) \qquad X_t = x + \frac{1}{\beta}\left(v - V_t + \sigma B_t\right).$$

It is a Gaussian process with mean $E(X_t) = x + \beta^{-1}(1 - e^{-\beta t}) v$ and variance $D(X_t) = (\sigma^2/\beta^2)$ $t^2 + (\sigma^2/2\beta^3)(-3 + 4e^{-\beta t} - e^{-2\beta t})$. Thus, if $\beta \to \infty$ and $\sigma \to \infty$ so that $\hat{\sigma} := \sigma/\beta$ remains constant, then we have the following equilibrium relation

$$(1.8) \qquad \left(X_t\right)_{t\geq 0} \; \tilde{\longrightarrow} \; \left(x+\hat{\sigma}B_t\right)_{t\geq 0}$$

no matter what the initial velocity v is. This indicates that the Einstein-Smoluchowski (idealized) theory is a good approximation of the Ornstein-Uhlenbeck (Newtonian) theory for a Brownian particle under no influence of external force.

3. The Ornstein-Uhlenbeck velocity process (1.4) (started at $v=0$ for simplicity) can be obtained as a time-space change of standard Brownian motion:

$$(1.9) \qquad V_t = \frac{\sigma}{\sqrt{2\beta}}\, e^{-\beta t}\, B(e^{2\beta t}-1) \; .$$

If $V=(V_t)_{t\geq 0}$ is a stationary Ornstein-Uhlenbeck velocity process started at $v \sim N(0,\frac{\sigma^2}{2\beta})$ independently from B, then the following representation is valid:

$$(1.10) \qquad V_t = \frac{\sigma}{\sqrt{2\beta}}\, e^{-\beta t}\, B(e^{2\beta t}) \; .$$

These facts are well-known and can easily be verified by means of standard time-change techniques; they appear naturally in some *nonlinear optimal stopping* problems (see [10]).

4. Similarly to the fact that Brownian motion can be approximated by means of a discrete random walk, the Ornstein-Uhlenbeck velocity process can be approximated by an *Ehrenfest urn* model for diffusion of particles through a porous membrane (see [6] p.171-172).

5. The Ornstein-Uhlenbeck velocity process has found a large number of applications in modelling various other random phenomena. Perhaps the most notable lately are some models for *stochastic volatility* (see [8] p.154-155).

2. Description of the problem

Given an Ornstein-Uhlenbeck velocity process started at zero, our main objective in this note is to study the question on how to control the maximal value of this process taken up to a random instant of time by means of its terminal value.

Our aim is motivated by the fact that the Ornstein-Uhlenbeck velocity process is neither a martingale, nor a positive submartingale, for which such bounds are successfully established and play a significant role in probability theory (see [12] p.50-52), but nevertheless, due to its special relation to standard Brownian motion (Section 1), we feel that such a question is of interest and can be answered positively. Moreover, in this note we deal only with the Ornstein-Uhlenbeck *velocity* process, and yet this is another interesting aspect of the question above, not present in the martingale framework of standard Brownian motion B where the concept of velocity is meaningless. Thus, it is interesting to see to which extent the control of the maximal position of Brownian motion B by means of the terminal value carries over to a similar control of the maximal velocity; our first result below ((3.3)+(3.4)) shows that we have to pass from a polynomial in the former case (see [3] and [4]) to the exponential function of the terminal value in the latter; this dichotomy is then balanced in (3.12) by another appearance of the logarithm function (see (3.19)+(3.20)).

Our approach is based on the fact that the Ornstein-Uhlenbeck velocity process satisfies the strong Markov property. It would be interesting to see how this compares to any other approach which would be based more substantially on the Gaussian property too (see also [5]).

1. To formulate the above problem more precisely, let us assume we are given an Ornstein-Uhlenbeck velocity process $V = (V_t)_{t\geq 0}$ satisfying (1.1) and starting at some $v \in \mathbb{R}$, where $\beta > 0$, $\sigma > 0$ and $B = (B_t)_{t\geq 0}$ is a standard Brownian motion defined on $(\Omega, \mathcal{F}, P)$. Then

$$(2.1) \qquad\qquad |V| = (|V_t|)_{t\geq 0}$$

is a non-negative diffusion process having the same infinitesimal characteristics in $(0, \infty)$ as the diffusion V, and 0 as an instantaneously reflecting boundary point. Due to the symmetry of the process V, it is no restriction to assume that $v \geq 0$.

Associate with $|V|$ its maximum process

$$(2.2) \qquad\qquad S_t = \max_{0 \leq r \leq t} |V_t|$$

and note that $S_0 = |V_0| = v$ under P. Setting $S_t := S_t \vee s$ for $s \geq v$, the process S may start at any $s \geq v$, and to indicate that under P the process $(|V_t|, S_t)$ starts at (v, s), we write $P_{v,s}$ instead of P; thus $P_{v,s}(|V_0| = v, S_0 = s) = 1$ for any given $s \geq v \geq 0$.

2. Given a continuous map $v \mapsto c(v) > 0$, consider the optimal stopping problem with payoff

$$(2.3) \qquad\qquad \Delta(v, s) = \sup_{\tau} E_{v,s}\left(S_\tau - \int_0^\tau c(|V_t|)\, dt \right)$$

where the supremum is taken over all stopping times τ of V for which the integral has finite expectation. It follows from the general result of [11] that this problem has a solution (the payoff is finite and there exists an optimal stopping time), if and only if the following differential equation

$$(2.4) \qquad\qquad g'(s) = \frac{\sigma^2 L'(g(s))}{2 c(g(s)) (L(s) - L(g(s)))} \qquad (s \in \mathbb{R})$$

admits a maximal solution which stays strictly below the diagonal in $\mathbb{R}^2$ (the *maximality principle*). In this equation we denote by L the scale function of $|V|$; it is explicitly given by

$$(2.5) \qquad\qquad L(v) = \int_0^v e^{(\beta/\sigma^2) z^2}\, dz$$

for all $v \geq 0$. If $s \mapsto g_*(s)$ denote the maximal solution of (2.4), then the stopping time

$$(2.6) \qquad\qquad \tau_* = \inf\{ t > 0 : |V_t| \leq g_*(S_t) \}$$

is optimal for the problem (2.3), and the payoff is explicitly given by

$$(2.7) \qquad \Delta(v, s) = s_* + \int_0^v (L(v) - L(z)) c(z)\, m(dz) \ , \quad \text{if } 0 \leq v \leq s \leq s_*$$

$$= s + \int_{g_*(s)}^v (L(v) - L(z)) c(z)\, m(dz) \ , \quad \text{if } g_*(s) \leq v \leq s \text{ and } s \geq s_*$$

$$= s \ , \quad \text{if } 0 \leq v \leq g_*(s) \text{ and } s \geq s_*$$

where s_* is the zero point of $s \mapsto g_*(s)$. In (2.7) we denote by $m(dz)$ the *speed measure* of $|V|$; it is explicitly given by

$$(2.8) \qquad m(dz) = \frac{2\,dz}{\sigma^2 L'(z)} .$$

3. In view of our problem formulated above, the crucial question becomes how to choose the cost function $v \mapsto c(v)$ in the optimal stopping problem (2.3). The following considerations suggest how this can be done.

From (1.1) we read that the infinitesimal generator of V is given by

$$(2.9) \qquad \mathbb{L}_V = -\beta v \frac{\partial}{\partial v} + \frac{\sigma^2}{2} \frac{\partial^2}{\partial v^2} .$$

If $F(x) = G(x^2)$ is sufficiently smooth, then by Itô formula we find:

$$(2.10) \qquad F(V_t) = F(v) + \int_0^t (\mathbb{L}_V F)(V_r)\,dr + \int_0^t F'(V_r)\,dB_r .$$

By applying the optional sampling theorem (see [12] p.65) to the continuous local martingale $(M_t)_{t \geq 0} = (\int_0^t F'(V_r)\,dB_r)_{t \geq 0}$ in (2.10), we see that

$$(2.11) \qquad E_{v,s}\left(\int_0^\tau (\mathbb{L}_V F)(V_r)\,dr \right) = E_{v,s}(F(V_\tau)) - F(v)$$

for all stopping times τ of V for which the following condition is fulfilled:

$$(2.12) \qquad E_{v,s}\left(\int_0^\tau \left(F'(V_r) \right)^2 dr \right)^{1/2} < \infty .$$

Yet another sufficient condition for (2.11) is obtained similarly by passing to a localization sequence of stopping times for $(M_t)_{t \geq 0}$; this shows that (2.11) is satisfied, whenever $(F(V_{\tau \wedge t}))_{t \geq 0}$ is uniformly integrable, and $v \mapsto (\mathbb{L}_V F)(v)$ is non-negative for instance.

Motivated in part by (2.4), define the cost function in (2.3) by $c(v) = \kappa\,L'(v) = \kappa\,e^{(\beta/\sigma^2)v^2}$, where $\kappa > 0$ is arbitrary and not specified, and note that $F(v) = (\kappa/\beta)(e^{(\beta/\sigma^2)v^2} - 1)$ satisfies $\mathbb{L}_V F = c$. Thus (2.11) shows that

$$(2.13) \qquad E\left(\int_0^\tau e^{(\beta/\sigma^2)V_r^2}\,dr \right) = \frac{1}{\beta} E\left(e^{(\beta/\sigma^2)V_\tau^2} - 1 \right)$$

for all stopping times τ of V for which either of the conditions specified above holds; in particular, note that (2.13) holds for all bounded stopping times τ of V . Inserting $c(v) = \kappa\,e^{(\beta/\sigma^2)v^2}$ into (2.4), we see that this equation takes the following form:

$$(2.14) \qquad g'(s) = \frac{\sigma^2}{2\kappa \displaystyle\int_{g(s)}^s e^{(\beta/\sigma^2)z^2}\,dz} .$$

It is a nonlinear equation, and our problem in essence is reduced to detecting the maximal solution of this equation staying strictly below the diagonal in $I\!R^2$. This is the key argument in the proof of the main result which we state in the next section.

3. The result and proof

The main result of this note is contained in the following theorem. Observe that the simple estimate (3.2) stated below is considerably improved later on in (3.11), but yet we want to point it out for the useful role it plays in the derivation of (3.3), as well as to highlight technicalities of the final improvement (3.12). The first part of the theorem which we state now should be read together with the second part which is stated below; see also Corollary 3.2.

Theorem 3.1

(I): Let $V = (V_t)_{t\geq0}$ be the Ornstein-Uhlenbeck velocity process starting at zero and satisfying (1.1). Then the following inequality is satisfied:

$$(3.1) \qquad E\left(\max_{0\leq t\leq\tau} |V_t|\right) \leq s_*(\kappa; \beta, \sigma) + \kappa E\left(\int_0^\tau e^{(\beta/\sigma^2)V_r^2} dr\right)$$

for all stopping times τ of V and all $\kappa > 0$, where $s_*(\kappa; \beta, \sigma) > 0$ is the unique zero point of the maximal solution $s \mapsto g_*(s)$ of the equation (2.14) staying strictly below the diagonal in $I\!R^2$. This inequality is sharp, and equality can be attained for each $\kappa > 0$ at $\tau_*(\kappa)$ from (2.6).

The following estimate is valid (see (3.11) below):

$$(3.2) \qquad s_*(\kappa; \beta, \sigma) \leq \frac{\sigma^2}{2\kappa}$$

for all $\beta > 0$, where equality is attained if $\beta \downarrow 0$. In particular, this yields (see (3.12) below)

$$(3.3) \qquad E\left(\max_{0\leq t\leq\tau} |V_t|\right) \leq \sigma\sqrt{2E\left(\int_0^\tau e^{(\beta/\sigma^2)V_r^2} dr\right)}$$

for all stopping times τ of V. The constant $\sigma\sqrt{2}$ is best possible in this inequality.

(II) : The following identity holds in either (3.1) or (3.3):

$$(3.4) \qquad E\left(\int_0^\tau e^{(\beta/\sigma^2)V_r^2} dr\right) = \frac{1}{\beta} E\left(e^{(\beta/\sigma^2)V_\tau^2} - 1\right)$$

for all stopping times τ of V satisfying either

$$(3.5) \qquad E\left(\int_0^\tau V_r^2 e^{2(\beta/\sigma^2)V_r^2} dr\right)^{1/2} < \infty$$

or that the process $(e^{(\beta/\sigma^2)V_{\tau\wedge t}^2})_{t\geq0}$ is uniformly integrable, the latter condition being more general. In particular, the condition (3.5) is satisfied, and thus (3.4) holds as well, whenever the stopping time τ is bounded.

Proof. (I): Motivated by our considerations in the previous section, consider the optimal stopping problem (2.3) with the cost function $c(v) = \kappa e^{(\beta/\sigma^2)v^2}$ where $\kappa > 0$. Then we know that this problem has a solution, if and only if the differential equation (2.14) admits a maximal solution which stays strictly below the diagonal in $I\!R^2$. By using a simple comparison argument we will now prove that such a solution exists.

1. Consider the following equation:

$$(3.6) \qquad h'(s) = \frac{\sigma^2}{2\kappa\,(s-h(s))} \qquad (s \in I\!R) \; .$$

Observe that $h_*(s) = s - \sigma^2/2\kappa$ solves this equation, and moreover it is easily verified that $s \mapsto h_*(s)$ is the maximal solution of (3.6) which stays strictly below the diagonal in $I\!R^2$. To do so, pass to the inverse function $t \mapsto h^{-1}(t)$ in (3.6), and note that this equation becomes linear, and thus admits the general solution in closed form which is found easily.

2. A simple comparison of (2.14) and (3.6) shows that

$$(3.7) \qquad g'(s) \leq h'(s)$$

for all $s \in I\!R$. Thus, if we consider (2.14) as the initial value problem with $g(\sigma^2/2\kappa) = 0$, and use Picard's theorem step-by-step to the right from $\sigma^2/2\kappa$, we obtain a solution $s \mapsto g_1(s)$ of (2.14) satisfying $g_1(\sigma^2/2\kappa) = 0$ and staying strictly below $s \mapsto h_*(s)$ on $[\sigma^2/2\kappa, \infty)$; the solution $s \mapsto g_1(s)$ cannot hit $s \mapsto h_*(s)$ at any $s > \sigma^2/2\kappa$, since this would violate (3.7); observe also that $g_1'(\sigma^2/2\kappa) \leq 1$. Thus, in particular, the solution $s \mapsto g_1(s)$ stays strictly below the diagonal on $[\sigma^2/2\kappa, \infty)$.

3. We now claim that this solution extends to $(-\infty, \sigma^2/2\kappa)$ so that there it also stays below the diagonal. This is again seen by applying Picard's theorem step-by-step to the left from $\sigma^2/2\kappa$. By doing so we obtain a solution $s \mapsto g_1(s)$ on $(-\infty, \sigma^2/2\kappa)$ which stays between the diagonal and $s \mapsto h_*(s)$; it cannot hit the diagonal, since this would violate the fact of (2.14) that $g_1'(s+) = +\infty$ if $g_1(s+) = s$, neither it can hit $s \mapsto h_*(s)$, since this would violate (3.7). (From (2.14) it is also clear that $g_1(s)$ approaches the diagonal very rapidly as $s \to -\infty$.)

4. In this way we have proved that the equation (2.14) admits a global solution $s \mapsto g_1(s)$ which stays strictly below the diagonal in $I\!R^2$. Passing then to the equivalent integral formulation of (2.14), and using the monotone convergence theorem, we may conclude that there exists a maximal solution $s \mapsto g_*(s)$ of (2.14) which stays strictly below the diagonal in $I\!R^2$. The existence of the maximal solution implies that the problem (2.3) has a solution; from (2.7) with $v = s = 0$ we find that (3.1) holds. Observe that $s \mapsto g_*(s)$ depends on κ , β and σ ; therefore we write $s_*(\kappa; \beta, \sigma)$ to denote its unique zero point. Finally, since $g_1(\sigma^2/2\kappa) = 0$, by the maximality property of $s \mapsto g_*(s)$ we see that (3.2) holds. Inserting this into (3.1), and minimising the right-hand side over all $\kappa > 0$, we obtain (3.3) by easy calculations. Observe also that equality in (3.2) is attained if we let $\beta \downarrow 0$.

(II): These facts follow from our considerations about (2.11)-(2.13) in the previous section. The proof is complete. $\qquad\square$

Remarks: 1. Variational ideas applied in the theorem above are generally well-known. In a study of similar maximal inequalities for Bessel processes they were used very effectively in [2]. Similar ideas are also used in [3] and [4].

2. By (1.5) we can compute that

$$(3.8) \qquad E\left(e^{(\beta/\sigma^2)V_t^2}\right) = e^{\beta t}$$

for all t . Inserting this via (3.4) into the right-hand side of (3.3), we obtain the following estimate:

$$(3.9) \qquad E\left(\max_{0 \le r \le t} |V_r|\right) \le \sigma \sqrt{\frac{2}{\beta}\left(e^{\beta t} - 1\right)}$$

for all t . Letting $\beta \downarrow 0$, and using that $V_t \to \sigma B_t$, we see that (3.9) reduces to the inequality $E(S_t) \le \sqrt{2t}$ where $S_t = \max_{0 \le r \le t} |B_r|$; recall that $E(|B_t|) = \sqrt{(2/\pi)\,t}$.

3. Similarly, if we let $\beta \downarrow 0$ in (3.3), we obtain the following inequality:

$$(3.10) \qquad E\left(\max_{0 \le r \le \tau} |B_r|\right) \le \sqrt{2E(\tau)}$$

for all stopping times τ of B . This inequality is known to be sharp (see [2]).

4. While for standard Brownian motion B we have $B_t \sim N(0,t)$ and therefore the variance of B_t equals t and tends to infinity as $t \to \infty$, for the Ornstein-Uhlenbeck velocity process V satisfying (1.1) we have (1.5) and thus the variance of V_t remains bounded by the constant $\sigma^2/2\beta$. Thus it is not surprising that to control the maximum of the process we have to pass from a polynomial in the former case (recall results of [3] and [4]) to the exponential function of the terminal value in the latter (see also Remark 8 below). It may also be observed that for $V_\infty \sim N(0, \sigma^2/2\beta)$ we have $E(e^{\lambda V_\infty^2}) < \infty$ if and only if $\lambda < \beta/\sigma^2$; this is in agreement with the apparent fact that the left-hand side of (3.1) or (3.3) tends to infinity as τ does so.

5. In view of (1.9) and the law of iterated logarithm for Brownian motion, we see that the estimate (3.9) is much too crude; we expect the left-hand side in (3.9) to increase at least as slow as $\sqrt{\log(t)}$ as $t \to \infty$. The reason for this lack of precision comes from the estimate (3.2). There we expect $s_*(\kappa; \beta, \sigma)$ to increase to infinity at least as slow as $\log(1/\kappa)$ or even $\sqrt{\log(1/\kappa)}$ when $\kappa \downarrow 0$. For this reason we continue by obtaining a sharper estimate of $s_*(\kappa; \beta, \sigma)$ which is aimed to match the law-of-iterated-logarithm prediction.

Theorem 3.1 (continued)
The following refinement of (3.2) is valid (see (3.23) below):

$$(3.11) \qquad s_*(\kappa; \beta, \sigma) \le \frac{\sigma}{\sqrt{\beta}}\,\Psi^{-1}\left(\frac{\sqrt{\beta}\,\sigma}{2\kappa}\right)$$

for all $\kappa > 0$, $\beta > 0$ and $\sigma > 0$, where $\Psi(x) = \int_0^x e^{z^2} dz$. This estimate is sharp.
Consequently, the following refinement of (3.3) is valid (see (3.19) and (3.20) below):

$$(3.12) \qquad E\left(\max_{0 \le t \le \tau} |V_t|\right) \le \frac{\sigma}{\sqrt{\beta}}\left(\left(\Psi^{-1} \circ \rho^{-1}\right)\left(\beta E(I_\tau)\right) + \frac{1}{2}\,\frac{\beta E(I_\tau)}{\rho^{-1}\left(\beta E(I_\tau)\right)}\right)$$

for all stopping times τ of V , where $\rho(x) = 2x^2 \exp\left(-(\Psi^{-1}(x))^2\right)$ and $I_\tau = \int_0^\tau e^{(\beta/\sigma^2)V_r^2}\,dr$. This inequality is sharp.

Proof. We shall establish (3.11) by applying the following trick which will enable us to extend the simple argument used in the proof above when considering the initial value problem (2.14) with $g(\sigma^2/2\kappa) = 0$. It was observed above that the solution $s \mapsto g_1(s)$ satisfies $g_1'(\sigma^2/2\kappa) \leq 1$.

On the other hand, if we compute the second derivative of any function $s \mapsto g(s)$ satisfying (2.14), we find that $g''(s) \leq 0$ if and only if

$$(3.13) \qquad L'(s) - L'\big(g(s)\big)g'(s) \geq 0$$

where $L'(s) = e^{(\beta/\sigma^2)s^2}$. In particular, we see that $g'(s) \leq 1$ will imply that $g''(s) \leq 0$ whenever s satisfies $g(s) < s$. This fact combined with our observation above that $g_1'(\sigma^2/2\kappa) \leq 1$ motivates us to consider the initial value problem (2.14) with $g(s_*) = 0$, and find the smallest positive $s_* \leq \sigma^2/2\kappa$ satisfying this property with $g'(s_*) \leq 1$. In other words, we shall be looking for the smallest positive $s_* \leq \sigma^2/2\kappa$ satisfying

$$(3.14) \qquad \frac{\sigma^2}{2\kappa} \leq \int_0^{s_*} e^{(\beta/\sigma^2)z^2}\, dz \ .$$

The solution $s \mapsto g_2(s)$ obtained from (2.14) will then satisfy $g_2(s_*) = 0$, and clearly this solution will stay below the straight line $s \mapsto s - s_*$ for all $s > s_*$; otherwise $g_2''(s) \leq 0$ for some $s > s_*$ would be violated. In particular, the solution $s \mapsto g_2(s)$ stays below the diagonal in $\mathbb{R}^2$, and thus we may conclude that $s_*(\kappa; \beta, \sigma) \leq s_*$. This now can be quantitatively expressed by (3.11) if one makes use of (3.14).

Inserting (3.11) into (3.1), and minimising the right-hand side over all $\kappa > 0$, we obtain (3.12). This verification is somewhat lengthy, but quite straightforward. The proof is complete. $\qquad \square$

6. To obtain a better understanding of (3.12), the following estimates show useful:

$$(3.15) \qquad \frac{e^{x^2}-1}{2x} \leq \int_0^x e^{z^2}\, dz \leq \frac{e^{x^2}}{x} \qquad (\forall x > 0) \ .$$

By using these estimates it is possible to verify through elementary calculations that

$$(3.16) \qquad \lim_{x \to \infty} \frac{(\Psi^{-1} \circ \rho^{-1})(x)}{\sqrt{\log(x)}} = 1$$

$$(3.17) \qquad \frac{x}{\rho^{-1}(x)} \leq 2 \qquad (\forall x > 0)$$

$$(3.18) \qquad \lim_{x \to 0} \frac{x}{\rho^{-1}(x)} = \lim_{x \to \infty} \frac{x}{\rho^{-1}(x)} = 0 \ .$$

Observe also that the map $x \mapsto \rho(x)$ is strictly increasing on $[0, \infty)$.

Corollary 3.2

Let $V = (V_t)_{t \geq 0}$ be the Ornstein-Uhlenbeck velocity process starting at zero and satisfying (1.1). Then there exists a universal constant $C > 0$ such that (see Remark 9 below)

$$(3.19) \qquad E\left(\max_{0 \leq t \leq \tau} |V_t| \right) \leq C \frac{\sigma}{\sqrt{\beta}} \sqrt{\log\left(1 + \beta E(I_\tau)\right)}$$

for all stopping times τ of V , where $I_\tau = \int_0^\tau e^{(\beta/\sigma^2)V_r^2}\,dr$. In particular, we have

(3.20)
$$E\left(\max_{0\le t\le\tau}|V_t|\right) \le C\,\frac{\sigma}{\sqrt{\beta}}\,\sqrt{\log E\left(e^{(\beta/\sigma^2)V_\tau^2}\right)}$$

for all stopping times τ of V satisfying either (3.5) or that the process $(e^{(\beta/\sigma^2)V_{\tau\wedge t}^2})_{t\ge0}$ is uniformly integrable.

Proof. From (3.12) with (3.16) and (3.17) we have

(3.21)
$$E\left(\max_{0\le t\le\tau}|V_t|\right) \le (1+\varepsilon)\,\frac{\sigma}{\sqrt{\beta}}\left(\sqrt{\log\left(\beta E(I_\tau)\right)}+1\right)$$

whenever $E(I_\tau) \ge M_\varepsilon$ (large enough). Since $1+\sqrt{\log(\beta x)} \le A\sqrt{\log(1+\beta x)}$ for $\beta x \ge M_A$ (large enough) where $A > 1$, from (3.21) we get

(3.22)
$$E\left(\max_{0\le t\le\tau}|V_t|\right) \le A\,(1+\varepsilon)\,\frac{\sigma}{\sqrt{\beta}}\left(\sqrt{\log\left(1+\beta E(I_\tau)\right)}\right)$$

whenever $E(I_\tau) \ge M_\varepsilon\vee M_A$. Finally, since $\sqrt{\beta x} \le B\sqrt{\log(1+\beta x)}$ for $0\le\beta x\le M_\varepsilon\vee M_A$ for some $B>0$ large enough, combining (3.3) and (3.22) we obtain (3.19) with $C = A(1+\varepsilon)\vee B\sqrt{2}$. The inequality (3.20) follows from (3.19) and Part II of Theorem 3.1. The proof is complete. $\square$

7. Observe in (3.11) that $\Psi^{-1}(x) \le x$ for all $x \ge 0$; thus the right-hand side in (3.11) is always smaller than the right-hand side in (3.2); therefore the right-hand side in (3.12) is always smaller than the right-hand side in (3.3). In fact, it is possible to verify by (3.15) that

(3.23)
$$\Psi^{-1}(x) \le (1+\varepsilon)\sqrt{\log(x)}$$

for all $x \ge x_\varepsilon$ (large enough); thus the improvement (3.12) upon (3.3) is substantial for large $E(I_\tau)$; this is expressed in a more readable way through the appearance of the logarithm function in (3.19); note also from the proof above that (3.19) has been obtained by merging (3.12) and (3.3) together; observe from the facts just pointed out, however, that no matter what value for C in (3.19) is found the best possible in the proof above, the inequality (3.12) will always give at least as good bound, although not as elegant in appearance.

8. Note by Jensen's inequality that $\log E(e^{(\beta/\sigma^2)V_\tau^2}) \ge (\beta/\sigma^2)\,E(V_\tau^2)$; since the variance of V_t remains bounded over all t , and the left-hand side in (3.20) tends to infinity when τ does so, we see that $\log E(e^{(\beta/\sigma^2)V_\tau^2})$ in (3.20) cannot generally be replaced by $(\beta/\sigma^2)\,E(V_\tau^2)$; observe, however, that the inequality (3.20) may be formally viewed as such an estimate if we neglect the expectation sign; for instance, if V_τ is identically constant, then (3.20) says that the maximum of $|V_t|$ over $0\le t\le\tau$ is controlled by the terminal value $|V_\tau|$ whenever the uniform integrability condition is fulfilled; in view of the specific form of the drift term in (1.1) which always directs the process to the origin, it does not come quite with surprise.

9. By using a different method it is possible to prove that the second expectation sign in (3.19) can be pulled out in front of the square-root and logarithm sign (see [5]); in view of Jensen's inequal-

ity this bound is better; moreover, it is possible to prove that this inequality is two-sided; although it is difficult to compute these bounds explicitly, the knowledge of their existence helps to grasp the real nature of the error when passing to the terminal-value bound in (3.12) or (3.20); observe, however, that no such maneuver of the expectation sign can be applied directly to (3.12) or (3.20).

References

[1] DOOB, J. L. (1942). The Brownian movement and stochastic equations. *Ann. of Math.* 43 (351-369).

[2] DUBINS, L. E. SHEPP, L. A. *and* SHIRYAEV, A. N. (1993). Optimal stopping rules and maximal inequalities for Bessel processes. *Theory Probab. Appl.* 38 (226-261).

[3] GRAVERSEN, S. E. *and* PESKIR, G. (1995). On Doob's maximal inequality for Brownian motion. *Research Report* No. 337, *Dept. Theoret. Statist. Aarhus* (13 pp). *Stochastic Process. Appl.* 69, 1997 (111-125).

[4] GRAVERSEN, S. E. *and* PESKIR, G. (1996). Optimal stopping in the $L \log L$-inequality of Hardy and Littlewood. *Research Report* No. 360, *Dept. Theoret. Statist. Aarhus* (12 pp). *Bull. London Math. Soc.* 30, 1998 (171-181).

[5] GRAVERSEN, S. E. *and* PESKIR, G. (1998). Maximal inequalities for the Ornstein-Uhlenbeck process. *Research Report* No. 393, *Dept. Theoret. Statist. Aarhus* (8 pp).

[6] KARLIN, S. *and* TAYLOR, H. M. (1981). *A Second Course in Stochastic Processes.* Academic Press.

[7] LANGEVIN, P. (1908). Sur la theorie du mouvement brownien. *C. R. Acad. Sci. Paris* 146 (530-533).

[8] MUSIELA, M. *and* RUTKOWSKI, M. (1997). *Martingale Methods in Financial Modelling.* Springer-Verlag.

[9] NELSON, E. (1967). *Dynamical Theories of Brownian Motion.* Princeton Univ. Press.

[10] PEDERSEN, J. L. *and* PESKIR, G. (1997). Solving non-linear optimal stopping problems by the method of time-change. *Research Report* No. 390, *Dept. Theoret. Statist. Aarhus* (24 pp).

[11] PESKIR, G. (1997). Optimal stopping of the maximum process: The maximality principle. *Research Report* No. 377, *Dept. Theoret. Statist. Aarhus* (30 pp). To appear in *Ann. Probab.*

[12] REVUZ, D. *and* YOR, M. (1994). *Continuous Martingales and Brownian Motion.* Springer-Verlag.

[13] UHLENBECK, G. E. *and* ORNSTEIN, L. S. (1930). On the theory of Brownian motion. *Physical Review* 36 (823-841).

CHEBYSHEV POLYNOMIALS WITH INTEGER COEFFICIENTS *

IGOR E. PRITSKER[†]

Department of Mathematics, Case Western Reserve University, 10900 Euclid Avenue, Cleveland, Ohio 44106-7058, U.S.A.

Dedicated to Professor R. S. Varga on the occasion of his seventieth birthday.

Abstract. We study the asymptotic structure of polynomials with integer coefficients and smallest uniform norms on an interval of the real line. Introducing methods of the weighted potential theory into this problem, we improve the bounds for the multiplicities of some factors of the integer Chebyshev polynomials.

1. Introduction

Let $\mathcal{P}_n(\mathbb{C})$ and $\mathcal{P}_n(\mathbb{Z})$ be the sets of algebraic polynomials of degree at most n, respectively with complex and with integer coefficients. Define the uniform norm on the interval $[a, b] \subset \mathbb{R}$ by

$$\|f\|_{[a,b]} := \max_{x \in [a,b]} |f(x)|.$$

It is very well known that the *Chebyshev polynomial*

$$T_n(x) := 2^{1-n} \cos(n \arccos x)$$

is a monic polynomial of degree n, which minimizes the uniform norm on $[-1, 1]$ in the class of all *monic* polynomials from $\mathcal{P}_n(\mathbb{C})$ (see [2], [12] and

---

1991 AMS Subject Classification. Primary 30C10, 11C08; Secondary 31A05, 31A15.

Key words and phrases. Chebyshev polynomials; Integer Chebyshev constant; Integer transfinite diameter; Multiple factors; Asymptotics; Potentials; Weighted polynomials.

[†]Research supported in part by the National Science Foundation grant DMS-9707359.

T.M. Rassias and H.M. Srivastava (eds.), Analytic and Geometric Inequalities and Applications, 335–348.
© *1999 Kluwer Academic Publishers.*

[15]). The case of an arbitrary interval $[a, b] \subset \mathbb{R}$ can be reduced to that of $[-1, 1]$ by a change of variable. Thus we immediately obtain that

$$t_n(x) := \left(\frac{b-a}{2}\right)^n T_n\left(\frac{2x-a-b}{b-a}\right)$$

is a monic polynomial with the smallest uniform norm on $[a, b]$ among all *monic* polynomials from $\mathcal{P}_n(\mathbb{C})$. Clearly,

$$\|t_n\|_{[a,b]} = 2\left(\frac{b-a}{4}\right)^n, \qquad n \in \mathbb{N}, \tag{1.1}$$

and the *Chebyshev constant* for $[a, b]$ is given by

$$cheb([a, b]) := \lim_{n\to\infty} \|t_n\|_{[a,b]}^{1/n} = \frac{b-a}{4}. \tag{1.2}$$

Chebyshev polynomials and Chebyshev constant represent very classical topics in analysis. These ideas have applications in many areas of mathematics, see [2], [12] and [15]. We remark that the Chebyshev constant of a compact set in $\mathbb{C}$ is equal to its transfinite diameter and to its logarithmic capacity (cf. [18, pp. 71-75] for the general definitions and a discussion).

A corresponding minimization problem in the class of polynomials with integer coefficients $\mathcal{P}_n(\mathbb{Z})$ also has a long and interesting history, surveyed in [14, Ch. 10] and [3]. An *integer Chebyshev polynomial* $q_n \in \mathcal{P}_n(\mathbb{Z})$ is defined in this case as follows:

$$\|q_n\|_{[a,b]} = \inf_{0\neq p_n \in \mathcal{P}_n(\mathbb{Z})} \|p_n\|_{[a,b]}, \tag{1.3}$$

where the inf in (1.3) is taken over all polynomials from $\mathcal{P}_n(\mathbb{Z})$, which are not identically zero. Further, one can define the *integer Chebyshev constant* (integer transfinite diameter) for $[a, b]$ similarly to (1.2)

$$inch([a, b]) := \lim_{n\to\infty} \|q_n\|_{[a,b]}^{1/n}. \tag{1.4}$$

It is not difficult to see that the above limit exists (cf. [14, Ch. 10] or [3]). Observe from (1.1) and (1.3) that if $b - a \geq 4$ then $q_n(x) \equiv 1$ for any $n \in \mathbb{N}$ and $inch([a, b]) = 1$. However, if $b - a < 4$ then

$$\frac{b - a}{4} = cheb([a, b]) \leq inch([a, b]). \tag{1.5}$$

On the other hand, the results of Hilbert [10] and Fekete [5] imply that

$$inch([a, b]) \leq \sqrt{\frac{b - a}{4}} \tag{1.6}$$

(see [3]). The exact value of the integer Chebyshev constant and an explicit (or even asymptotic) form of the integer Chebyshev polynomials is not known for any $[a, b]$ with $b - a < 4$. Perhaps the most studied case, due to the interest in the distribution of prime numbers, is the case of $[0, 1]$ (cf. [14, Ch. 10], [3] and [4]). The best known bounds for $inch([0, 1])$ are as follows:

$$0.42072638 < inch([0, 1]) \leq 0.42347945. \tag{1.7}$$

The lower bound in (1.7) is obtained with the help of the Gorshkov-Wirsing polynomials (see [14, Ch. 10]). It was believed to be the precise value of $inch([0, 1])$, but Borwein and Erdélyi [3] recently showed that there must be a *strict* inequality on the left of (1.7). The upper bound can be found from the very definition of the integer Chebyshev constant in (1.3)-(1.4), using various optimization techniques. Although this is by no means straightforward, both theoretically and practically, this nevertheless becomes more accessible for computations with growing power of modern computers. Thus, the upper bound in (1.7) has recently been improved several times (cf. [3], [6], [7] and [9]). The value in (1.7) is taken from [9], and to our knowledge is the best computed upper bound.

2. Asymptotic structure of the integer Chebyshev polynomials

We are interested in the asymptotic structure of the polynomials $Q_n \in \mathcal{P}_n(\mathbb{Z})$ satisfying

$$\|Q_n\|_{[0,1]} = \inf_{0 \neq p_n \in \mathcal{P}_n(\mathbb{Z})} \|p_n\|_{[0,1]}, \quad n \in \mathbb{N}. \tag{2.1}$$

This problem was originally proposed by A. O. Gelfond (cf. [1]). It is known that the polynomials Q_n satisfying (2.1) have factors that tend to repeat and to increase in power as $n \to \infty$ (see [14, Ch. 10] and [3] for a discussion). In particular, Aparicio (cf. Theorem 3 in [1]) showed that if $\{Q_n\}_{n=1}^{\infty} \subset \mathcal{P}_n(\mathbb{Z})$ satisfy (2.1), then

$$Q_n(x) = (x(1-x))^{[\alpha_1 n]}(2x-1)^{[\alpha_2 n]}(5x^2 - 5x + 1)^{[\alpha_3 n]}R_n(x), \quad \text{as } n \to \infty, \tag{2.2}$$

where

$$\alpha_1 \geq 0.1456, \quad \alpha_2 \geq 0.0166 \quad \text{and} \quad \alpha_3 \geq 0.0037, \tag{2.3}$$

and $R_n \in \mathcal{P}_n(\mathbb{Z})$, $n \in \mathbb{N}$. Borwein and Erdélyi proved that

$$\alpha_1 \geq 0.26 \tag{2.4}$$

in (2.2) (see Theorem 3.1 of [3]). Flammang, Rhin and Smyth [8] recently generalized the ideas of [3] and obtained the following lower bounds

$$\alpha_1 \geq 0.264151, \quad \alpha_2 \geq 0.021963 \quad \text{and} \quad \alpha_3 \geq 0.005285. \tag{2.5}$$

They also considered six additional factors of $Q_n(x)$ and studied other intervals (cf. [8] for the details).

We use the methods of the weighted potential theory, developed during the last two decades, to study the integer Chebyshev problem and to improve the bounds for α_1 and α_2. A complete account on the weighted potential theory is contained in [16].

Theorem 2.1. *Let* $\{Q_n\}_{n=1}^{\infty}$ *be a sequence of polynomials with integer coefficients satisfying (2.1). Then (2.2) holds with*

$$0.2961 \leq \alpha_1 \leq 0.3634 \quad and \quad 0.0952 \leq \alpha_2 \leq 0.1767. \tag{2.6}$$

Furthermore, the pair (α_1, α_2) *must belong to the region G pictured below in Figure 1, which is determined by (3.25), (3.26) and (3.28).*

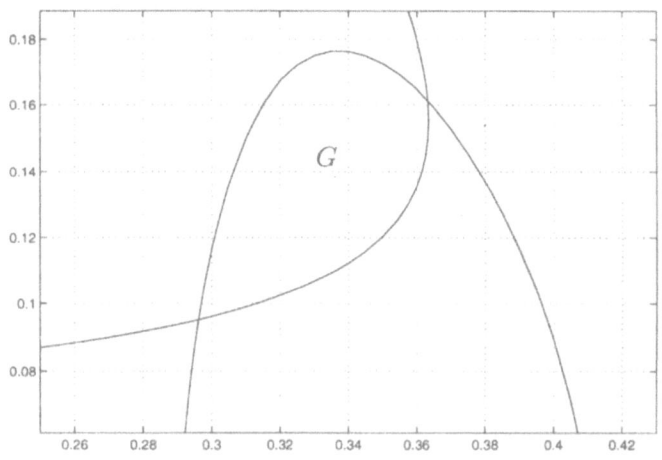

Figure 1: Region G for α_1 and α_2.

Note that (2.6) also gives the upper bounds for α_1 and α_2. Moreover, we believe that the methods introduced here can produce bounds for α_3 and for the multiplicities of other factors of the integer Chebyshev polynomials (see [3], [8] and [9] for lists of such factors). The proof of Theorem 2.1 is presented in Section 3, after the development of necessary techniques.

3. Weighted polynomials and weighted potentials

Using the idea of symmetry and the change of variable $x(1-x) \to x$, one can see that

$$(inch([0,1]))^2 = inch([0,1/4]). \tag{3.1}$$

Furthermore, we have by Lemmas 1-2 of [9] that

$$Q_{2k}(x) = q_k(x(1-x)) \tag{3.2}$$

and

$$Q_{2k+1}(x) = (1-2x)q_k(x(1-x)), \tag{3.3}$$

where

$$\lim_{k \to \infty} \|q_k\|_{[0,1/4]}^{1/k} = inch([0,1/4]). \tag{3.4}$$

Hence we can study the sequences $\{q_n\}_{n=0}^{\infty} \subset \mathcal{P}_n(\mathbb{Z})$ satisfying (3.4) instead of considering the original sequence $\{Q_n\}_{n=0}^{\infty}$ satisfying (2.1), which is more convenient for technical reasons. Note that the factors $x(1-x)$ and $(1-2x)^2$ for $Q_{2k}(x)$ are transformed into the factors x and $(1-4x)$ for $q_k(x)$, under the change of variable $x(1-x) \to x$. On writing

$$q_n(x) = x^{k_1(n)}(1-4x)^{k_2(n)}r_{n-k_1(n)-k_2(n)}(x), \quad n \in \mathbb{N}, \tag{3.5}$$

where $r_{n-k_1(n)-k_2(n)}(0) \neq 0$ and $r_{n-k_1(n)-k_2(n)}(1/4) \neq 0$, we can assume that the following limits exist:

$$\lim_{n \to \infty} \frac{k_1(n)}{n} =: u \quad \text{and} \quad \lim_{n \to \infty} \frac{k_2(n)}{n} =: v \tag{3.6}$$

(passing to subsequences, if necessary). It follows that in the study of the n-th root behavior of (3.5) one may equivalently consider

$$\left(x^{\frac{u}{1-u-v}}(1-4x)^{\frac{v}{1-u-v}}\right)^{n-k_1(n)-k_2(n)} r_{n-k_1(n)-k_2(n)}(x), \quad x \in [0,1/4], \quad (3.7)$$

as $n \to \infty$, where (3.7) is the so-called weighted polynomial with varying weight $(w(x))^{n-k_1(n)-k_2(n)}$ and where

$$w(x) := x^{\frac{u}{1-u-v}}(1-4x)^{\frac{v}{1-u-v}}, \quad x \in [0,1/4] \qquad (3.8)$$

(see [16]). Clearly, the uniform norms of both (3.5) and (3.7) on $[0,1/4]$ cannot be attained at the endpoints. Furthermore, if $u > 0$ and $v > 0$ then these uniform norms "live" on an interval $[a,b] \subset (0,1/4)$. This type of problem for the Jacobi weights (see Example IV.1.17 in [16, p. 206]) has first been considered in [13], [11] and [17], where the sharp values for a and b were found. We follow the modern and general approach to the problem via the weighted potential theory, described in [16].

For $[a,b] \subset \mathbb{R}$, let $\Omega := \bar{\mathbb{C}} \setminus [a,b]$ and let $g_\Omega(z,p)$ be the Green function of Ω with pole at $p \in \Omega$ (cf. [18, p. 14]). Consider the following natural extension for $w(x)$ of (3.8):

$$w(z) := |z|^{\frac{u}{1-u-v}}|1-4z|^{\frac{v}{1-u-v}}, \quad z \in \mathbb{C}, \qquad (3.9)$$

where $u \geq 0$, $v \geq 0$ and $u+v \leq 1$.

Lemma 3.1. *Suppose that $P_n \in \mathcal{P}_n(\mathbb{C})$ and w is defined by (3.9). Then there exists an interval $[a,b] \subset [0,1/4]$ with*

$$a := (u^2 - v^2 - \sqrt{\Delta} + 1)/8 \text{ and } b := (u^2 - v^2 + \sqrt{\Delta} + 1)/8, \qquad (3.10)$$

where $\Delta := (1-(u+v)^2)(1-(u-v)^2)$, and a continuous in $\mathbb{C}$ and harmonic in $\mathbb{C} \setminus [a,b]$ function

$$\begin{aligned}
h(z) \quad &:= \quad (g_\Omega(z,\infty) - u(\log|z| + g_\Omega(z,0)) \\
&- \quad v(\log|4z-1| + g_\Omega(z,1/4)))/(1-u-v), \qquad (3.11)
\end{aligned}$$

such that

$$|P_n(z)| \le \|w^n P_n\|_{[a,b]} e^{nh(z)}, \qquad z \in \mathbb{C}. \tag{3.12}$$

Moreover,

$$\|w^n P_n(z)\|_{[0,1/4]} = \|w^n P_n\|_{[a,b]}. \tag{3.13}$$

Proof. It follows from Theorem III.2.1 of [16] that

$$|P_n(z)| \le \|w^n P_n\|_{S_w} \exp(n(F_w - U^{\mu_w}(z))), \qquad z \in \mathbb{C}, \tag{3.14}$$

where μ_w is a positive unit Borel measure with the support $S_w \subset [0, 1/4]$, which is the solution of the weighted energy problem for the weight w of (3.9) on $[0, 1/4]$, considered in Section I.1 of [16]. Here, $U^{\mu_w}(z)$ is the logarithmic potential of μ_w

$$U^{\mu_w}(z) := \int \log \frac{1}{|z - t|} d\mu_w(t), \tag{3.15}$$

and F_w is the *modified Robin constant* for w. Note that the weight w of (3.9) is just a special case of the Jacobi weights of Example IV.1.17 in [16]. Thus our problem on the interval $[0, 1/4]$ is easily reduced to that on the interval $[-1, 1]$ considered there, with the help of the change of variable $x \to (x+1)/8$. We obtain from Example IV.1.17 of [16] that $S_w = [a, b]$, with a and b given by (3.10) (see (1.27) and (1.28) in [16, p. 207]). Equation (3.13) now follows from Corollary III.2.6 of [16]. Also, Theorem I.1.3 of [16] yields that

$$F_w - U^{\mu_w}(z) = -\log w(z), \tag{3.16}$$

for quasi every $z \in [a, b]$ (i.e., with the exception of a set of zero capacity). Hence $F_w - U^{\mu_w}(z) - h(z)$ is a harmonic function in Ω such that

$$F_w - U^{\mu_w}(z) - h(z) = 0$$

for quasi every $z \in [a, b] = \partial \Omega$ by (3.11), (3.9), (3.16) and the basic properties of Green functions (see [18, p. 14]). Using the uniqueness theorem for the solution of the Dirichlet problem in Ω (cf. Theorem III.28 and its Corollary in [18]), we conclude that

$$F_w - U^{\mu w}(z) \equiv h(z), \quad z \in \mathbb{C}.$$

Thus (3.12) follows from (3.14). ∎

Proof of Theorem 2.1. Suppose that $\{q_n\}_{n=0}^{\infty}$ is a sequence of polynomials with integer coefficients, satisfying (3.2)-(3.4). We also assume that (3.5)-(3.6) hold for this sequence, as before. It follows from (3.6) that

$$\lim_{n \to \infty} |z|^{\frac{k_1(n)}{n-k_1(n)-k_2(n)}} |1 - 4z|^{\frac{k_2(n)}{n-k_1(n)-k_2(n)}} = w(z), \tag{3.17}$$

where $w(z)$ is given by (3.9) and where the above convergence is uniform on compact subsets of $\mathbb{C}$. Since $r_{n-k_1(n)-k_2(n)}(0) \neq 0$ by (3.5), we obtain from (3.12) that

$$
\begin{aligned}
1 &\leq |r_{n-k_1(n)-k_2(n)}(0)| \\
&\leq \|w^{n-k_1(n)-k_2(n)} r_{n-k_2(n)-k_2(n)}\|_{[a,b]} e^{(n-k_1(n)-k_2(n))h(0)} \\
&\leq \|w^{n-k_1(n)-k_2(n)} r_{n-k_1(n)-k_2(n)}\|_{[0,1/4]} e^{(n-k_1(n)-k_2(n))h(0)}.
\end{aligned}
$$

Extracting the n-th root in the above inequality, passing to the limit as $n \to \infty$ and, using (3.4)-(3.6) and (3.17), we arrive at

$$1 \leq inch([0, 1/4]) e^{(1-u-v)h(0)}. \tag{3.18}$$

A similar argument applied in the case $z = 1/4$ gives that

$$\left(\frac{1}{4}\right)^{n-k_1(n)-k_2(n)} \leq \left| r_{n-k_1(n)-k_2(n)} \left(\frac{1}{4}\right) \right|$$

and that

$$4^{u+v-1} \le inch([0, 1/4])e^{(1-u-v)h(1/4)}. \tag{3.19}$$

The Green functions in the definition of $h(z)$ in (3.11) can be found explicitly, by using the conformal mappings of Ω onto the exterior of the unit disk $D' := \{w : |w| > 1\}$. Indeed, introducing these conformal mappings by

$$\Phi_\infty(z) := \frac{2z - a - b + 2\sqrt{(z-a)(z-b)}}{b-a}, \quad z \in \Omega, \tag{3.20}$$

$$\Phi_0(z) := \frac{2z^{-1} - b^{-1} - a^{-1} + 2\sqrt{(z^{-1} - b^{-1})(z^{-1} - a^{-1})}}{b^{-1} - a^{-1}}, \quad z \in \Omega, \tag{3.21}$$

and

$$\Phi_{1/4}(z) := \frac{2(z - 1/4)^{-1} - (b - 1/4)^{-1} - (a - 1/4)^{-1}}{(b - 1/4)^{-1} - (a - 1/4)^{-1}} + $$
$$\frac{2\sqrt{((z - 1/4)^{-1} - (a - 1/4)^{-1})((z - 1/4)^{-1} - (b - 1/4)^{-1})}}{(b - 1/4)^{-1} - (a - 1/4)^{-1}}, \quad z \in \Omega, \tag{3.22}$$

we observe that

$$\Phi_\infty(\infty) = \infty, \quad \Phi_0(0) = \infty \quad \text{and} \quad \Phi_{1/4}(1/4) = \infty. \tag{3.23}$$

Hence

$$g_\Omega(z, \infty) = \log|\Phi_\infty(z)|, \quad g_\Omega(z, 0) = \log|\Phi_0(z)|$$
$$\text{and} \quad g_\Omega(z, 1/4) = \log|\Phi_{1/4}(z)|, \quad z \in \Omega, \tag{3.24}$$

by Theorem I.17 of [18, p. 18]. Taking (3.24) into account, we rewrite (3.18) as

$$1 \le inch([0, 1/4])|\Phi_\infty(0)| \left(\lim_{z \to 0} |z\Phi_0(z)| \right)^{-u} |\Phi_{1/4}(0)|^{-v} \tag{3.25}$$

and (3.19) as

$$4^{2v-1} \leq inch([0,1/4])|\Phi_\infty(1/4)||\Phi_0(1/4)|^{-u} \left(\lim_{z \to 1/4} |z - 1/4||\Phi_{1/4}(z)| \right)^{-v}.$$
(3.26)

Thus $u \in [0,1]$ and $v \in [0,1]$ must satisfy the inequalities (3.25) and (3.26). Applying (3.1) and the upper bound of (1.7) in (3.25)-(3.26), we obtain the region H of Figure 2 below and the following bounds for u and v:

$$0.5923 \leq u \leq 0.7268 \quad \text{and} \quad 0.0952 \leq v \leq 0.1767. \tag{3.27}$$

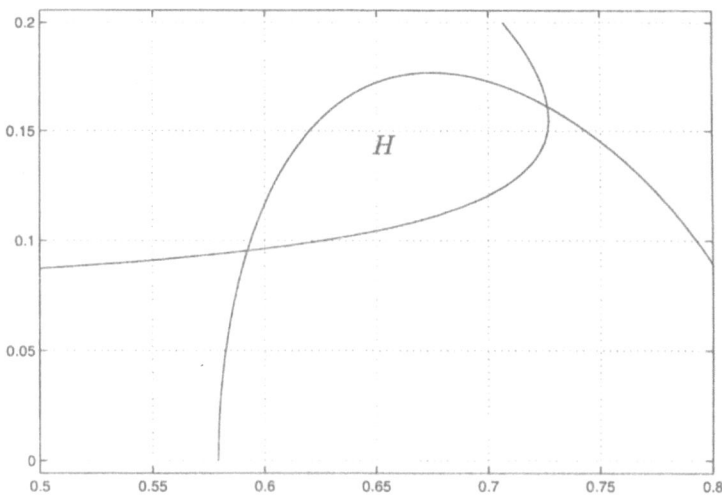

Figure 2: Region H for u and v.

Using (3.2) - (3.4), we conclude that (2.2) holds with

$$\alpha_1 = u/2 \quad \text{and} \quad \alpha_2 = v, \tag{3.28}$$

so that Theorem 2.1 follows from the results for u and v. ∎

Remark. As a consequence of the above proof and (3.13), we have that

$$inch([0, 1/4]) = inch([A, B]),$$

where

$$A := \inf_{(u,v) \in H} a(u, v) \approx 0.089 \quad \text{and} \quad B := \sup_{(u,v) \in H} b(u, v) \approx 0.247,$$

with $a(u, v)$ and $b(u, v)$ as in (3.10).

References

[1] Emiliano Aparicio Bernardo, *On the asymptotic structure of the polynomials of minimal diophantic deviation from zero*, J. Approx. Theory **55** (1988), 270-278.

[2] P. Borwein and T. Erdélyi, *Polynomials and Polynomial Inequalities*, Springer-Verlag, New York, 1995.

[3] P. Borwein and T. Erdélyi, *The integer Chebyshev problem*, Math. Comp. **65** (1996), 661-681.

[4] G. V. Chudnovsky, *Number theoretic applications of polynomials with rational coefficients defined by extremality conditions*, Arithmetic and Geometry, vol. I (M. Artin and J. Tate, eds.), pp. 61-105, Birkhäuser, Boston, 1983.

[5] M. Fekete, *Über die Verteilung der Wurzeln bei gewissen algebraischen Gleichungen mit ganzzahligen Koeffizienten*, Math. Zeit. **17** (1923), 228-249.

[6] V. Flammang, *Sur la longueur des entiers algébriques totalement positifs*, J. Number Theory **54** (1995), 60-72.

[7] V. Flammang, *Sur le diamètre transfini entier d'un intervalle à extrémités rationnelles*, Ann. Inst. Fourier Gren. **45** (1995), 779-793.

[8] V. Flammang, G. Rhin and C. J. Smyth, *The integer transfinite diameter of intervals and totally real algebraic integers*, J. Théor. Nombres Bordeaux **9** (1997), 137-168.

[9] L. Habsieger and B. Salvy, *On integer Chebyshev polynomials*, Math. Comp. **66** (1997), 763-770.

[10] D. Hilbert, *Ein Beitrag zur Theorie des Legendreschen Polynoms*, Acta Math. **18** (1894), 155-159.

[11] M. Lachance, E. B. Saff and R. S. Varga, *Bounds for incomplete polynomials vanishing at both endpoints of an interval*. In "Constructive approaches to Mathematical Models", pp. 421-437, Academic Press, New York, 1979.

[12] G. V. Milovanović, D. S. Mitrinović and Th. M. Rassias, *Topics in Polynomials: Extremal Problems, Inequalities, Zeros*, World Scientific, Singapore, 1994.

[13] D. S. Moak, E. B. Saff and R. S. Varga, *On the zeros of Jacobi polynomials* $P_n^{(\alpha_n,\beta_n)}(x)$, Trans. Amer. Math. Soc. **249** (1979), 159-162.

[14] H. L. Montgomery, *Ten Lectures on the Interface Between Analytic Number Theory and Harmonic Analysis*, CBMS, vol. 84, Amer. Math. Soc., Providence, R.I., 1994.

[15] T. J. Rivlin, *Chebyshev Polynomials*, John Wiley & Sons, New York, 1990.

[16] E. B. Saff and V. Totik, *Logarithmic Potentials with External Fields*, Springer-Verlag, Berlin, 1997.

[17] E. B. Saff, J. L. Ullman and R. S. Varga, *Incomplete polynomials: An electrostatics approach*. In "Approximation Theory III", pp. 769-782, Academic Press, San Diego, 1980.

[18] M. Tsuji, *Potential Theory in Modern Function Theory*, Chelsea Publ. Co., New York, 1975.

DISTORTION INEQUALITIES FOR ANALYTIC AND UNIVALENT FUNCTIONS ASSOCIATED WITH CERTAIN FRACTIONAL CALCULUS AND OTHER LINEAR OPERATORS

H.M. SRIVASTAVA
Department of Mathematics and Statistics
University of Victoria
Victoria, British Columbia V8W 3P4
Canada
E-Mail: harimsri@math.uvic.ca

Abstract. By applying certain operators of fractional calculus (that is, fractional integral and fractional derivative), the author presents a systematic investigation of several generalizations of various growth-and-distortion type inequalities for some novel classes of analytic and univalent functions. These general inequalities are shown to stem naturally from some recent conjectures and theorems in Geometric Function Theory.

1. Introduction, Definitions, and Preliminaries

Denote by $\mathcal{A}$ the class of functions $f(z)$ *normalized*, as usual in Geometric Function Theory, by

$$f(0) = f'(0) - 1 = 0 \tag{1.1}$$

so that

$$f(z) = z + \sum_{n=1}^{\infty} a_n z^n, \tag{1.2}$$

which are *analytic* in the *open* unit disk

$$\mathcal{U} := \{z : z \in \mathbb{C} \quad \text{and} \quad |z| < 1\}. \tag{1.3}$$

1991 *Mathematics Subject Classification.* Primary 26A33, 30C45; Secondary 33C20.
Key words and phrases. Analytic functions, univalent functions, fractional calculus (fractional integral and fractional derivative), growth-and-distortion type inequalities, Geometric Function Theory, Koebe function, Löwner's result, removable singularities, Gauss hypergeometric function, generalized hypergeometric function, Euler's transformation, Chu-Vandermonde summation theorem, Hadamard product (or convolution), Ruscheweyh derivatives, Clausenian hypergeometric function.

T.M. Rassias and H.M. Srivastava (eds.), Analytic and Geometric Inequalities and Applications, 349–374.
© 1999 *Kluwer Academic Publishers.*

Also let S denote the class of all functions in $\mathcal{A}$ which are *univalent* in $\mathcal{U}$. We denote by $S^*(\alpha)$ and $\mathcal{K}(\alpha)$ the subclasses of S consisting of all functions which are, respectively, *starlike of order* α and *convex of order* α in $\mathcal{U}$ $(0 \leq \alpha < 1)$, that is,

$$S^*(\alpha) := \left\{ f : f \in S \quad \text{and} \quad \Re\left(\frac{z\,f'(z)}{f(z)}\right) > \alpha \qquad (0 \leq \alpha < 1;\ z \in \mathcal{U}) \right\} \quad (1.4)$$

and

$$\mathcal{K}(\alpha) := \left\{ f : f \in S \quad \text{and} \quad \Re\left(1 + \frac{z\,f''(z)}{f'(z)}\right) > \alpha \qquad (0 \leq \alpha < 1;\ z \in \mathcal{U}) \right\}. \quad (1.5)$$

It follows readily from the definitions (1.4) and (1.5) that

$$f(z) \in \mathcal{K}(\alpha) \Leftrightarrow z\,f'(z) \in S^*(\alpha) \qquad (0 \leq \alpha < 1), \quad (1.6)$$

whose special case, when $\alpha = 0$, is the familiar Alexander theorem (*cf.*, *e.g.*, Duren [6, p. 43, Theorem 2.12]). We note also that

$$\mathcal{K}(\alpha) \subset S^*(\alpha) \subset S \qquad (0 \leq \alpha < 1), \quad (1.7)$$

$$S^*(\alpha) \subseteq S^*(0) =: S^* \qquad (0 \leq \alpha < 1), \quad (1.8)$$

and

$$\mathcal{K}(\alpha) \subseteq \mathcal{K}(0) =: \mathcal{K} \qquad (0 \leq \alpha < 1), \quad (1.9)$$

where S^* denotes the class of all functions in $\mathcal{A}$ which are *starlike* (*with respect to the origin*) in $\mathcal{U}$.

In statements like those involved in the definitions (1.4) and (1.5), and in analogous situations throughout this paper, it should be understood that functions such as (see also Srivastava and Owa [28])

$$\frac{z\,f'(z)}{f(z)} \quad \text{and} \quad \frac{z\,f''(z)}{f'(z)},$$

which have *removable singularities* at $z = 0$, have had these singularities removed.

For functions $f(z)$ of the form (1.2) and belonging to the classes $S^*(\alpha)$ and $\mathcal{K}(\alpha)$ defined by Equations (1.4) and (1.5), respectively, it is well-known that (*cf.* Robertson [20])

$$f \in S^*(\alpha) \Rightarrow |a_n| \leq \frac{\prod\limits_{j=2}^{n}(j - 2\alpha)}{(n-1)!} \qquad (n \in \mathbb{N} \setminus \{1\}) \quad (1.10)$$

and

$$f \in \mathcal{K}(\alpha) \Rightarrow |a_n| \leq \frac{\prod\limits_{j=2}^{n}(j - 2\alpha)}{n!} \qquad (n \in \mathbb{N} \setminus \{1\}), \quad (1.11)$$

where, as usual,

$$\mathbb{N} := \{1, 2, 3, \cdots\} \quad \text{and} \quad \mathbb{N}_0 := \mathbb{N} \cup \{0\}. \tag{1.12}$$

In view of the *second* inclusion relation in (1.7), the special case of the assertion (1.10) when $\alpha = 0$ is substantially weaker than de Branges' result (*cf.* [5]):

$$f \in \mathcal{S} \, (\supset \mathcal{S}^*) \Rightarrow |a_n| \le n \quad (n \in \mathbb{N} \setminus \{1\}), \tag{1.13}$$

where the equality holds true for all integers $n \ge 2$ only if $f(z)$ is any rotation of the (Koebe) function $K(z)$ defined by

$$K(z) := \frac{z}{(1-z)^2} = \sum_{n=1}^{\infty} n \, z^n \quad (z \in \mathcal{U}). \tag{1.14}$$

On the other hand, in its special case when $\alpha = 0$, the assertion (1.11) immediately yields Löwner's result (*cf.* [9]):

$$f \in \mathcal{K} \Rightarrow |a_n| \le 1 \quad (n \in \mathbb{N} \setminus \{1\}), \tag{1.15}$$

where the equality holds true for all integers $n \ge 2$ only if $f(z)$ is any rotation of the (Löwner) function $L(z)$ defined by

$$L(z) := \frac{z}{1-z} = \sum_{n=1}^{\infty} z^n \quad (z \in \mathcal{U}). \tag{1.16}$$

By applying the assertions (1.13) and (1.15), it is fairly straightforward to derive the following familiar growth-and-distortion type inequalities for the function classes $\mathcal{S}$ and $\mathcal{K}$, respectively (*cf.*, *e.g.*, Goodman [8] and Duren [6]).

Theorem A. *If the function $f(z)$ is in the class $\mathcal{S}$, then*

$$|f^{(n)}(z)| \le \frac{n! \, (n + |z|)}{(1 - |z|)^{n+2}} \quad (z \in \mathcal{U}; \; n \in \mathbb{N}_0), \tag{1.17}$$

where the equality holds true for the Koebe function $K(z)$ given by (1.14).

Theorem B. *If the function $f(z)$ is in the class $\mathcal{K}$, then*

$$|f^{(n)}(z)| \le \frac{n!}{(1 - |z|)^{n+1}} \quad (z \in \mathcal{U}; \; n \in \mathbb{N}_0), \tag{1.18}$$

where the equality holds true for the Löwner function $L(z)$ given by (1.16).

Owa *et al.* ([13] and [16]) conjectured the existence of growth and distortion inequalities of the types (1.17) and (1.18) involving the fractional derivative operator $D_z^{n+\lambda}$ ($n \in \mathbb{N}_0$; $0 \le \lambda < 1$) given by Definition 2 below. These conjectures were subsequently shown to be false by Cho *et al.* [3] and Chen *et al.* [1], who did indeed prove generalizations of Theorem A and Theorem B, respectively, in terms of fractional derivatives of analytic and univalent functions. Recently, many authors (including, for example, Owa [12], Choi [4], Srivastava [26], and Chen *et al.* [2]) presented a number of growth-and-distortion type inequalities relevant to (or extending) the works of Cho *et al.* [3] and Chen *et al.* [1]. The main object of this paper is to present a systematic investigation of various generalizations of these growth-and-distortion type inequalities involving some novel classes of analytic functions in $\mathcal{U}$. We also show how the main inequalities presented in this paper are related to those given by the aforementioned and other workers on the subject. Some recent conjectures and theorems, relevant to the work presented here, are also considered briefly.

2. Operators of Fractional Calculus and Their Applications Involving the Classes $\mathcal{S}$ and $\mathcal{K}$

Our present investigation shall make use of certain operators of *fractional calculus* (that is, *fractional integral* and *fractional derivative*). From among the numerous operators of fractional claculus (which have indeed been studied in the mathematical literature in one context or the other), we first choose to recall here the fractional calculus operators given by Definition 1 and Definition 2 below (*cf.* Owa [10]; see also Owa and Srivastava [15]):

Definition 1 (Fractional Integral Operator). The *fractional integral of order* λ is defined, for a function $f(z)$, by

$$D_z^{-\lambda} f(z) := \frac{1}{\Gamma(\lambda)} \int_0^z \frac{f(\zeta)}{(z-\zeta)^{1-\lambda}} \, d\zeta \qquad (\lambda > 0), \tag{2.1}$$

where $f(z)$ is an analytic function in a simply-connected region of the z-plane containing the origin, and the multiplicity of $(z-\zeta)^{\lambda-1}$ is removed by requiring $\log(z-\zeta)$ to be *real* when $z-\zeta > 0$.

Definition 2 (Fractional Derivative Operator). The *fractional derivative of order* λ is defined, for a function $f(z)$, by

$$D_z^{\lambda} f(z) := \begin{cases} \dfrac{1}{\Gamma(1-\lambda)} \dfrac{d}{dz} \displaystyle\int_0^z \frac{f(\zeta)}{(z-\zeta)^{\lambda}} \, d\zeta & (0 \leq \lambda < 1) \\[3mm] \dfrac{d^n}{dz^n} D_z^{\lambda-n} f(z) & (n \leq \lambda < n+1; \; n \in \mathbb{N}), \end{cases} \tag{2.2}$$

where $f(z)$ is constrained, and the multiplicity of $(z-\zeta)^{-\lambda}$ is removed, as in Definition 1.

We shall also require, in our investigation of the function classes stemming essentially from the classes $\mathcal{S}$ and $\mathcal{K}$ defined above, the familiar *Gaussian case*

$$\ell - 1 = m = 1$$

and other important cases of the *generalized hypergeometric function* ${}_\ell F_m$ given by

Definition 3 (Generalized Hypergeometric Function). Let λ_j $(j = 1, \cdots, \ell)$ and μ_j $(j = 1, \cdots, m)$ be complex numbers such that

$$\mu_j \neq 0, -1, -2, \cdots \qquad (j = 1, \cdots, m).$$

Then the *generalized hypergeometric function* ${}_\ell F_m(z)$ is defined by

$$
\begin{aligned}
{}_\ell F_m(z) &\equiv {}_\ell F_m(\lambda_1, \cdots, \lambda_\ell; \ \mu_1, \cdots, \mu_m; \ z) \\
&= {}_\ell F_m \begin{bmatrix} \lambda_1, \cdots, \lambda_\ell; \\ \mu_1, \cdots, \mu_m; \end{bmatrix} z \\
&:= \sum_{n=0}^{\infty} \frac{(\lambda_1)_n \cdots (\lambda_\ell)_n}{(\mu_1)_n \cdots (\mu_m)_n} \frac{z^n}{n!} \qquad (\ell \leq m + 1),
\end{aligned}
\tag{2.3}
$$

where $(\lambda)_n$ denotes the *Pochhammer symbol* defined, in terms of the Gamma functions, by

$$
(\lambda)_n := \frac{\Gamma(\lambda + n)}{\Gamma(\lambda)} = \begin{cases} 1 & (n = 0; \ \lambda \neq 0) \\ \lambda(\lambda + 1) \cdots (\lambda + n - 1) & (n \in \mathbb{N}). \end{cases}
\tag{2.4}
$$

We observe in passing that

$$z \, {}_\ell F_m(\lambda_1, \cdots, \lambda_\ell; \ \mu_1, \cdots, \mu_m; \ z) \in \mathcal{A},
\tag{2.5}$$

since the ${}_\ell F_m$ series in (2.3) converges absolutely for (*cf.*, *e.g.*, Slater [23, Chapter 2])

(i) $|z| < \infty$ if $\ell < m + 1$;

(ii) $z \in \mathcal{U}$ if $\ell = m + 1$;

(iii) $z \in \partial \mathcal{U} := \{z : z \in \mathbb{C} \text{ and } |z| = 1\}$ if $\ell = m + 1$, provided further that

$$\Re \left(\sum_{j=1}^{m} \mu_j - \sum_{j=1}^{\ell} \lambda_j \right) > 0,
\tag{2.6}$$

unless (of course) the series terminates.

In terms of the Gaussian case

$$\ell - 1 = m = 1$$

of the generalized hypergeometric function (2.3), we now recall the definitions of some useful generalizations of the fractional calculus operators given by Definition 1 and Definition 2 above.

Definition 4 (Generalized Fractional Integral Operator). Under the hypotheses of Definition 1, the *generalized fractional integral of order* λ is defined, for a function $f(z)$, by

$$I_{0,z}^{\lambda,\mu,\nu} f(z) := \frac{z^{-\lambda-\mu}}{\Gamma(\lambda)} \int_0^z (z-\zeta)^{\lambda-1} \, {}_2F_1\left(\lambda+\mu, -\nu; \lambda; 1-\frac{\zeta}{z}\right) f(\zeta) \, d\zeta \tag{2.7}$$

$$(\lambda > 0; \; \kappa > \max\{0, \; \mu - \nu\} - 1),$$

provided further that

$$f(z) = O(|z|^\kappa) \qquad (z \to 0). \tag{2.8}$$

It follows readily from Definition 1 and Definition 4 that

$$D_z^{-\lambda} f(z) = I_{0,z}^{\lambda,-\lambda,\nu} f(z) \qquad (\lambda > 0). \tag{2.9}$$

Furthermore, since

$$_2F_1(a,b;b;z) = {}_1F_0(a;\text{---};z) = (1-z)^{-a} \qquad (z \in \mathcal{U}), \tag{2.10}$$

we have the relationship:

$$I_{0,z}^{\lambda,\mu,-\lambda} f(z) = D_z^{-\lambda} z^{-\lambda-\mu} f(z) \qquad (\lambda > 0). \tag{2.11}$$

The operator $I_{0,z}^{\lambda,\mu,\nu}$ is a *generalization* of the fractional integral operator which was studied by Saigo [22] and applied subsequently by Srivastava and Saigo [29] in solving various boundary value problems involving the Euler-Darboux equation:

$$\frac{\partial^2 u}{\partial x \partial y} - \frac{1}{x-y}\left(\beta \frac{\partial u}{\partial x} - \alpha \frac{\partial u}{\partial y}\right) = 0 \tag{2.12}$$

$$(\alpha > 0; \; \beta > 0; \; \alpha + \beta < 1).$$

Definition 5 (Generalized Fractional Derivative Operator). Under the hypotheses of Definition 2, the *generalized fractional derivative of order* λ is defined, for a function $f(z)$, by

$$J_{0,z}^{\lambda,\mu,\nu} f(z) := \begin{cases} \dfrac{1}{\Gamma(1-\lambda)} \dfrac{d}{dz}\left\{z^{\lambda-\mu} \displaystyle\int_0^z (z-\zeta)^{-\lambda}\right. \\ \qquad \left. \cdot\, {}_2F_1\left(\mu-\lambda, -\nu; 1-\lambda; 1-\dfrac{\zeta}{z}\right) f(\zeta) \, d\zeta\right\} & (0 \leq \lambda < 1) \\[2mm] \dfrac{d^n}{dz^n} J_{0,z}^{\lambda-n,\mu,\nu} f(z) & (n \leq \lambda < n+1; \; n \in \mathbb{N}) \end{cases}$$

$$\tag{2.13}$$

$$(\kappa > \max\{0, \; \mu - \nu - 1\} - 1),$$

where κ is given, as before, by the order estimate (2.8).

It follows at once from Definition 5 that

$$J_{0,z}^{\lambda,\lambda,\nu} f(z) = D_z^\lambda f(z) \qquad (0 \le \lambda < 1), \tag{2.14}$$

where the fractional calculus operator D_z^λ is, in fact, given by Definitions 1 and 2 for all values of λ (see, *e.g.*, Srivastava and Owa [27, p. 343]). Furthermore, in terms of the Gamma functions, we have (*cf.* [14] and [30])

$$J_{0,z}^{\lambda,\mu,\nu} z^\rho = \frac{\Gamma(\rho+1)\Gamma(\rho-\mu+\nu+2)}{\Gamma(\rho-\mu+1)\Gamma(\rho-\lambda+\nu+2)} z^{\rho-\mu} \tag{2.15}$$

$$(0 \le \lambda < 1; \ \rho > \max\{0, \ \mu - \nu - 1\} - 1).$$

We now recall the conjectures of Owa *et al.* ([13] and [16]), which were referred to in Section 1, involving the fractional derivative operator $D_z^{n+\lambda}$ ($n \in \mathbb{N}_0$; $0 \le \lambda < 1$) given by Definition 2.

Conjecture 1 (Owa *et al.* [13, p. 88]). *If the function $f(z)$ is in the class $\mathcal{S}$, then*

$$|D_z^{n+\lambda} f(z)| \le \frac{(n+\lambda+|z|)\,\Gamma(n+\lambda+1)}{(1-|z|)^{n+\lambda+2}} \tag{2.16}$$

$$(z \in \mathcal{U}; \ n \in \mathbb{N}_0; \ 0 \le \lambda < 1),$$

where the equality holds true for the Koebe function $K(z)$ given by (1.14).

Conjecture 2 (Owa and Srivastava [16, p. 225]). *If the function $f(z)$ is in the class $\mathcal{K}$, then*

$$|D_z^{n+\lambda} f(z)| \le \frac{\Gamma(n+\lambda+1)}{(1-|z|)^{n+\lambda+1}} \qquad (z \in \mathcal{U}; \ n \in \mathbb{N}; \ 0 \le \lambda < 1), \tag{2.17}$$

where the equality holds true for the Löwner function $L(z)$ given by (1.16).

For $\lambda = 0$, Conjecture 1 and Conjecture 2 can indeed be validated by means of Theorem A and Theorem B, respectively. With a view to examining and investigating the validity of Conjecture 1 and Conjecture 2 when the parameter λ is constrained by $0 < \lambda < 1$, we first state a generalization of Theorem A in terms of fractional derivatives, which is contained in (*cf.* Cho *et al.* [3]; see also Owa [12])

Theorem 1. *If the function $f(z)$ is in the class $\mathcal{S}$, then*

$$|D_z^{n+\lambda} f(z)| \le M_n(\lambda; |z|) + \frac{n!\,|z|^{-\lambda}}{\Gamma(2-\lambda)} (1-|z|)^{-n-\lambda-2}$$

$$\cdot \left\{ n(1-\lambda)(1-|z|) \,{}_2F_1 \left[\begin{array}{c} -n-\lambda, \ -\lambda; \\ 1-\lambda; \end{array} |z| \right] \right. \tag{2.18}$$

$$\left. + (n+1)|z| \,{}_2F_1 \left[\begin{array}{c} -n-\lambda, \ -\lambda; \\ 2-\lambda; \end{array} |z| \right] \right\}$$

$$(0 < |z| < 1; \; n \in \mathbb{N}_0; \; 0 \leq \lambda < 1),$$

where

$$M_n(\lambda; |z|) := \sum_{k=1}^{n-1} \frac{k \cdot k!}{|\Gamma(k - n - \lambda + 1)|} |z|^{k-n-\lambda}, \tag{2.19}$$

the sum being assumed to be nil for $n = 0$ and $n = 1$.

Remark 1. Since [*cf.* Equation (2.19)]

$$M_n(0; |z|) = 0 \qquad (n \in \mathbb{N}_0; \; 0 < |z| < 1), \tag{2.20}$$

and in view of the fact that *each* of the Gaussian hypergeometric functions occurring in (2.18) reduces to its first term 1 when $\lambda = 0$, Theorem 1 does yield the assertion (1.17) of Theorem A in the special case when $\lambda = 0$.

Remark 2. By virtue of the known identity (*cf.*, *e.g.*, Srivastava [25, p. 39, Equation (6)]):

$$_{\ell}F_m \left[\begin{array}{c} \beta_1 + n, \; \alpha_2, \cdots, \alpha_\ell; \\ \beta_1, \cdots, \beta_m; \end{array} z \right] = \sum_{j=0}^{n} \binom{n}{j} \frac{(\alpha_2)_j \cdots (\alpha_\ell)_j}{(\beta_1)_j \cdots (\beta_m)_j}$$

$$\cdot z^j \; _{\ell-1}F_{m-1} \left[\begin{array}{c} \alpha_2 + j, \cdots, \alpha_\ell + j; \\ \beta_2 + j, \cdots, \beta_m + j; \end{array} z \right], \tag{2.21}$$

it is not difficult to rewrite the assertion (2.18) of Theorem 1 in the following *equivalent* forms:

$$|D_z^{n+\lambda} f(z)| \leq M_n(\lambda; |z|) + \frac{n! \, n |z|^{-\lambda}}{\Gamma(1-\lambda)} \; _3F_2 \left[\begin{array}{c} n+1, \; n+1, \; 1; \\ n, \; 1-\lambda; \end{array} |z| \right] \tag{2.22}$$

$$(0 < |z| < 1; \; n \in \mathbb{N}; \; 0 \leq \lambda < 1),$$

and

$$|D_z^{n+\lambda} f(z)| \leq M_n(\lambda; |z|) + \frac{n! \, n |z|^{-\lambda}}{\Gamma(1-\lambda)}$$

$$\cdot \left\{ _2F_1 \left[\begin{array}{c} n+1, \; 1; \\ 1-\lambda; \end{array} |z| \right] + \frac{n+1}{n(1-\lambda)} |z| \; _2F_1 \left[\begin{array}{c} n+2, \; 2; \\ 2-\lambda; \end{array} |z| \right] \right\} \tag{2.23}$$

$$(0 < |z| < 1; \; n \in \mathbb{N}; \; 0 \leq \lambda < 1),$$

each of which extends Theorem A to hold true for *fractional derivatives.*

We have already observed how Theorem 1 reduces to the inequality (1.17) in the special case when $\lambda = 0$. In its special case when $\lambda = 0$, the inequality (2.22) would reduce to (1.17) by virtue of (2.20) *and* the fact that

$$
{}_3F_2\left[\begin{matrix} n+1,\ n+1,\ 1; \\ n,\ 1-\lambda; \end{matrix}\ |z|\right]_{\lambda=0} = {}_2F_1\left[\begin{matrix} n+1,\ n+1; \\ n; \end{matrix}\ |z|\right]
$$

$$
= (1-|z|)^{-n-2}\ {}_2F_1\left[\begin{matrix} -1,\ -1; \\ n; \end{matrix}\ |z|\right] \tag{2.24}
$$

$$
= (1-|z|)^{-n-2}\left(1+\frac{|z|}{n}\right) \qquad (z \in \mathcal{U};\ n \in \mathbb{N}),
$$

where we have also made use of Euler's transformation (*cf.* Erdélyi *et al.* [7, p. 64, Equation 2.1.4 (23)]):

$$
{}_2F_1\left[\begin{matrix} \alpha,\ \beta; \\ \gamma; \end{matrix}\ z\right] = (1-z)^{\gamma-\alpha-\beta}\ {}_2F_1\left[\begin{matrix} \gamma-\alpha,\ \gamma-\beta; \\ \gamma; \end{matrix}\ z\right] \qquad (z \in \mathcal{U}). \tag{2.25}
$$

Similarly, since [*cf.* Equation (2.10)]

$$
{}_2F_1\left[\begin{matrix} n+1,\ 1; \\ 1-\lambda; \end{matrix}\ |z|\right]_{\lambda=0} = (1-|z|)^{-n-1} \qquad (z \in \mathcal{U};\ n \in \mathbb{N}_0) \tag{2.26}
$$

and

$$
{}_2F_1\left[\begin{matrix} n+2,\ 2; \\ 2-\lambda; \end{matrix}\ |z|\right]_{\lambda=0} = (1-|z|)^{-n-2} \qquad (z \in \mathcal{U};\ n \in \mathbb{N}_0), \tag{2.27}
$$

the special case of the assertion (2.23) when $\lambda = 0$ would also readily yield the inequality (1.17) in view of (2.20). The upper bounds in (2.18), (2.22), and (2.23) differ rather markedly from the *conjectured* upper bound in (2.16).

Next we state a generalization of Theorem B in terms of fractional derivatives, which is given by (*cf.* Chen *et al.* [1])

Theorem 2. *If the function $f(z)$ is in the class $\mathcal{K}$, then*

$$
|D_z^{n+\lambda} f(z)| \leq N_n(\lambda;|z|) + \frac{n!\,|z|^{-\lambda}}{\Gamma(1-\lambda)}(1-|z|)^{-n-\lambda-1}
$$

$$
\cdot\ {}_2F_1\left[\begin{matrix} -n-\lambda,\ -\lambda; \\ 1-\lambda; \end{matrix}\ |z|\right] \tag{2.28}
$$

$$
(0 < |z| < 1;\ n \in \mathbb{N}_0;\ 0 \leq \lambda < 1),
$$

where

$$
N_n(\lambda;|z|) := \sum_{k=1}^{n-1} \frac{k!}{|\Gamma(k-n-\lambda+1)|}\,|z|^{k-n-\lambda}, \tag{2.29}
$$

the sum being assumed to be nil for $n = 0$ and $n = 1$.

Remark 3. By appealing to Euler's transformation (2.25), we can easily rewrite the assertion (2.28) of Theorem 2 in the following *equivalent* form:

$$|D_z^{n+\lambda} f(z)| \leq N_n(\lambda; |z|) + \frac{n! \, |z|^{-\lambda}}{\Gamma(1-\lambda)} \, {}_2F_1 \left[\begin{matrix} n+1, \; 1; \\ 1-\lambda; \end{matrix} \, |z| \right] \qquad (2.30)$$

$$(0 < |z| < 1; \; n \in \mathbb{N}_0; \; 0 \leq \lambda < 1),$$

where $N_n(\lambda; |z|)$ is given by (2.29).

Remark 4. Since [*cf.* Equation (2.29)]

$$N_n(0, |z|) = 0 \qquad (n \in \mathbb{N}_0; \; 0 < |z| < 1), \qquad (2.31)$$

and since the Gaussian hypergeometric function occurring in (2.28) reduces (when $\lambda = 0$) to its first term 1, Theorem 2 does yield the assertion (1.18) of Theorem B in the special case when $\lambda = 0$. Furthermore, in its special case when $\lambda = 0$, the inequality (2.30) would also reduce to the assertion (1.18) of Theorem B by virtue of (2.31) *and* the fact that [*cf.* Equation (2.10)]

$$ {}_2F_1 \left[\begin{matrix} n+1, \; 1; \\ 1-\lambda; \end{matrix} \, |z| \right]_{\lambda=0} = \; {}_1F_0 \left[\begin{matrix} n+1; \\ \text{---}; \end{matrix} \, |z| \right] \qquad (2.32)$$

$$= (1 - |z|)^{-n-1} \qquad (z \in \mathcal{U}; \; n \in \mathbb{N}_0),$$

which incidentally remains valid even when n is replaced by an arbitrary (real or complex) parameter ν.

Thus, in addition to the assertion (2.28) of Theorem 2, the inequality (2.30) provides an *equivalent* upper bound for

$$|D_z^{n+\lambda} f(z)| \qquad (0 < |z| < 1; \; n \in \mathbb{N}_0; \; 0 \leq \lambda < 1), \qquad (2.33)$$

each extending the known inequality (1.18) to fractional derivatives. Furthermore, the upper bounds in (2.28) and (2.30) differ rather markedly from the *conjectured* upper bound in (2.17).

3. Distortion Inequalities Involving Fractional Derivatives of Functions in the Classes $\mathcal{S}^*(\alpha)$ and $\mathcal{K}(\alpha)$

By applying the assertions (1.10) and (1.11), instead of (1.13) and (1.15), it is not difficult to prove Theorem 3 and Theorem 4 below (*cf.* Owa [12]).

Theorem 3. *If the function $f(z)$ is in the class $S^*(\alpha)$, then*

$$|D_z^{n+\lambda} f(z)| \leq P_n^{(\alpha)}(\lambda; |z|) + \frac{(2-2\alpha)_n \, |z|^{1-\lambda}}{\Gamma(3-\lambda)} (1-|z|)^{2\alpha-\lambda-n-2}$$

$$\cdot \left\{ (n+1)(2-\lambda)(1-|z|) \, {}_2F_1 \left[\begin{matrix} 2\alpha-\lambda-n, \ 1-\lambda; \\ 2-\lambda; \end{matrix} |z| \right] \right. \tag{3.1}$$

$$\left. + (n-2\alpha+2)|z| \, {}_2F_1 \left[\begin{matrix} 2\alpha-\lambda-n, \ 1-\lambda; \\ 3-\lambda; \end{matrix} |z| \right] \right\}$$

$$(0 < |z| < 1; \ n \in \mathbb{N}_0; \ 0 \leq \lambda < 1; \ 0 \leq \alpha < 1),$$

where

$$P_n^{(\alpha)}(\lambda; |z|) := \sum_{k=0}^{n-1} \frac{(k+1)(2-2\alpha)_k}{|\Gamma(k-n-\lambda+2)|} \, |z|^{k-n-\lambda+1}, \tag{3.2}$$

the sum being assumed to be nil for $n = 0$.

Theorem 4. *If the function $f(z)$ is in the class $\mathcal{K}(\alpha)$, then*

$$|D_z^{n+\lambda} f(z)| \leq Q_n^{(\alpha)}(\lambda; |z|) + \frac{(2-2\alpha)_n \, |z|^{1-\lambda}}{\Gamma(2-\lambda)} (1-|z|)^{2\alpha-\lambda-n-1} \tag{3.3}$$

$$\cdot {}_2F_1 \left[\begin{matrix} 2\alpha-\lambda-n, \ 1-\lambda; \\ 2-\lambda; \end{matrix} |z| \right]$$

$$(0 < |z| < 1; \ n \in \mathbb{N}_0; \ 0 \leq \lambda < 1; \ 0 \leq \alpha < 1),$$

where

$$Q_n^{(\alpha)}(\lambda; |z|) := \sum_{k=0}^{n-1} \frac{(2-2\alpha)_k}{|\Gamma(k-n-\lambda+2)|} \, |z|^{k-n-\lambda+1}, \tag{3.4}$$

the sum being assumed to be nil for $n = 0$.

It is easily observed from the definitions (3.2) and (3.4) (*with $\alpha = 0$*) that

$$P_n^{(0)}(\lambda; |z|) = \sum_{k=0}^{n-1} \frac{(k+1) \cdot (k+1)!}{|\Gamma(k-n-\lambda+2)|} \, |z|^{k-n-\lambda+1}$$

$$= \sum_{k=1}^{n} \frac{k \cdot k!}{|\Gamma(k-n-\lambda+1)|} \, |z|^{k-n-\lambda} \tag{3.5}$$

$$= M_n(\lambda; |z|) + \frac{n \cdot n!}{\Gamma(1-\lambda)} \, |z|^{-\lambda}$$

and

$$Q_n^{(0)}(\lambda; |z|) = \sum_{k=0}^{n-1} \frac{(k+1)!}{|\Gamma(k-n-\lambda+2)|} |z|^{k-n-\lambda+1}$$

$$= \sum_{k=1}^{n} \frac{k!}{|\Gamma(k-n-\lambda+1)|} |z|^{k-n-\lambda} \tag{3.6}$$

$$= N_n(\lambda; |z|) + \frac{n!}{\Gamma(1-\lambda)} |z|^{-\lambda},$$

where $M_n(\lambda; |z|)$ and $N_n(\lambda; |z|)$ are defined by (2.19) and (2.29), respectively.

Thus, by making use of the relationships (3.5) and (3.6), it is not difficult to show that the special cases of Theorem 3 and Theorem 4 when $\alpha = 0$ are already contained in Theorem 1 and Theorem 2, respectively. The details involved in these derivations may be left as an exercise for the interested reader.

4. Inequalities Associated with a General Class of Analytic Functions

By appealing appropriately to the familiar Euler transformation [cf. Equation (2.25)]:

$$_2F_1(a, b; c; z) = (1-z)^{c-a-b} \,_2F_1(c-a, c-b; c; z) \tag{4.1}$$

$$(|\arg(1-z)| \leq \pi - \epsilon \quad (0 < \epsilon < \pi); \quad c \neq 0, -1, -2, \cdots)$$

as well as the Chu-Vandermonde summation theorem [23, p. 243, Equation (III.4)]:

$$_2F_1(-n, b; c; 1) = \frac{(c-b)_n}{(c)_n} \tag{4.2}$$

$$(n \in \mathbb{N}_0; \ c \neq 0, -1, -2, \cdots),$$

where $(\lambda)_n$ is defined, as before, by (2.4), it is not difficult to verify that

$$_2F_1(-n, \xi; \eta; \omega) > 0 \quad (n \in \mathbb{N}_0; \ \eta > \max\{0, \xi\}; \ 0 \leq \omega \leq 1). \tag{4.3}$$

This last positivity result (4.3) leads us naturally to a novel subclass (given by Definition 6 below) of the general class $\mathcal{A}(p)$ consisting of functions $f(z)$ of the form:

$$f(z) = \sum_{n=p}^{\infty} a_n z^n \quad (p \in \mathbb{N}; \ a_p \neq 0), \tag{4.4}$$

which are analytic in the open unit disk $\mathcal{U}$. Clearly, we have [cf. Equation (1.2)]

$$\mathcal{A}(1) = \mathcal{A}, \tag{4.5}$$

where it is understood that $a_1 = 1$.

Definition 6. A function $f(z)$ given by (4.4) and belonging to the class $\mathcal{A}(p)$ is said to be in the class $\mathcal{H}_p(\omega; \xi, \eta)$ if there exist real numbers ξ, η, and ω such that

$$f \in \mathcal{H}_p(\omega; \xi, \eta) \Rightarrow |a_n| \leq {}_2F_1(1 - n, \xi; \eta; \omega) \tag{4.6}$$

$$(n = p, p + 1, p + 2, \cdots; \; p \in \mathbb{N}; \; \eta > \max\{0, \xi\}; \; 0 \leq \omega \leq 1),$$

where ${}_2F_1$ is the Gauss hypergeometric function corresponding to the case

$$\ell - 1 = m = 1$$

of Definition 3.

In view of the Chu-Vandermonde summation theorem (4.2), it is easily verified that

$$\mathcal{H}_p(1; \eta - \xi, \eta) = \mathcal{S}_p(\xi, \eta) \qquad (p \in \mathbb{N}; \; \xi, \eta \in \mathbb{R}_+), \tag{4.7}$$

where $\mathcal{S}_p(\xi, \eta)$ denotes the subclass of $\mathcal{A}(p)$ considered recently by Srivastava [26]. In particular, it is not difficult to observe from Definition 6, and from (1.10), (1.11), and (1.13), that the class $\mathcal{S}_p(\xi, \eta)$ [that is, the class $\mathcal{H}_p(1; \eta - \xi, \eta)$] is analogous to

(i) the class $\mathcal{S}^*(\alpha)$ when

$$p = 1, \; \xi = 2 - 2\alpha, \quad \text{and} \quad \eta = 1 \qquad (0 \leq \alpha < 1); \tag{4.8}$$

(ii) the class $\mathcal{K}(\alpha)$ when

$$p = 1, \; \xi = 2 - 2\alpha, \quad \text{and} \quad \eta = 2 \qquad (0 \leq \alpha < 1), \tag{4.9}$$

and

(iii) the familiar class $\mathcal{S}$ itself when

$$p = 1, \; \xi = 2, \quad \text{and} \quad \eta = 1, \tag{4.10}$$

it being understood, just as with (4.5), that $a_1 = 1$ in (4.4) *with $p = 1$.*

By appealing to the definition (4.4) and the assertion (4.6), we now prove a class of distortion inequalities contained in (*cf.* Chen *et al.* [2])

Theorem 5. *If the function $f(z)$ is in the class $\mathcal{H}_p(\omega; \xi, \eta)$, then*

$$\left| J_{0,z}^{n+\lambda,\mu,\nu} f(z) \right| \leq \Theta_n^{(\xi,\eta;\omega)}(\lambda, \mu, \nu; |z|)$$

$$+ \sum_{k=0}^{\infty} \frac{(k + n + p)! \, \Gamma(k + n + p - \mu + \nu + 2)}{\Gamma(k + p - \mu + 1) \, \Gamma(k + n + p - \lambda + \nu + 2)} \tag{4.11}$$

$$\cdot {}_2F_1(1 - k - n - p, \xi; \eta; \omega) |z|^{k+p-\mu}$$

$$\left(0 < |z| < 1; \; n \in \mathbb{N}_0; \; p \in \mathbb{N}; \; 0 \leq \lambda < \min\{1, \nu + 2\}; \right.$$

$$\left. \mu < \min\{1, \nu + 2\}; \; \eta > \max\{0, \xi\}; \; 0 \leq \omega \leq 1 \right),$$

where

$$\Theta_n^{(\xi,\eta;\omega)}(\lambda,\mu,\nu;|z|) := \sum_{k=0}^{n-1} {}_2F_1(1-k-p,\xi;\eta;\omega)$$

$$\cdot \frac{(k+p)!\,\Gamma(k+p-\mu+\nu+2)}{|\Gamma(k+p-n-\mu+1)|\,\Gamma(k+p-\lambda+\nu+2)}\,|z|^{k+p-n-\mu},$$

(4.12)

the sum being assumed to be nil for $n = 0$.

Proof. First of all, by Definition 5 as well as the generalized fractional derivative formula (2.15), we find from (4.4) that

$$J_{0,z}^{n+\lambda,\mu,\nu}\,f(z) = \frac{d^n}{dz^n}\,J_{0,z}^{\lambda,\mu,\nu}\sum_{k=p}^{\infty} a_k\,z^k$$

$$= \sum_{k=p}^{\infty} a_k\,\frac{k!\,\Gamma(k-\mu+\nu+2)}{\Gamma(k-n-\mu+1)\,\Gamma(k-\lambda+\nu+2)}\,z^{k-n-\mu}$$

(4.13)

$$\left(0 < |z| < 1;\; n \in \mathbb{N}_0;\; p \in \mathbb{N};\; 0 \le \lambda < \min\{1,\nu+2\};\right.$$

$$\left.\mu < \min\{1,\nu+2\}\right).$$

Since $f \in \mathcal{H}_p(\omega;\xi,\eta)$, we can apply the assertion (4.6). Equation (4.13) thus yields

$$\left|J_{0,z}^{n+\lambda,\mu,\nu}\,f(z)\right| \le \sum_{k=p}^{\infty} {}_2F_1(1-k,\xi;\eta;\omega)$$

$$\cdot \frac{k!\,\Gamma(k-\lambda+\nu+2)}{|\Gamma(k-n-\mu+1)|\,\Gamma(k-\lambda+\nu+2)}\,|z|^{k-n-\mu}$$

(4.14)

$$= \Theta_n^{(\xi,\eta;\omega)}(\lambda,\mu,\nu;|z|) + \sum_{k=n}^{\infty} {}_2F_1(1-k-p,\xi;\eta;\omega)$$

$$\cdot \frac{(k+p)!\,\Gamma(k+p-\mu+\nu+2)}{|\Gamma(k+p-n-\mu+1)|\,\Gamma(k+p-\lambda+\nu+2)}\,|z|^{k+p-n-\mu}$$

$$\left(0 < |z| < 1;\; n \in \mathbb{N}_0;\; p \in \mathbb{N};\; 0 \le \lambda < \min\{1,\nu+2\};\right.$$

$$\left.\mu < \min\{1,\nu+2\};\; \eta > \max\{0,\xi\};\; 0 \le \omega \le 1\right),$$

where $\Theta_n^{(\xi,\eta;\omega)}(\lambda,\mu,\nu;|z|)$ is given by (4.12).

Finally, by an obvious shift of the index of summation, the last sum in (4.14) is rewritten precisely as the sum occurring in the assertion (4.11). This evidently completes the proof of Theorem 5.

By applying the relationship (2.14) between the fractional derivative operators $J_{0,z}^{\lambda,\mu,\nu}$ and D_z^{λ}, the special case of Theorem 1 when $\mu = \lambda$ immediately yields

Theorem 6. *If the function $f(z)$ is in the class $\mathcal{H}_p(\omega; \xi, \eta)$, then*

$$\left|D_z^{n+\lambda} f(z)\right| \leq \Phi_n^{(\xi,\eta;\omega)}(\lambda; |z|)$$

$$+ \sum_{k=0}^{\infty} \frac{(k+n+p)!}{\Gamma(k+p-\lambda+1)} \, {}_2F_1(1-k-n-p, \xi; \eta; \omega) |z|^{k+p-\lambda} \qquad (4.15)$$

$$(0 < |z| < 1; \; n \in \mathbb{N}_0; \; p \in \mathbb{N}; \; 0 \leq \lambda < 1; \; \eta > \max\{0, \xi\}; \; 0 \leq \omega \leq 1),$$

where

$$\Phi_n^{(\xi,\eta;\omega)}(\lambda; |z|) := \sum_{k=0}^{n-1} {}_2F_1(1-k-p, \xi; \eta; \omega)$$

$$\cdot \frac{(k+p)!}{|\Gamma(k+p-n-\lambda+1)|} \, |z|^{k+p-n-\lambda}, \qquad (4.16)$$

the sum being assumed to be nil for $n = 0$.

It seems to be worthwhile to record here yet another set of growth-and-distortion type inequalities for the class $\mathcal{H}_p(\omega; \xi, \eta)$, which unifies and extends a number of known results (including, for example, the assertions of Theorem A and Theorem B). Indeed, by setting $\lambda = 0$ in Theorem 6 (or, alternatively, by setting $\mu = \lambda = 0$ in Theorem 5), we obtain

Theorem 7. *If the function $f(z)$ is in the class $\mathcal{H}_p(\omega; \xi, \eta)$, then*

$$\left|f^{(n)}(z)\right| \leq \Psi_n^{(\xi,\eta;\omega)}(|z|)$$

$$+ \sum_{k=0}^{\infty} \binom{k+n+p}{n} n! \, {}_2F_1(1-k-n-p, \xi; \eta; \omega) |z|^{k+p} \qquad (4.17)$$

$$(z \in \mathcal{U}; \; n \in \mathbb{N}_0; \; p \in \mathbb{N}; \; \eta > \max\{0, \xi\}; \; 0 \leq \omega \leq 1),$$

where

$$\Psi_n^{(\xi,\eta;\omega)}(|z|) := \sum_{k=0}^{n-1} \binom{k+p}{n} n! \, {}_2F_1(1-k-n-p, \xi; \eta; \omega) |z|^{k+p-n}, \qquad (4.18)$$

the sum being assumed to be nil for $n = 0$.

Various known upper bounds for the fractional derivatives of functions belonging to such simpler classes of analytic functions as

$$\mathcal{S}_p(\xi, \eta), \; \mathcal{S}^*(\alpha), \; \mathcal{K}(\alpha), \; \mathcal{S}, \quad \text{and} \quad \mathcal{K},$$

given earlier by (for example) Srivastava [26], Owa [12], Choi [4], Cho *et al.* [3], and Chen *et al.* [1], would follow from one or the other of Theorems 5, 6, and

7 above by suitably specializing the parameters involved. We choose to present here the growth-and-distortion type inequalities for the class $S_p(\xi, \eta)$ given by the relationship (4.7).

Theorem 8. *If the function $f(z)$ is in the class $S_p(\xi, \eta)$, then*

$$\left| J_{0,z}^{n+\lambda,\mu,\nu} f(z) \right| \le X_n^{(\xi,\eta)}(\lambda, \mu, \nu; |z|)$$

$$+ \frac{(\xi)_{n+p-1}}{(\eta)_{n+p-1}} \frac{(n+p)!\,\Gamma(n+p-\mu+\nu+2)}{\Gamma(p-\mu+1)\,\Gamma(n+p-\lambda+\nu+2)} \qquad (4.19)$$

$$\cdot |z|^{p-\mu} {}_4F_3 \left[\begin{matrix} \xi+n+p-1,\ n+p+1,\ n+p-\mu+\nu+2,\ 1; \\ \eta+n+p-1,\ p-\mu+1,\ n+p-\lambda+\nu+2; \end{matrix} |z| \right]$$

$$\left(0 < |z| < 1;\ n \in \mathbb{N}_0;\ p \in \mathbb{N};\ 0 \le \lambda < \min\{1, \nu+2\}; \right.$$

$$\left. \mu < \min\{1, \nu+2\};\ \xi, \eta \in \mathbb{R}_+ \right),$$

where

$$X_n^{(\xi,\eta)}(\lambda, \mu, \nu; |z|) := \sum_{k=0}^{n-1} \frac{(\xi)_{k+p-1}}{(\eta)_{k+p-1}}$$

$$\cdot \frac{(k+p)!\,\Gamma(k+p-\mu+\nu+2)}{|\Gamma(k+p-n-\mu+1)|\,\Gamma(k+p-\lambda+\nu+2)} |z|^{k+p-n-\mu}, \qquad (4.20)$$

the sum being assumed to be nil for $n = 0$.

In view of the relationship (2.14), Theorem 8 would reduce fairly readily to Theorem 9 below by setting

$$\mu = \lambda \qquad (0 \le \lambda < 1).$$

Theorem 9. *If the function $f(z)$ is in the class $S_p(\xi, \eta)$, then*

$$|D_z^{n+\lambda} f(z)| \le R_n^{(\xi,\eta)}(\lambda; |z|) + \frac{(\xi)_{n+p-1}}{(\eta)_{n+p-1}} \frac{(n+p)!}{\Gamma(p-\lambda+1)} |z|^{p-\lambda} \qquad (4.21)$$

$$\cdot {}_3F_2 \left[\begin{matrix} \xi+n+p-1,\ n+p+1,\ 1; \\ \eta+n+p-1,\ p-\lambda+1; \end{matrix} |z| \right]$$

$$(0 < |z| < 1;\ n \in \mathbb{N}_0;\ p \in \mathbb{N};\ 0 \le \lambda < 1;\ \xi, \eta \in \mathbb{R}_+),$$

where

$$R_n^{(\xi,\eta)}(\lambda; |z|) := \sum_{k=0}^{n-1} \frac{(\xi)_{k+p-1}}{(\eta)_{k+p-1}} \frac{(k+p)!}{|\Gamma(k+p-n-\lambda+1)|} |z|^{k+p-n-\lambda}, \qquad (4.22)$$

the sum being assumed to be nil for $n = 0$.

In its special case when

$$\xi = \eta + m \qquad (m \in \mathbb{N}), \tag{4.23}$$

the assertion (4.21) of Theorem 9 can be expressed in terms of the Gaussian hypergeometric function by appealing to the reduction formula (2.21). We thus obtain

Theorem 10. *If the function* $f(z)$ *is in the class* $\mathcal{S}_p(\eta + m, \eta)$, *then*

$$|D_z^{n+\lambda} f(z)| \leq S_n^{(\eta)}(m; \lambda; |z|) + \frac{(\eta + n + p - 1)_m}{(\eta)_m} \frac{(n + p)!}{\Gamma(p - \lambda + 1)} |z|^{p - \lambda}$$

$$\cdot \sum_{j=0}^{m} \binom{m}{j} \frac{(n + p + 1)_j \, (1)_j}{(\eta + n + p - 1)_j \, (p - \lambda + 1)_j} |z|^j \tag{4.24}$$

$$\cdot \, {}_2F_1 \left[\begin{matrix} n + p + j + 1, \ j + 1; \\ p - \lambda + j + 1; \end{matrix} \ |z| \right]$$

$$\left(0 < |z| < 1; \ n \in \mathbb{N}_0; \ m, p \in \mathbb{N}; \ 0 \leq \lambda < 1; \ \eta \in \mathbb{R}_+ \right),$$

where

$$S_n^{(\eta)}(m; \lambda; |z|) := \sum_{k=0}^{n-1} \frac{(\eta + k + p - 1)_m}{(\eta)_m}$$

$$\cdot \frac{(k + p)!}{|\Gamma(k + p - n - \lambda + 1)|} |z|^{k + p - n - \lambda}, \tag{4.25}$$

the sum being assumed to be nil for $n = 0$.

By virtue of the definition (2.4), it is not difficult to verify from (4.22) and (4.25) that

$$S_n^{(\eta)}(m; \lambda; |z|) = R_n^{(\eta + m, \eta)}(\lambda; |z|). \tag{4.26}$$

Yet another special case of Theorem 8, when the parameters ξ and η are constrained by means of the relationship (4.23), is worthy of mention here. In this special case, the assertion (4.19) of Theorem 8 can be expressed in terms of the Clausenian hypergeometric function, and we are thus led to

Theorem 11. *If the function* $f(z)$ *is in the class* $\mathcal{S}_p(\eta + m, \eta)$, *then*

$$\left| J_{0,z}^{n+\lambda,\mu,\nu} f(z) \right| \leq Y_n^{(\eta)}(m; \lambda, \mu, \nu; |z|)$$

$$+ \frac{(\eta + n + p - 1)_m}{(\eta)_m} \frac{(n + p)! \, \Gamma(n + p - \mu + \nu + 2)}{\Gamma(p - \mu + 1) \, \Gamma(n + p - \lambda + \nu + 2)} \tag{4.27}$$

$$\cdot |z|^{p-\mu} \sum_{j=0}^{m} \binom{m}{j} \frac{(n+p+1)_j (n+p-\mu+\nu+2)_j (1)_j}{(\eta+n+p-1)_j (p-\mu+1)_j (n+p-\lambda+\nu+2)_j} |z|^j$$

$$\cdot {}_3F_2 \left[\begin{matrix} n+p+j+1, \ n+p+j-\mu+\nu+2, \ j+1; \\ p-\mu+j+1, \ n+p+j-\lambda+\nu+2; \end{matrix} |z| \right]$$

$$\left(0 < |z| < 1; \ n \in \mathbb{N}_0; \ m, p \in \mathbb{N}; \ 0 \le \lambda < \min\{1, \nu+2\}; \right.$$

$$\left. \mu < \min\{1, \nu+2\}; \ \eta \in \mathbb{R}_+ \right),$$

where

$$Y_n^{(\eta)}(m; \lambda, \mu, \nu; |z|) := \sum_{k=0}^{n-1} \frac{(\eta+k+p-1)_m}{(\eta)_m}$$

$$\cdot \frac{(k+p)! \, \Gamma(k+p-\mu+\nu+2)}{|\Gamma(k+p-n-\mu+1)| \, \Gamma(k+p-\lambda+\nu+2)} |z|^{k+p-n-\mu}, \tag{4.28}$$

the sum being assumed to be nil for $n = 0$.

By comparing the definitions (4.20) and (4.28), it is easily verified that

$$Y_n^{(\eta)}(m; \lambda, \mu, \nu; |z|) = X_n^{(\eta+m, n)}(\lambda, \mu, \nu; |z|). \tag{4.29}$$

Finally, we record a distortion inequality for the class $\mathcal{S}_p(\xi, \eta)$, which may be deduced from Theorem 9 by setting $\lambda = 0$ (or, alternatively, from Theorem 8 by setting $\mu = \lambda = 0$). We thus obtain

Theorem 12. *If the function $f(z)$ is in the class $\mathcal{S}_p(\xi, \eta)$, then*

$$|f^{(n)}(z)| \le Z_n^{(\xi, \eta)}(|z|) + \frac{(\xi)_{n+p-1}}{(\eta)_{n+p-1}} \frac{(n+p)!}{p!} |z|^p$$

$$\cdot {}_3F_2 \left[\begin{matrix} \xi+n+p-1, \ n+p+1, \ 1; \\ \eta+n+p-1, \ p+1; \end{matrix} |z| \right] \tag{4.30}$$

$$(z \in \mathcal{U}; \ n \in \mathbb{N}_0; \ p \in \mathbb{N}; \ \xi, \eta \in \mathbb{R}_+),$$

where

$$Z_n^{(\xi, \eta)}(|z|) := \sum_{k=0}^{n-1} \frac{(\xi)_{k+p-1}}{(\eta)_{k+p-1}} \frac{(k+p)!}{(k+p-n)!} |z|^{k+p-n}, \tag{4.31}$$

the sum being assumed to be nil for $n = 0$.

Evidently, when

$$n \ge p \qquad (n \in \mathbb{N}; \ p \in \mathbb{N}), \tag{4.32}$$

the sum in (4.31) would be taken from $k = n - p$ to $k = n - 1$.

5. Distortion Inequalities Involving Ruscheweyh Derivatives

For $f_j(z) \in \mathcal{A}$ given by

$$f_j(z) = z + \sum_{k=2}^{\infty} a_{k,j} \, z^k \qquad (j = 1, 2), \tag{5.1}$$

the Hadamard product (or convolution) $(f_1 * f_2)(z)$ of $f_1(z)$ and $f_2(z)$ is defined by

$$(f_1 * f_2)(z) := z + \sum_{k=2}^{\infty} a_{k,1} \, a_{k,2} \, z^k. \tag{5.2}$$

Using the convolution (5.2), Ruscheweyh [21] introduced what is now referred to as the Ruscheweyh derivative $\mathcal{D}^\lambda f(z)$ of order λ of $f(z) \in \mathcal{A}$ by

$$\mathcal{D}^\lambda f(z) := \frac{z}{(1 - z)^{\lambda+1}} * f(z) \qquad (\lambda > -1). \tag{5.3}$$

It follows from (5.3) that

$$\mathcal{D}^0 f(z) = f(z), \quad \mathcal{D}^1 f(z) = z \, f'(z), \tag{5.4}$$

and, in general,

$$\mathcal{D}^n f(z) = \frac{z \left(z^{n-1} f(z)\right)^{(n)}}{n!} \qquad (n \in \mathbb{N}_0). \tag{5.5}$$

Furthermore, we have

$$\mathcal{D}^\lambda f(z) = z + \sum_{k=2}^{\infty} C(\lambda, k) \, a_k \, z^k, \tag{5.6}$$

where, for convenience,

$$C(\lambda, k) := \frac{\prod_{j=1}^{k-1} (j + \lambda)}{(k - 1)!} = \binom{\lambda + k - 1}{k - 1} \qquad (k \in \mathbb{N} \setminus \{1\}). \tag{5.7}$$

Our first distortion inequality involving Ruscheweyh derivatives is contained in (cf. Owa and Srivastava [17]; see also Owa [11]).

Theorem 13. *If a function $f(z)$ belongs to the class $\mathcal{S}^*(\alpha)$, then*

$$|\mathcal{D}^\lambda f(z)| \le M(n, \lambda, \alpha; |z|)$$
$$+ \frac{(1 + \lambda)_{n-1}(2 - 2\alpha)_{n-1}}{\{(n - 1)!\}^2} |z|^n \, {}_3F_2 \left(\begin{matrix} n + \lambda, \ n + 1 - 2\alpha, \ 1; \\ n, \ n; \end{matrix} |z| \right) \tag{5.8}$$

$$(z \in \mathcal{U}; \ n \in \mathbb{N} \setminus \{1, 2\}),$$

where

$$M(n, \lambda, \alpha; |z|) := |z| + \sum_{k=2}^{n-1} \frac{(1+\lambda)_{k-1}(2-2\alpha)_{k-1}}{\{(k-1)!\}^2} |z|^k. \tag{5.9}$$

Proof. We begin by noting from the definition (5.7) that

$$C(\lambda, k) = \frac{\prod_{j=1}^{k-1}(j+\lambda)}{(k-1)!} = \frac{(1+\lambda)_{k-1}}{(k-1)!}, \tag{5.10}$$

and

$$|a_k| \leq \frac{\prod_{j=2}^{k}(j-2\alpha)}{(k-1)!} = \frac{(2-2\alpha)_{k-1}}{(k-1)!} \qquad (f \in \mathcal{S}^*(\alpha)). \tag{5.11}$$

It follows from (5.10) and (5.11) that

$$|\mathcal{D}^\lambda f(z)| \leq |z| + \sum_{k=2}^{\infty} C(\lambda, k) |a_k| |z|^k$$

$$\leq |z| + \sum_{k=2}^{\infty} \frac{(1+\lambda)_{k-1}(2-2\alpha)_{k-1}}{\{(k-1)!\}^2} |z|^k$$

$$= \left\{ |z| + \sum_{k=2}^{n-1} \frac{(1+\lambda)_{k-1}(2-2\alpha)_{k-1}}{\{(k-1)!\}^2} |z|^k \right\} \tag{5.12}$$

$$+ \sum_{k=n}^{\infty} \frac{(1+\lambda)_{k-1}(2-2\alpha)_{k-1}}{\{(k-1)!\}^2} |z|^k$$

$$= M(n, \lambda, \alpha; |z|) + \sum_{k=0}^{\infty} \frac{(1+\lambda)_{k+n-1}(2-2\alpha)_{k+n-1}}{\{(k+n-1)!\}^2} |z|^{k+n}.$$

Since

$$(1+\lambda)_{k+n-1} = (1+\lambda)_{n-1}(n+\lambda)_k, \tag{5.13}$$

$$(2-2\alpha)_{k+n-1} = (2-2\alpha)_{n-1}(n+1-2\alpha)_k, \tag{5.14}$$

and

$$(k+n-1)! = (n-1)!(n)_k, \tag{5.15}$$

we see that

$$|\mathcal{D}^\lambda f(z)| \leq M(n, \lambda, \alpha; |z|)$$

$$+ \frac{(1+\lambda)_{n-1}(2-2\alpha)_{n-1}}{\{(n-1)!\}^2} |z|^n \sum_{k=0}^{\infty} \frac{(n+\lambda)_k(n+1-2\alpha)_k}{\{(n)_k\}^2} |z|^k$$

$$(5.16)$$

$$= M(n, \lambda, \alpha; |z|)$$

$$+ \frac{(1+\lambda)_{n-1}(2-2\alpha)_{n-1}}{\{(n-1)!\}^2} |z|^n \, {}_3F_2 \left(\begin{matrix} n+\lambda, \ n+1-2\alpha, \ 1; \\ n, \ n; \end{matrix} |z| \right),$$

which is precisely the assertion (5.8) of Theorem 13, $M(n, \lambda, \alpha; |z|)$ being given by (5.9).

Theorem 14. *If a function $f(z)$ belongs to the class $\mathcal{S}^*(\alpha)$, then*

$$|\mathcal{D}^m f(z)| \leq M(n, m, \alpha; |z|) + \frac{(1+m)_{n-1}(2-2\alpha)_{n-1}}{\{(n-1)!\}^2} |z|^n \qquad (5.17)$$

$$\cdot \left\{ \sum_{k=0}^{m} \binom{m}{k} \frac{(n+1-2\alpha)_k(1)_k}{\{(n)_k\}^2} \frac{|z|^k}{(1-|z|)^{k+2-2\alpha}} \, {}_2F_1 \left(\begin{matrix} 2\alpha-1, \ n-1; \\ n+k; \end{matrix} |z| \right) \right\},$$

$$(z \in \mathcal{U}; \ m \in \mathbb{N}),$$

where $M(n, m, \alpha; |z|)$ is defined by (5.9) with, of course, $\lambda = m$.

Proof. In view of the known hypergeometric identities (2.21) and (2.25), the Clausenian hypergeometric function occurring in (5.8) can be simplified, when $\lambda = m$ ($m \in \mathbb{N}$), as follows:

$${}_3F_2 \left(\begin{matrix} n+m, \ n+1-2\alpha, \ 1; \\ n, \ n; \end{matrix} |a| \right)$$

$$= \sum_{k=0}^{m} \binom{m}{k} \frac{(n+1-2\alpha)_k(1)_k}{\{(n)_k\}^2} |z|^k \, {}_2F_1 \left(\begin{matrix} n+1+k-2\alpha, \ 1+k; \\ n+k; \end{matrix} |z| \right) \quad (5.18)$$

$$= \sum_{k=0}^{m} \binom{m}{k} \frac{(n+1-2\alpha)_k(1)_k}{\{(n)_k\}^2} \frac{|z|^k}{(1-|z|)^{k+2-2\alpha}} \, {}_2F_1 \left(\begin{matrix} 2\alpha-1, \ n-1; \\ n+k; \end{matrix} |z| \right),$$

which yields the assertion (5.17) of Theorem 14.

Setting $\alpha = \frac{1}{2}$ in Theorem 13, we obtain

Theorem 15. *If a function $f(z)$ belongs to the class $\mathcal{S}^* \left(\frac{1}{2} \right)$, then*

$$|\mathcal{D}^\lambda f(z)| \leq M \left(n, \lambda, \frac{1}{2}; |z| \right)$$

$$+ \frac{(1+\lambda)_{n-1}}{(n-1)!} \frac{|z|^n}{(1-|z|)^{1+\lambda}} \, {}_2F_1 \left(\begin{matrix} -\lambda, \ n-1; \\ n; \end{matrix} |z| \right) \quad (z \in \mathcal{U}), \qquad (5.19)$$

where

$$M\left(n, \lambda, \tfrac{1}{2}; |z|\right) = |z| + \sum_{k=2}^{n-1} \frac{(1+\lambda)_{k-1}}{(k-1)!} |z|^k. \tag{5.20}$$

Further, taking $\alpha = 0$ in Theorem 13, we have

Theorem 16. *If a function $f(z)$ belongs to the class $\mathcal{S}^*$, then*

$$|\mathcal{D}^\lambda f(z)| \leq M(n, \lambda, 0; |z|)$$

$$+ \frac{n(1+\lambda)_{n-1}}{(n-1)!} |z|^n \left\{ \frac{1}{(1-|z|)^{1+\lambda}} \,_2F_1\left(\begin{matrix} -\lambda, \ n-1; \\ n; \end{matrix} |z| \right) \right. \tag{5.21}$$

$$+ \left. \frac{n+\lambda}{n^2} \frac{|z|}{(1-|z|)^{2+\lambda}} \,_2F_1\left(\begin{matrix} -\lambda, \ n-1; \\ n+1; \end{matrix} |z| \right) \right\} \quad (z \in \mathcal{U}),$$

where

$$M(n, \lambda, 0; |z|) = |z| + \sum_{k=1}^{n-1} \frac{k(1+\lambda)_{k-1}}{(k-1)!} |z|^k. \tag{5.22}$$

For the Ruscheweyh derivatives of convex functions belonging to the class $\mathcal{K}(\alpha)$, we can similarly obtain Theorems 17 to 20 below:

Theorem 17. *If a function $f(z)$ belongs to the class $\mathcal{K}(\alpha)$, then*

$$|\mathcal{D}^\lambda f(z)| \leq N(n, \lambda, \alpha; |z|)$$

$$+ \frac{(1+\lambda)_{n-1}(2-2\alpha)_{n-1}}{n!\,(n-1)!} |z|^n \,_3F_2\left(\begin{matrix} \lambda+n, \ n+1-2\alpha, \ 1; \\ n, \ n+1; \end{matrix} |z| \right) \tag{5.23}$$

$$(z \in \mathcal{U}; \ n \in \mathbb{N} \setminus \{1,2\}),$$

where

$$N(n, \lambda, \alpha; |z|) = |z| + \sum_{k=2}^{n-1} \frac{(1+\lambda)_{k-1}(2-2\alpha)_{k-1}}{k!\,(k-1)!} |z|^k. \tag{5.24}$$

Theorem 18. *If a function $f(z)$ belongs to the class $\mathcal{K}(\alpha)$, then*

$$|\mathcal{D}^m f(z)| \leq N(n, m, \alpha; |z|) + \frac{(1+m)_{n-1}(2-2\alpha)_{n-1}}{n!\,(n-1)!} \tag{5.25}$$

$$\cdot \left\{ \sum_{k=0}^{m} \binom{m}{k} \frac{(n+1-2\alpha)_k (1)_k}{(n)_k (n+1)_k} \frac{|z|^k}{(1-|z|)^{k+1-2\alpha}} \,_2F_1\left(\begin{matrix} 2\alpha, \ n; \\ n+k+1; \end{matrix} |z| \right) \right\}$$

$$(z \in \mathcal{U}; \ m \in \mathbb{N}),$$

where $N(n, m, \alpha; |z|)$ is defined by (5.24) with, of course, $\lambda = m$.

Theorem 19. *If a function $f(z)$ belongs to the class $\mathcal{K}\left(\frac{1}{2}\right)$, then*

$$|\mathcal{D}^\lambda f(z)| \leq N\left(n, \lambda, \tfrac{1}{2}; |z|\right)$$

$$+ \frac{(1+\lambda)_{n-1}}{n!} \frac{|z|^n}{(1-|z|)^\lambda} \, {}_2F_1\left(\begin{matrix} 1-\lambda, \ n; \\ n+1; \end{matrix} |z|\right) \quad (z \in \mathcal{U}),$$

$$(5.26)$$

where

$$N\left(n, \lambda, \tfrac{1}{2}; |z|\right) = |z| + \sum_{k=2}^{n-1} \frac{(1+\lambda)_{k-1}}{k!} |z|^k. \tag{5.27}$$

Theorem 20. *If a function $f(z)$ belongs to the class $\mathcal{K}$, then*

$$|\mathcal{D}^\lambda f(z)| \leq N(n, \lambda, 0; |z|)$$

$$+ \frac{(1+\lambda)_{n-1}}{(n-1)!} \frac{|z|^n}{(1-|z|)^{1+\lambda}} \, {}_2F_1\left(\begin{matrix} -\lambda, \ n-1; \\ n; \end{matrix} |z|\right) \quad (z \in \mathcal{U}),$$

$$(5.28)$$

where

$$N(n, \lambda, 0; |z|) = |z| + \sum_{k=2}^{n-1} \frac{(1+\lambda)_{k-1}}{(k-1)!} |z|^k. \tag{5.29}$$

6. Further Growth-and-Distortion Type Inequalities: Concluding Remarks and Observations

First of all, by applying the fractional derivative formula (2.15) *directly* (or by specializing Theorem 5 and Theorem 8 in accordance with (4.7) and (4.10) appropriately), we can prove a generalization of Theorem 1, which is contained in

Theorem 21. *If the function $f(z)$ is in the class $\mathcal{S}$, then*

$$\left| J_{0,z}^{n+\lambda,\mu,\nu} f(z) \right| \leq U_n(\lambda, \mu, \nu; |z|) + \frac{n! \, |z|^{-\mu} \, \Gamma(n-\mu+\nu+3)}{\Gamma(2-\mu)\,\Gamma(n-\lambda+\nu+3)}$$

$$\cdot \left\{ \frac{n(1-\mu)(n-\lambda+\nu+2)}{n-\mu+\nu+2} \, {}_3F_2\left[\begin{matrix} n+1, \ n-\mu+\nu+2, \ 1; \\ 1-\mu, \ n-\lambda+\nu+2; \end{matrix} |z|\right] \right.$$

$$\left. + (n+1)|z| \, {}_3F_2\left[\begin{matrix} n+2, \ n-\mu+\nu+3, \ 2; \\ 2-\mu, \ n-\lambda+\nu+3; \end{matrix} |z|\right] \right\}$$

$$(6.1)$$

$$\left(0 < |z| < 1; \ n \in \mathbb{N}_0; \ 0 \leq \lambda < \min\{1, \nu+2\}; \ \mu < \min\{1, \nu+2\}\right),$$

where

$$U_n(\lambda, \mu, \nu; |z|) := \sum_{k=1}^{n-1} \frac{k \cdot k! \, \Gamma(k - \mu + \nu + 2)}{|\Gamma(k - \mu - n + 1)| \, \Gamma(k - \lambda + \nu + 2)} \, |z|^{k - \mu - n}, \qquad (6.2)$$

the sum being assumed to be nil for $n = 0$ and $n = 1$.

In a similar manner, a generalization of Theorem 2 in terms of the fractional derivative operator $J_{0,z}^{\lambda, \mu, \nu}$ is given by

Theorem 22. *If the function $f(z)$ is in the class $\mathcal{K}$, then*

$$\left| J_{0,z}^{n+\lambda, \mu, \nu} \, f(z) \right| \leq V_n(\lambda, \mu, \nu; |z|) + \frac{n! \, \Gamma(n - \mu + \nu + 2)}{\Gamma(1 - \mu) \, \Gamma(n - \lambda + \nu + 2)} \, |z|^{-\mu} \qquad (6.3)$$

$$\cdot {}_3F_2 \left[\begin{matrix} n + 1, \ n - \mu + \nu + 2, \ 1; \\ 1 - \mu, \ n - \lambda + \nu + 2; \end{matrix} \, |z| \right]$$

$$(0 < |z| < 1; \ n \in \mathbb{N}_0; \ 0 \leq \lambda < \min\{1, \nu + 2\}; \ \mu < \min\{1, \nu + 2\}),$$

where

$$V_n(\lambda, \mu, \nu; |z|) := \sum_{k=1}^{n-1} \frac{k! \, \Gamma(k - \mu + \nu + 2)}{|\Gamma(k - \mu - n + 1)| \, \Gamma(k - \lambda + \nu + 2)} \, |z|^{k - \mu - n}, \qquad (6.4)$$

the sum being assumed to be nil for $n = 0$ and $n = 1$.

Numerous further extensions of the growth-and-distortion type inequalities contained in Theorems 1 to 22 of this and the preceding sections can indeed be given for *multivalent* functions in classes analogous (amongst others) to $\mathcal{H}_p(\omega; \xi, \eta)$, with missing coefficients or with fixed (finitely many) coefficients, and so on. And then there are several families of integral, convolution, or other linear operators which are capable of providing interesting generalizations of the operators D_z^λ, $J_{0,z}^{\lambda, \mu, \nu}$, and $\mathcal{D}^\lambda$ used in our present investigation. Each of these general operators can be applied *instead*. However, these extensions are fairly straightforward, and we choose to omit the details involved.

The recent paper of Choi [4] is motivated also by (and provides generalizations of) the earlier works of Chen *et al.* [1] and Cho *et al.* [3]. In fact, the *main* results of Choi [4] are stated as two theorems in his paper: Theorem 1 of Choi [4] is a *mild* generalization of Theorem 21 of this section; Theorem 2 of Choi [4], on the other hand, is a *mild* generalization of Theorem 22 of this section. These generalizations by Choi [4] are accomplished by introducing an additional (seemingly inconsequential) parameter m which Raina [18] had introduced earlier in the familiar Definition 5 *without* paying any attention to the resulting infinitely many singularities of the integrand in $\mathcal{U}$. Similar remarks would apply also to a very recent conjecture by Raina and Kalia [19, p. 13].

Finally, we do not see any significance or usefulness of the aforementioned parameter m in the familiar fractional calculus operators (that is, the case $m = 1$) whose study was initiated by Saigo [22] and others (*cf.*, *e.g.*, Srivastava *et al.* [30], Owa *et al.* [14], and Sohi [24]), except *possibly* for the sole purpose of *further* generalizing whatever is known for these already sufficiently general fractional calculus operators involving the Gaussian hypergeometric function in their kernel.

Acknowledgments

The present investigation was supported, in part, by the *Natural Sciences and Engineering Research Council of Canada* under Grant OGP0007353.

References

[1] M.-P. Chen, H.M. Srivastava, and C.-S. Yu, A note on a conjecture involving fractional derivatives of convex functions, *J. Frac. Calc.* 5(1994), 81–85.

[2] M.-P. Chen, H.M. Srivastava, and C.-S. Yu, Some operators of fractional calculus and their applications involving a novel class of analytic functions, *Appl. Math. Comput.* 91(1998), 285–296.

[3] N.E. Cho, S. Owa, and H.M. Srivastava, Some remarks on a conjectured upper bound for the fractional derivative of univalent functions, *Internat. J. Math. Statist. Sci.* 2(1993), 117–125.

[4] J.H. Choi, Distortion properties of some univalent and convex functions involving a generalized fractional derivative operator, *Internat. J. Math. Statist. Sci.* 5(1996), 161–177.

[5] L. de Branges, A proof of the Bieberbach conjecture, *Acta Math.* 154(1985), 137–152.

[6] P.L. Duren, *Univalent Functions*, Grundlehren der Mathematischen Wissenschaften, Bd. 259, Springer-Verlag, New York, Berlin, Heidelberg, and Tokyo, 1983.

[7] A. Erdélyi, W. Magnus, F. Oberhettinger, and F.G. Tricomi, *Higher Transcendental Functions*, Vol. I, McGraw-Hill Book Company, New York, Toronto, and London, 1953.

[8] A.W. Goodman, *Univalent Functions*, Vol. I, Polygonal Publishing House, Washington, New Jersey, 1983.

[9] K. Löwner, Untersuchungen über die Verzerrung bei donformen Abbildungen des Einheitskreises $|z| < 1$, die durch Funktionen mit nicht verschwindender Ableitung geliefert werden, *Ber. Verh. Sächs. Ges. Wiss. Leipzig* 69(1917), 89–106.

[10] S. Owa, On the distortion theorems. I, *Kyungpook Math. J.* 18(1978), 53–59.

[11] S. Owa, A distortion theorem for Ruscheweyh derivatives, *Far East J. Math. Sci.* 2(1994), 137–142.

[12] S. Owa, Fractional calculus of analytic functions, in *Transform Methods and Special Functions* (Proceedings of the First International Workshop held at Bankya [Sofia] on August 12–17, 1994) (P. Rusev, I. Dimovski, and V. Kiryakova, Editors), Science Culture Technology Publisying Company, Singapore, 1995, 213–219.

[13] S. Owa, K. Nishimoto, S.K. Lee, and N.E. Cho, A note on certain fractional operator, *Bull. Calcutta Math. Soc.* 83(1991), 87–90.

[14] S. Owa, M. Saigo, and H.M. Srivastava, Some characterization theorems for starlike and convex functions involving a certain fractional integral operator, *J. Math. Anal. Appl.* 140(1989), 419–426.

[15] S. Owa and H.M. Srivastava, Univalent and starlike generalized hypergeometric functions, *Canad. J. Math.* **39**(1987), 1057–1077.

[16] S. Owa and H.M. Srivastava, A distortion theorem and a related conjecture involving fractional derivatives of convex functions, in *Univalent Functions, Fractional Calculus, and Their Applications* (H.M. Srivastava and S. Owa, Editors), Halsted Press (Ellis Horwood Limited, Chichester), John Wiley and Sons, New York, Chichester, Brisbane, and Toronto, 1989, 219–228.

[17] S. Owa and H.M. Srivastava, Distortion inequalities for Ruscheweyh derivatives, *Math. Inequal. Appl.* **1**(1998), 239–246.

[18] R.K. Raina, On certain classes of analytic functions and applications to fractional calculus operators, *Integral Transform. Spec. Funct.* **5**(1997), 247–260.

[19] R.K. Raina and R.N. Kalia, On a conjecture giving upper bound for certain fractional derivative operator of convex functions, *Math. Sci. Res. Hot-Line* **1**(2) (1997), 12–13.

[20] M.S. Robertson, On the theory of univalent functions, *Ann. of Math.* **37**(1936), 374–408.

[21] St. Ruscheweyh, New criteria for univalent functions, *Proc. Amer. Math. Soc.* **49**(1975), 109–115.

[22] M. Saigo, A remark on integral operators involving the Gauss hypergeometric functions, *Math. Rep. College General Ed. Kyushu Univ.* **11**(1978), 135–145.

[23] L.J. Slater, *Generalized Hypergeometric Functions*, Cambridge University Press, Cambridge, London, and New York, 1966.

[24] N.S. Sohi, Distortion theorems involving certain operators of fractional calculus on a class of *p*-valent functions, in *Fractional Calculus and Its Applications* (Proceedings of the International Conference held at the Nihon University Center in Tokyo on May 29 – June 1, 1989) (K. Nishimoto, Editor), College of Engineering (Nihon University), Koriyama, 1990, 245–252.

[25] H.M. Srivastava, Generalized hypergeometric functions with integral parameter differences, *Nederl. Akad. Wetensch. Indag. Math.* **35**(1973), 38–40.

[26] H.M. Srivastava, Fractional calculus and its applications in analytic function theory, in *Proceedings of the International Conference on Analysis* (Gyongsan; December 11–14, 1996) (Y.C. Kim, Editor), Yeungnam University, Gyongsan, 1996, 1–25.

[27] H.M. Srivastava and S. Owa (Editors), *Univalent Functions, Fractional Calculus, and Their Applications*, Halsted Press (Ellis Horwood Limited, Chichester), John Wiley and Sons, New York, Chichester, Brisbane, and Toronto, 1989.

[28] H.M. Srivastava and S. Owa (Editors), *Current Topics in Analytic Function Theory*, World Scientific Publishing Company, Singapore, New Jersey, London, and Hong Kong, 1992.

[29] H.M. Srivastava and M. Saigo. Multiplication of fractional calculus operators and boundary value problems involving the Euler-Darboux equation, *J. Math. Anal. Appl.* **121**(1987), 325–369.

[30] H.M. Srivastava, M. Saigo, and S. Owa, A class of distortion theorems involving certain operators of fractional calculus, *J. Math. Anal. Appl.* **131**(1988), 412–420.

INDEX

375